GLOBAL STUDIES

AFRICA

TENTH EDITION

Dr. F. Jeffress Ramsay

Dr. Wayne Edge

OTHER BOOKS IN THE GLOBAL STUDIES SERIES
- China
- Europe
- India and South Asia
- Japan and the Pacific Rim
- Latin America
- The Middle East
- Russia, the Eurasian Republics, and
  Central/Eastern Europe

McGraw-Hill/Dushkin Company
530 Old Whitfield Street, Guilford, Connecticut 06437
Visit us on the Internet—http://www.dushkin.com

Staff

Jeffrey L. Hahn	*Vice President/Publisher*
Theodore Knight, Ph.D.	*Managing Editor*
Brenda S. Filley	*Director of Production*
Lisa Clyde Nielsen	*Series Developmental Editor*
Roberta Monaco	*Managing Developmental Editor*
Charles Vitelli	*Designer*
Robin Zarnetske	*Permissions Editor*
Marie Lazauskas	*Permissions Assistant*
Lisa Holmes-Doebrick	*Senior Program Coordinator*
Michael Campbell	*Graphics*
Eldis Lima	*Graphics/Cover Design*
Juliana Arbo	*Typesetting Supervisor/Co-designer*
Jocelyn Proto	*Typesetter*
Cynthia Powers	*Typesetter*

Sources for Statistical Reports

U.S. State Department *Background Notes* (2003)

C.I.A. *World Factbook* (2002)

World Bank *World Development Reports* (2002/2003)

UN *Population and Vital Statistics Reports* (2002/2003)

World Statistics in Brief (2002)

The Statesman's Yearbook (2003)

Population Reference Bureau *World Population Data Sheet* (2002)

The World Almanac (2003)

The Economist Intelligence Unit (2003)

Copyright

Cataloging in Publication Data
Main entry under title: Global Studies: Africa. 10th ed.
     1. Africa—History—1960–.
I. Title:Africa. II. Ramsey, F. Jeffress, *comp.*; Edge, Wayne, *comp.*
ISBN 0–07–284713–1   960.3   91–71258   ISSN 1098-3880

Tenth Edition

Printed in the United States of America   1234567890BAHBAH54   Printed on Recycled Paper

AFRICA

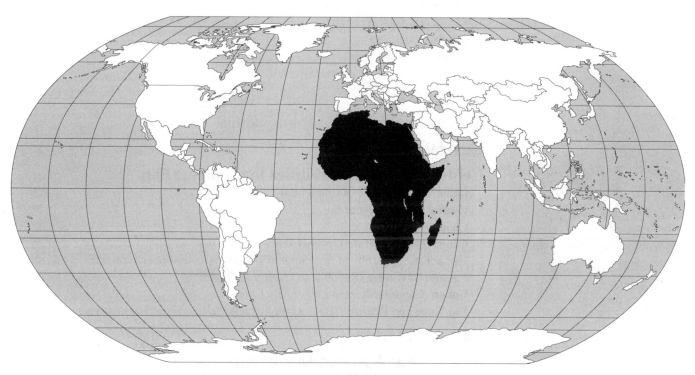

AUTHOR/EDITORS

Dr. F. Jeffress Ramsay

Dr. F. Jeffress ("Jeff") Ramsay, the author/editor of *Global Studies: Africa*, obtained his Ph.D. in African history from Boston University. He has extensive experience in both secondary and tertiary education in the United States and in Botswana, where he is currently the principal of Legae Academy in Gaborone. Dr. Ramsay writes regularly for Botswana newspapers as well as popular and academic periodicals; he is the chairperson of the Botswana Media Consultative Council; and he has been involved in the development of regional museums. A recent recipient of Botswana's Presidential Order of Honour for his varied contributions to the country, Dr. Ramsay is a Botswana citizen.

Along with Barry Morton and Themba Mgadla, Dr. Ramsay is the coauthor of *Building a Nation, a History of Botswana from 1800–1910* (Longman Botswana, 1996); he is the coeditor with Fred Morton of *Birth of Botswana, a History of the Bechuanaland Protectorate from 1910–1966* (Longman Botswana, 1987); he coauthored *A Historical Dictionary of Botswana* (Scarecrow Press, 3rd ed., 1997), with Barry Morton and Fred Morton; he collaborated with Barry Morton on *The Making of a President, Sir Ketumile Masire's Early Years* (Pula Publishing, 1996) and on *Comrade Fish, Memories of a Freedom Fighter on the Botswana Underground* (Pula Publishing, 2000); and, along with Lucey Clarke, he coauthored *New Three Year I. C. Social Studies Revision Notes* (Tasalls, 1997).

Dr. Wayne Edge

Dr. Wayne Edge is a longtime resident of Botswana. Having obtained his Ph.D. in political science at the University of Delaware, he was until recently a lecturer at the University of Botswana. He has also taught at the University of the Virgin Islands, Lincoln University, the University of Delaware, and the School for International Training, Botswana Program. Dr. Edge's other publications include authoring *The Autobiography of Motsamai Mpho* (Lepopo Publishers, 1996); and coediting, with M. Lekorewe, as well as contributing to *Botswana: Politics and Society* (J. L. Van Schaik Publishers, 1998).

Contents

West Africa Map

West Africa: Seeking Unity in Diversity

Articles from the World Press

Using Global Studies: Africa

THE GLOBAL STUDIES SERIES

The Global Studies series was created to help readers acquire a basic knowledge and understanding of the regions and countries in the world. Each volume provides a foundation of information—geographic, cultural, economic, political, historical, artistic, and religious—that will allow readers to better assess the current and future problems within these countries and regions and to comprehend how events there might affect their own well-being. In short, these volumes present the background information necessary to respond to the realities of our global age.

Each of the volumes in the Global Studies series is crafted under the careful direction of an author/editor—an expert in the area under study. The author/editors teach and conduct research and have traveled extensively through the regions about which they are writing.

In this *Global Studies: Africa* edition, the author/editor has written an introductory essay on the continent as a whole, several regional essays, and country reports for each of the countries included.

MAJOR FEATURES OF THE GLOBAL STUDIES SERIES

The Global Studies volumes are organized to provide concise information on the regions and countries within those areas under study. The major sections and features of the books are described here.

Regional Essays

For *Global Studies: Africa,* the author/editor has written several essays focusing on the religious, cultural, sociopolitical, and economic differences and similarities of the countries and peoples in the various regions of Africa. Regional maps accompany the essays.

Country Reports

Concise reports are written for each of the countries within the region under study. These reports are the heart of each Global Studies volume. *Global Studies: Africa, Tenth Edition,* contains 48 country reports.

The country reports are composed of five standard elements. Each report contains a detailed map visually positioning the country among its neighboring states; a summary of statistical information; a current essay providing important historical, geographical, political, cultural, and economic information; a historical timeline, offering a convenient visual survey of a few key historical events; and four "graphic indicators," with summary statements about the country in terms of development, freedom, health/welfare, and achievements.

A Note on the Statistical Reports

The statistical information provided for each country has been drawn from a wide range of sources. (The most frequently referenced are listed on page 2.) Every effort has been made to provide the most current and accurate information available. However, sometimes the information cited by these sources differs to some extent; and, all too often, the most current information available for some countries is somewhat dated. Aside from these occasional difficulties, the statistical summary of each country is generally quite complete and up to date. Care should be taken, however, in using these statistics (or, for that matter, any published statistics) in making hard comparisons among countries. We have also provided comparable statistics for the United States and Canada, which can be found on pages 8 and 9.

World Press Articles

Within each Global Studies volume is reprinted a number of articles carefully selected by our editorial staff and the author/editor from a broad range of international periodicals and newspapers. The articles have been chosen for currency, interest, and their differing perspectives on the subject countries. There are 12 articles in *Global Studies: Africa, Tenth Edition.*

The articles section is preceded by an annotated table of contents. This resource offers a brief summary of each article.

WWW Sites

An extensive annotated list of selected World Wide Web sites can be found on the facing page 7 in this edition of *Global Studies: Africa.* In addition, the URL addresses for country-specific Web sites are provided on the statistics page of most countries. All of the Web site addresses were correct and operational at press time. Instructors and students alike are urged to refer to those sites often to enhance their understanding of the region and to keep up with current events.

Glossary, Bibliography, Index

At the back of each Global Studies volume, readers will find a glossary of terms and abbreviations, which provides a quick reference to the specialized vocabulary of the area under study and to the standard abbreviations used throughout the volume.

Following the glossary is a bibliography, which lists general works, national histories, and current-events publications and periodicals that provide regular coverage on Africa.

The index at the end of the volume is an accurate reference to the contents of the volume. Readers seeking specific information and citations should consult this standard index.

Currency and Usefulness

Global Studies: Africa, like the other Global Studies volumes, is intended to provide the most current and useful information available necessary to understand the events that are shaping the cultures of the region today.

This volume is revised on a regular basis. The statistics are updated, regional essays and country reports revised, and world press articles replaced. In order to accomplish this task, we turn to our author/editor, our advisory boards, and—hopefully—to you, the users of this volume. Your comments are more than welcome. If you have an idea that you think will make the next edition more useful, an article or bit of information that will make it more current, or a general comment on its organization, content, or features that you would like to share with us, please send it in for serious consideration.

Selected World Wide Web Sites for Africa

(Some Web sites continually change their structure and content, so the information listed here may not always be available. Check our Web site at: http://www.dushkin.com/online/ —Ed.)

GENERAL SITES

BBC World Service
http://www.bbc.co.uk/worldservice/index.htm

The BBC, one of the world's most successful radio networks, provides the latest news from around the world, including news from almost all of the African countries.

C-SPAN ONLINE
http://www.c-span.org

Access C-SPAN International on the Web for International Programming Highlights and archived C-SPAN programs.

International Network Information Center at University of Texas
http://inic.utexas.edu

This gateway has pointers to international sites, including Africa, as well as African Studies Resources.

I-Trade International Trade Resources & Data Exchange
http://www.i-trade.com

Monthly exchange-rate data, U.S. Global Trade Outlook, and recent World Fact Book statistical demographic and geographic data for 180-plus countries can be found on this Web site.

Penn Library: Resources by Subject
http://www.library.upenn.edu/resources/subject/subject.html

This vast site is rich in links to information about African studies, including demography and population.

ReliefWeb
http://www.reliefweb.int

The UN's Department of Humanitarian Affairs clearinghouse for international humanitarian emergencies presents daily news updates, including Reuters, VOA, PANA.

Social Science Information Gateway (SOSIG)
http://sosig.esrc.bris.ac.uk

The Economic and Social Research Council [ESRC] project catalogs 22 subjects and lists developing countries' URL addresses.

United Nations System
http://www.unsystem.org

The official Web site for the United Nations system of organizations can be found here. Everything is listed alphabetically, and examples include UNICC and the Food and Agriculture Organization.

UN Development Programme (UNDP)
http://www.undp.org

Publications and current information on world poverty, Mission Statement, UN Development Fund for Women, and more are available on this Web site. Be sure to see the Poverty Clock.

UN Environmental Programme (UNEP)
http://www.unep.org

This UNEP official site provides information on UN environmental programs, products, services, and events. A search engine is also available.

U.S. Agency for International Development (USAID)
http://www.usaid.gov/regions/afr/

The U.S. policy regarding assistance to African countries is presented at this site.

U.S. Central Intelligence Agency (CIA)
http://www.odci.gov

This site includes publications of the CIA, such as the World Fact Book, Factbook on Intelligence, Handbook of International Economic Statistics, and CIA maps.

U.S. Department of State
http://www.state.gov/countries/

Organized alphabetically, data on human rights issues, international organizations, and country reports as well as other data are available here.

World Bank Group
http://www.worldbank.org/html/extdr/regions.htm

News (press releases, summary of new projects, speeches), publications, topics in development, and reports on countries and regions can be accessed on this Web site. Links to other financial organizations are also provided.

World Health Organization (WHO)
http://www.who.int/en

Maintained by WHO's headquarters in Geneva, Switzerland, this site uses the Excite search engine to conduct keyword searches.

World Trade Organization (WTO)
http://www.wto.org

WTO's Web site topics include information on world trade systems, data on textiles, intellectual property rights, legal frameworks, trade and environmental policies, recent agreements, and other issues.

GENERAL AFRICA SITES

Africa News Web Site: Crisis in the Great Lakes Region
http://www.africanews.org/specials/greatlakes.html

African News Web Site on Great Lakes (Rwanda, Burundi, Zaire, and Kenya, Tanzania, Uganda) can be found here with frequent updates and good links to other sites. It is possible to order e-mail crisis updates here.

African Policy Information Center (APIC)
http://www.africapolicy.org

Developed by the Washington Office on Africa to widen policy debate in the United States on African issues, this Web site includes special topic briefs, regular reports, and documents on African politics.

Africa: South of the Sahara
http://www-sul.stanford.edu/depts/ssrg/africa/guide.html

On this site, Topics and Regions link headings will lead to a wealth of information.

African Studies WWW (U.Penn)
http://www.sas.upenn.edu/African_Studies/AS.html

This Web site provides facts about each African country, which includes news, statistics, and links to other Web sites.

AllAfrica Global Media
http://allafrica.com

From this page, explore African news by region or country. Topics covered include conflict and security; economy, business, and finance; environment and sustainable development; health; and human rights, plus many more.

Library of Congress Country Studies
http://lcweb2.loc.gov/frd/cs/cshome.html#toc

Of the 71 countries that are covered in the continuing series of books available at this Web site, at least a dozen of them are in Africa.

South African Government Index
http://www.polity.org.za/gnuindex.html

This official site includes links to government agencies, data on structures of government, and links to detailed documents.

Weekly Mail & Guardian [Johannesburg]
http://www.mg.co.za

This free electronic daily South African newspaper includes archived back issues as well as links to other related sites on Africa.

World History Archives
http://www.hartford-hwp.com/archives/

Hartford Web Publishing offers historical archives for the continent of Africa as a whole as well as for all the regions and countries in Africa. See individual country report pages for additional Web sites.

See individual country report pages for additional Web sites.

The United States (United States of America)

GEOGRAPHY

Area in Square Miles (Kilometers):
3,717,792 (9,629,091) (about 1/2 the size of Russia)

Capital (Population): Washington, DC (3,997,000)

Environmental Concerns: air and water pollution; limited freshwater resources, desertification; loss of habitat; waste disposal; acid rain

Geographical Features: vast central plain, mountains in the west, hills and low mountains in the east; rugged mountains and broad river valleys in Alaska; volcanic topography in Hawaii

Climate: mostly temperate, but ranging from tropical to arctic

PEOPLE

Population

Total: 280,563,000

Annual Growth Rate: 0.89%

Rural/Urban Population Ratio: 24/76

Major Languages: predominantly English; a sizable Spanish-speaking minority; many others

Ethnic Makeup: 77% white; 13% black; 4% Asian; 6% Amerindian and others

Religions: 56% Protestant; 28% Roman Catholic; 2% Jewish; 4% others; 10% none or unaffiliated

Health

Life Expectancy at Birth: 74 years (male); 80 years (female)

Infant Mortality: 6.69/1,000 live births

Physicians Available: 1/365 people

HIV/AIDS Rate in Adults: 0.61%

Education

Adult Literacy Rate: 97% (official)

Compulsory (Ages): 7–16; free

COMMUNICATION

Telephones: 194,000,000 main lines

Daily Newspaper Circulation: 238/1,000 people

Televisions: 776/1,000 people

Internet Users: 165,750,000 (2002)

TRANSPORTATION

Highways in Miles (Kilometers): 3,906,960 (6,261,154)

Railroads in Miles (Kilometers): 149,161 (240,000)

Usable Airfields: 14,695

Motor Vehicles in Use: 206,000,000

GOVERNMENT

Type: federal republic

Independence Date: July 4, 1776

Head of State/Government: President George W. Bush is both head of state and head of government

Political Parties: Democratic Party; Republican Party; others of relatively minor political significance

Suffrage: universal at 18

MILITARY

Military Expenditures (% of GDP): 3.2%

Current Disputes: various boundary and territorial disputes; "war on terrorism"

ECONOMY

Per Capita Income/GDP: $36,300/$10.082 trillion

GDP Growth Rate: 0%

Inflation Rate: 3%

Unemployment Rate: 5.8%

Population Below Poverty Line: 13%

Natural Resources: many minerals and metals; petroleum; natural gas; timber; arable land

Agriculture: food grains; feed crops; fruits and vegetables; oil-bearing crops; livestock; dairy products

Industry: diversified in both capital and consumer-goods industries

Exports: $723 billion (primary partners Canada, Mexico, Japan)

Imports: $1.148 trillion (primary partners Canada, Mexico, Japan)

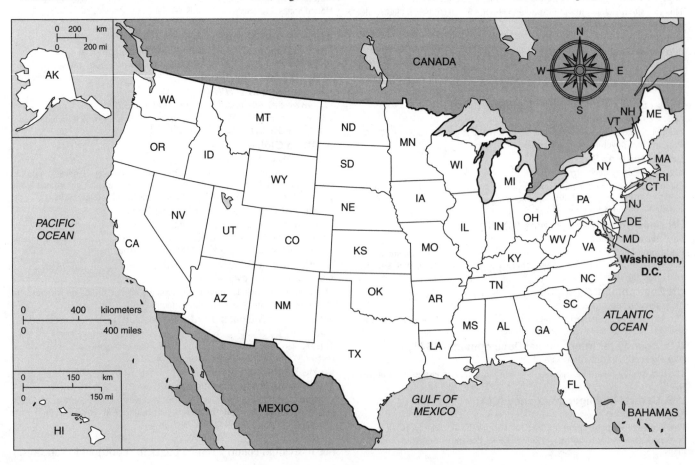

Canada

GEOGRAPHY

Area in Square Miles (Kilometers):
3,850,790 (9,976,140) (slightly larger than the United States)

Capital (Population): Ottawa (1,094,000)

Environmental Concerns: air and water pollution; acid rain; industrial damage to agriculture and forest productivity

Geographical Features: permafrost in the north; mountains in the west; central plains; lowlands in the southeast

Climate: varies from temperate to arctic

PEOPLE

Population

Total: 31,903,000

Annual Growth Rate: 0.96%

Rural/Urban Population Ratio: 23/77

Major Languages: both English and French are official

Ethnic Makeup: 28% British Isles origin; 23% French origin; 15% other European; 6% others; 2% indigenous; 26% mixed

Religions: 46% Roman Catholic; 36% Protestant; 18% others

Health

Life Expectancy at Birth: 76 years (male); 83 years (female)

Infant Mortality: 4.95/1,000 live births

Physicians Available: 1/534 people

HIV/AIDS Rate in Adults: 0.3%

Education

Adult Literacy Rate: 97%

Compulsory (Ages): primary school

COMMUNICATION

Telephones: 20,803,000 main lines

Daily Newspaper Circulation: 215/1,000 people

Televisions: 647/1,000 people

Internet Users: 16,840,000 (2002)

TRANSPORTATION

Highways in Miles (Kilometers): 559,240 (902,000)

Railroads in Miles (Kilometers): 22,320 (36,000)

Usable Airfields: 1,419

Motor Vehicles in Use: 16,800,000

GOVERNMENT

Type: confederation with parliamentary democracy

Independence Date: July 1, 1867

Head of State/Government: Queen Elizabeth II; Prime Minister Jean Chrétien

Political Parties: Progressive Conservative Party; Liberal Party; New Democratic Party; Bloc Québécois; Canadian Alliance

Suffrage: universal at 18

MILITARY

Military Expenditures (% of GDP): 1.1%

Current Disputes: maritime boundary disputes with the United States

ECONOMY

Currency ($U.S. equivalent): 1.46 Canadian dollars = $1

Per Capita Income/GDP: $27,700/$875 billion

GDP Growth Rate: 2%

Inflation Rate: 3%

Unemployment Rate: 7%

Labor Force by Occupation: 74% services; 15% manufacturing; 6% agriculture and others

Natural Resources: petroleum; natural gas; fish; minerals; cement; forestry products; wildlife; hydropower

Agriculture: grains; livestock; dairy products; potatoes; hogs; poultry and eggs; tobacco; fruits and vegetables

Industry: oil production and refining; natural-gas development; fish products; wood and paper products; chemicals; transportation equipment

Exports: $273.8 billion (primary partners United States, Japan, United Kingdom)

Imports: $238.3 billion (primary partners United States, European Union, Japan)

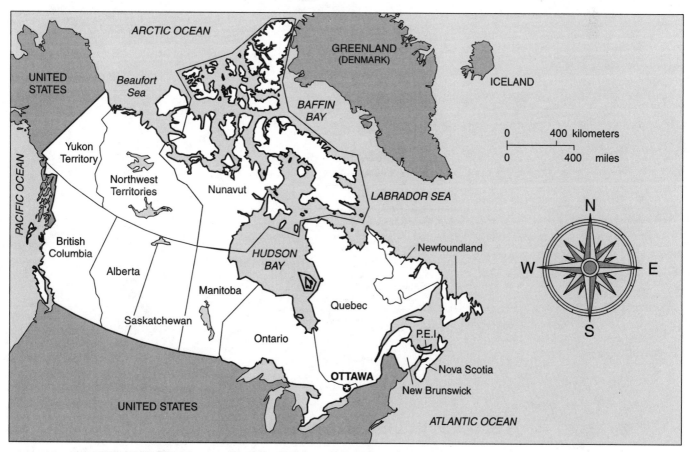

GLOBAL STUDIES

This map is provided to give you a graphic picture of where the countries of the world are located, the relationship they have with their region and neighbors, and their positions relative to major economic and political power blocs. We have focused on certain areas to illustrate these crowded regions more clearly. Africa is shaded for emphasis.

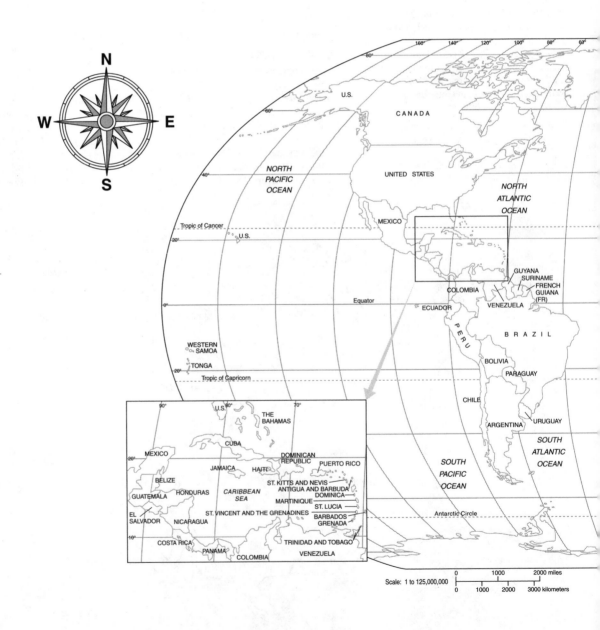

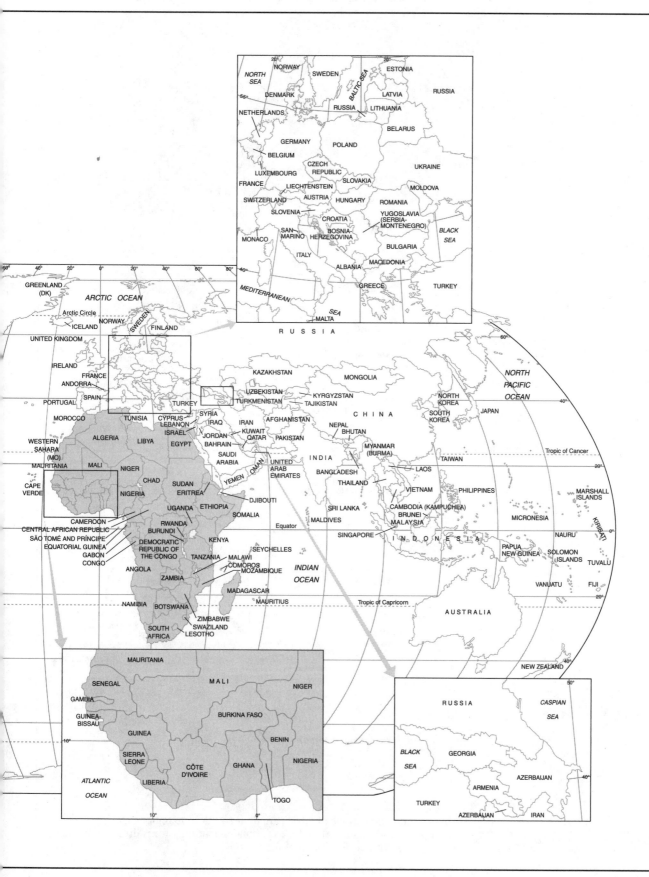

Africa

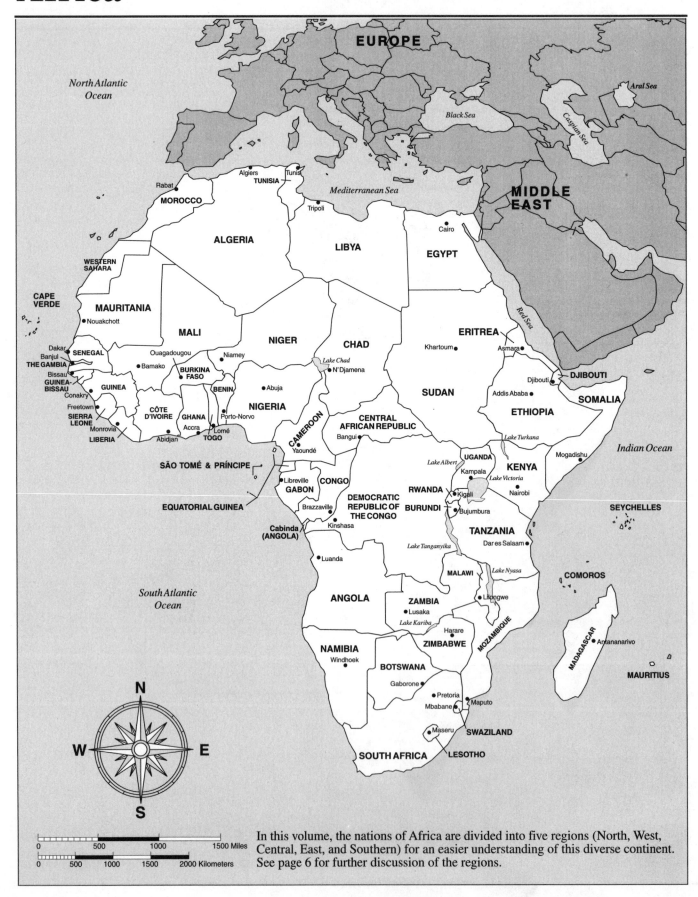

In this volume, the nations of Africa are divided into five regions (North, West, Central, East, and Southern) for an easier understanding of this diverse continent. See page 6 for further discussion of the regions.

Africa: Looking for a Renaissance

During the early 1960s, most of the African continent was liberated from colonial rule. Seventeen nations gained their independence in the great *Uhuru* ("Freedom") year of 1960 alone. The times were electric. In country after country, the banners of new states, whose leaders offered idealistic promises to remake the continent and thus the world, replaced the flags of various European states and the United Nations. Hopes were high, and even the most ambitious of goals seemed obtainable. Non-Africans also spoke of the resource-rich continent as being on the verge of a developmental takeoff. Some of the old racist myths about Africa were at last being questioned.

Yet four decades after the great freedom year, conditions throughout Africa are sobering rather than euphoric. For most Africans, independence has been a desperate struggle for survival rather than an exhilarating path to development. Nowadays Africa is often described in the global media as a "continent in crisis," a "region in turmoil," "on a precipice," and "suffering"—phrases that echo the sensationalist writings of nineteenth-century missionaries eager to convince others of the continent's need for "salvation." Unfortunately, the modern headlines are too often accurate, and certainly far more accurate than the mission tracts of yesteryear. Today, millions of Africans are indeed seeking some form of salvation from the grinding poverty, pestilence, and in many areas, wars that afflict their lives. Added to these miseries is the ongoing HIV/AIDS pandemic, which is devastating much of the continent; the majority of AIDS sufferers and fatalities have been African.

Perhaps the scale of Africa's maladies helps to explain why contemporary evangelists are so much more successful in swelling their congregations than were their counterparts in the past. It is certainly not for lack of competition; Africa is a continent of many, often overlapping faiths. In addition to Islam, Christianity, and other spiritual paths, Africans have embraced a myriad of secular ideologies: Marxism, African socialism, people's capitalism, structural adjustment, pan-Africanism, authenticity, nonracialism, the one-party state, and the multiparty state. The list is endless, but salvation seems ever more distant.

Africa's current circumstances are indeed difficult, yet it is also true that the postcolonial era has brought progress as well as problems. The goals so optimistically pronounced at independence have, for the most part, not been abandoned. Even when the states have faltered, the societies that they encompass have remained dynamic and adaptable to shifting opportunities. The support of strong families continues to allow most Africans to overcome enormous adversity. There are starving children in Africa today, but there are also many more in school uniforms studying to make their dreams a reality. Africa as a whole remains a dynamic, ever-changing continent that in recent years has seen much progress as well as instances of regression. For example, the use of mobile wireless telephones (cell phones) has spread across the continent like wildfire, bringing mass-communications capacity to millions for the first time. While most Africans remain on the other side of the "digital divide" in terms of their capacity to participate in the "global information age," the introduction of new information technologies is bringing change. Textile businesses in such diverse places as Mauritius, Lesotho, and Ghana use the new technologies to monitor their niche markets in Europe and North America on a daily basis. This trend will probably grow if trade access between the developed and developing world continues to be liberalized through the World Trade Organization (WTO) and other multilateral agreements such as the African Growth and Opportunity Act (AGOA) and the Cotonou Convention, which, respectively, have helped to open up U.S. and European Union markets to African goods.

It is also worth noting that most of what appears in the global media about Africa is but a snapshot of the continent in moments of catastrophic deprivation, war, and degradation. The global flow of information is dominated by a small number of media companies that must for the most part capture the eyes and ears of consumers in the world's wealthier societies to capture market share and sell advertising. Incremental progress in improving the quality of life for ordinary Africans is rarely perceived by these media companies as newsworthy. Thus, events such as the building of a new school, road, or water system that can dramatically transform the lives of people in a particular African locality are rarely of interest to the international broadcasters that increasingly influence what appears on the air waves of Africa itself as well as the rest of the world. Outsiders also often simply lose sight of the fact that Africa is a vast continent and not a country.

It would be naïve to downplay the serious challenges that face what is, in per capita economic terms, the world's least-developed continent. According to the 2002 United Nations Human Development Report, 39 of the world's 50 least developed countries are in Africa. But it is also worth appreciating that, with the passing of time, many of the ills that were inherited from colonialism have been remedied. In the process, new perspectives on how to move forward are being tested. The corruption and lack of capacity that characterize many African governments are mitigated by the resilience of local communities. Africans not only seek salvation from deprivation but also actively work for the delivery of a better tomorrow in which their children can enjoy the material benefits of being participants, rather than onlookers, in their continent's integration into the process of globalization.

A prerequisite for the realization of such dreams is the maintenance of peace. Without peace, there can be no prospect of freedom and development. In this context, recent progress in bringing an end to the armed conflicts in Angola, Sierra Leone, the Horn of Africa, and, perhaps most significantly, Congo, along with the consolidation of stability in such other areas as Mozambique and South Africa, have become sources of renewed hope.

If the momentum for peace can be maintained by the continent's growing ranks of legitimately elected leaders, ongoing talk among statespeople and intellectuals about the onset of an era of rebirth for the continent will appear less as a pipedream and more as a realistic possibility. Such a renaissance is predicated upon faith in the commitment of post–Cold War Africa's

WOULD YOU BELIEVE?

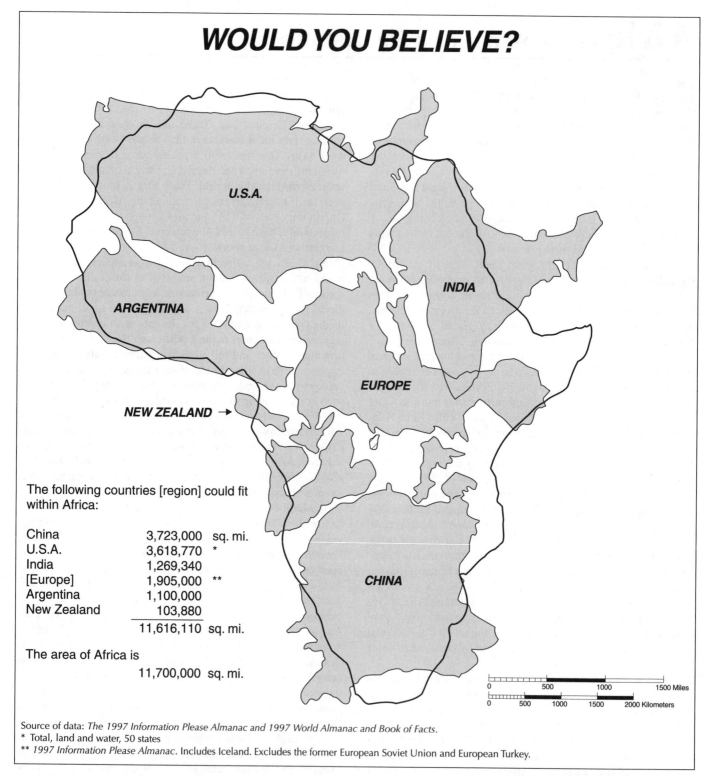

The following countries [region] could fit within Africa:

China	3,723,000	sq. mi.
U.S.A.	3,618,770	*
India	1,269,340	
[Europe]	1,905,000	**
Argentina	1,100,000	
New Zealand	103,880	
	11,616,110	sq. mi.

The area of Africa is

11,700,000 sq. mi.

Source of data: *The 1997 Information Please Almanac and 1997 World Almanac and Book of Facts.*
* Total, land and water, 50 states
** *1997 Information Please Almanac.* Includes Iceland. Excludes the former European Soviet Union and European Turkey.

capacity to resolve its internal problems in partnership with, but without undue interference from, nations outside Africa that have in the past contributed to the continent's instability.

A DIVERSE CONTINENT

Africa, which is almost four times the size of the United States (excluding Alaska), ranks just below Asia as the world's biggest continent. Well over one quarter of the membership of the United Nations consists of African states—53 in all. Such facts are worth noting, for even well-educated outsiders often lose sight of Africa's continental scope when they discuss its problems and prospects.

Not only is the African continent vast but, archaeology tells us, it was also the cradle of human civilization. (New evidence suggests that today's non-Africans are the descendants of Africans who moved out of the continent as recently as 80,000 years ago.) It should therefore not be surprising that the ways of life as well as genetic makeup of the 800 million or so contemporary inhabitants of Africa are characterized by extraordinary diversity. They speak more than 1,000 languages and live their lives according to a rich variety of household arrangements, kinship systems, and religious beliefs. The art and music styles of the continent are as varied as the people.

Given its diversity, it is not easy to generalize about Africa. For each statement, there is an exception. However, one aspect that is constant to all African societies is that they have always been changing, albeit in modern times at an ever-increasing rate. Cities have grown and people have moved back and forth between village and town, giving rise to new social groups, institutions, occupations, religions, and forms of communication that have made their mark in the countryside as well as in the urban centers. All Africans, whether they be urban computer programmers or hunter-gatherers living in the remote corners of the Kalahari Desert, have taken on new practices, interests, and burdens, yet they have retained their African identity. Uniquely African institutions, values, and histories underlie contemporary lifestyles throughout the continent.

Memories of past civilizations are a source of pride and community. The medieval Mali and Ghana empires, the glory of Pharaonic Egypt, the Fulani caliphate of northern Nigeria, the knights of Kanem and Bornu, the Great Zimbabwe, and the Kingdom of the Kongo, among others, are all remembered. The past is connected to the present through the generations and by ties to the land. In a continent where the majority of people are still farmers, land remains "the mother that never dies." It is valued for its fruits and because it is the place to which the ancestors came and were buried.

The art of personal relationships continues to be important. People typically live in large families. Children are considered precious, and large families are still desired for social as well as economic reasons. Elders are an important part of a household; nursing homes and retirement communities generally do not exist. People are not supposed to be loners. "I am because we are" remains a valued precept. In this age of nation-states, the "we" may refer to one's ethnic community, while obligations to one's extended family often take precedence over other loyalties.

Most Africans believe in a spiritual as well as a material world. The continent contains a rich variety of indigenous belief systems, which often coexist with the larger religions of Islam and various Christian sects. Many families believe that their lives are influenced by their ancestors. Africans from all walks of life will seek the services of professional "traditional" healers to explain an illness or suggest remedies for such things as sterility or bad fortune. But this pattern of behavior does not preclude one from turning to scientific medicine; all African governments face strong popular demands for better access to modern health-care facilities.

Islam has long been a strong force in Africa. Today that religion rivals Christianity as the fastest-growing faith on the continent. The followers of both religions often adapt their faiths to accommodate local traditions and values. Some people also join new religious movements and churches, such as the Brotherhood of the Cross and Star in Nigeria or the Church of Simon Kimbangu in Congo, that link Christian and indigenous beliefs with new ideas and rituals. Like other institutions in the towns and cities, the churches and mosques provide their followers with social networks.

Local art, like local religion, often reflects the influence of the changing world. An airplane is featured on a Nigerian gelede mask, the Apollo space mission inspires a Burkinabe carver, and an Ndebele dance wand is a beaded electric pole.

THE TROUBLED PRESENT

Some of the crises in Africa today threaten its peoples' traditional resiliency. The facts are grim: In material terms, the average African is poorer today than at independence, and it is predicted that poverty will only increase in the immediate future. Drought conditions in recent decades have led to food shortages across the continent. In the 1980s, widespread famine occurred in 22 African nations; the Food and Agriculture Organization (FAO) of the United Nations estimated that 70 percent of all Africans did not have enough to eat. An outpouring of assistance and relief efforts at the time saved as many as 35 million lives. Overall per capita food production in Africa dropped by 12 percent between 1961 and 1995. One factor in the decline has been the tendency of agricultural planners to ignore the fact that up to 70 percent of Africa's food crops are grown by women. It has also been estimated that up to 40 percent of the continent's food crops go uneaten as a result of inadequate transport and storage facilities. Although agricultural production rose modestly in the late 1990s, the food crisis continues. Since 1990, large parts of East and Southern Africa, in particular, have faced the prospect of renewed hunger. Other areas have become dependent on outside food aid. Declining commodity prices on world markets, explosive population growth, and recurring drought and locust infestations have often counterbalanced marginal advances in agricultural production through better incentives to farmers. Problems of climate irregularity, and obtaining and transporting needed goods and supplies require continued assistance and long-range planning. Wood, the average person's source of energy, grows scarcer every year, and most governments have had to contend with the rising cost of imported fuels.

Armed conflicts have devastated portions of Africa. Recent carnage in Angola, Djibouti, Sierra Leone, Somalia, and the Democratic Republic of the Congo (D.R.C., formerly Zaire), due to internal strife encouraged to greater or lesser degrees by outside forces, places them in a distinct class of suffering—a class that has also included (and that may yet again) Chad, Eritrea, Ethiopia, Mozambique, and Rwanda. By the end of 2002, civil war persisted in Burundi, Liberia, Sudan, Central African Republic, and northern Uganda, while threatening to ignite in long-stable Côte d'Ivoire. More than 3 million people have died in these countries over the past decade, while millions more have become refugees. Except for scattered enclaves, normal economic activities have been greatly disrupted or have ceased altogether.

Almost all African governments are deeply in debt. In 1991, the foreign debt owed by all the sub-Saharan African countries except South Africa already stood at about $175 billion. Although it is smaller in its absolute amount than that of Latin America, as a percentage of its economic output the continent's debt is the highest in the world and is rising swiftly. The combined gross national product (GNP) for the same countries,

MEASURING MISERY

United Nations Human Development Index, 2002*: Ranking of African States Among World Nations. Total Number of States Ranked: 173. African States Not Ranked: Liberia, Somalia

Rank	Country	Rank	Country
47	Seychelles	149	Djibouti
64	Libya	150	Uganda
67	Mauritius	151	Tanzania
100	Cape Verde	152	Mauritania
106	Algeria	153	Zambia
107	South Africa	154	Senegal
111	Equatorial Guinea	155	Democratic Republic of the Congo
115	Egypt	156	Côte d'Ivoire
117	Gabon	157	Eritrea
119	Sao Tome and Principe	158	Benin
122	Namibia	159	Guinea
123	Morocco	160	The Gambia
125	Swaziland	161	Angola
126	Botswana	162	Rwanda
128	Zimbabwe	163	Malawi
129	Ghana	164	Mali
132	Lesotho	165	Central African Republic
134	Kenya	166	Chad
135	Cameroon	167	Guinea-Bissau
136	Republic of Congo	168	Ethiopia
137	Comoros	169	Burkina Faso
139	Sudan	170	Mozambique
141	Togo	171	Burundi
147	Madagascar	172	Niger
148	Nigeria	173	Sierra Leone

*Standings among 173 countries, with the ranking 173 indicating the lowest development.

Source: UNDP, Human Development Report 2002.

questioning both the justice and practicality of these terms. In response there have been recent, though many would argue insufficient, moves by donor countries to arrange for debt relief in the poorest countries.

Another factor that helps to account for Africa's relative poverty is the low level of industrial output of all but a few of its countries. The decline of many commodity prices on the world market has further reduced national incomes. As a result, the foreign exchange needed to import food, machinery, fuel, and other goods is very limited in most African countries. In 1987, the continent's economy grew by only 0.8 percent, far below its population growth rate of about 3.2 percent. In the same year, cereal production declined 8 percent and overall agricultural production grew by only 0.5 percent. There has been some modest improvement in subsequent years. Recent estimates put the continent's economic growth rate at 1.5 percent—still the world's lowest, and far below that of the population growth rate. Perhaps more significant has been the high growth that has been recorded in those states, like Mozambique, that have managed to move away from civil war to governance based on democratic consensus.

But perhaps the greatest challenge facing Africa today is in the area of health. The HIV/AIDS pandemic has spread at an alarming rate over the past two decades, already claiming millions of lives and threatening millions more. Southern Africa has been especially hard hit in recent years, with more than one in five adults said to be HIV-positive in a number of countries. As a result, estimates of life expectancy have been declining dramatically. Botswana is a notable example. By the early 1990s, average overall life expectancy in the nation had risen to about 65 years, due to sustained public investment in providing universal access to primary health care. But the most recent estimate presented at the 2002 United Nations AIDS Conference in Spain suggest that average life expectancy in Botswana could drop to as little as 27 years by 2010. Put another way, unless the pandemic can be brought under control, it has been estimated that AIDS will claim the lives of one out of every two Batswana (as the people of Botswana are known) born in the new millennium. The battle against AIDS in Africa has been complicated by the existence of multiple strains of the HIV virus, of which the most virulent is currently concentrated in Southern Africa. The spread of HIV/AIDS has contributed to the resurgence of diseases such as tuberculosis. Mortality due to malaria—Africa's traditional scourge—has also been rising in many areas, as has cholera. Diseases afflicting livestock, such as rinderpest and foot-and-mouth disease, have also been making a comeback in certain regions.

THE EVOLUTION OF AFRICA'S ECONOMIES

Africa has seldom been rich, although it has vast resources, and some rulers and other elites have become very wealthy. In earlier centuries, the slave trade greatly contributed to limiting economic development in many African regions. During the period of European exploration and colonialism, Africa's involvement in the world economy greatly increased with the emergence of new forms of "legitimate" commerce. But colonial-era policies and practices assured that this development was of little long-term benefit to most of the continent's peoples.

whose total populations are in excess of 500 million, was less than $150 billion, a figure that represents only 1.2 percent of the global GNP and is about equal to the GNP of Belgium, a country of 10 million people. In Zambia, an extreme example, the per capita foreign debt theoretically owed by each of its citizens is nearly $1,000, more than twice its annual per capita income.

In order to obtain money to meet debts and pay for their running expenses, many African governments have been obliged to accept the stringent terms of global lending agencies, most notably the World Bank and the International Monetary Fund (IMF). These lending terms have led to great hardship, especially in urban areas, through austerity measures such as the abandonment of price controls on basic foodstuffs and the freezing of wages. Many African governments and experts are

THE AIDS PANDEMIC

Perhaps the greatest challenge currently facing Africa is the spread of HIV/AIDS. The statistics are chilling. According to the United Nations, of the 36 million people worldwide living with the HIV virus, some 24 million live in sub-Saharan Africa. Of the 5.3 million new infections estimated for the year 2000, 3.8 million were in Africa. AIDS is already the leading cause of death on the continent. In all, 2.4 million AIDS-related deaths were recorded for Africa in 2000, representing about 80 percent of the worldwide total.

AIDS-related fatalities have also resulted in a rapidly growing number of "AIDS orphans." According to Kingsley Amoako, executive secretary of the Economic Commission for Africa, more than 12 million children have been orphaned in Africa due to AIDS (out of the global estimate of just over 13 million). Speaking at a gathering of African leaders in November 2000, Amoako noted that, "Within the next 10 years, it is projected that there will be 40 million AIDS orphans in Africa.... The AIDS pandemic is undermining social and economic structures and reversing the fragile gains made since independence... in parts of Africa, AIDS is killing one in every three adults, making orphans out of every tenth child and decimating entire communities."

The worst-hit parts of the continent in recent years have been East and Southern Africa, with some countries having infection rates of more than a fifth of their adult populations. According to published figures, the most affected countries are Botswana, South Africa, and Zimbabwe, where it is currently estimated that one in every two people under age 15 could die from the disease.

Inevitably, the spread of HIV/AIDS is having a devastating impact on economic and social development. For example, it is estimated that in the next decade, South Africa's gross domestic product will be 17 percent lower than it would have been without the pandemic.

Amid the gloom there is, nonetheless, grounds for hope that the scourge can ultimately be overcome. According to a UN report issued at the end of 2000, some parts of the continent are finally seeing a decrease in new HIV cases. The report notes that this has resulted in a modest overall decrease in the total number of new HIV cases in Africa as a whole. The decrease has been partially attributed to the gradual success of prevention programs, especially in the East African countries.

Africa's ability to fight HIV/AIDS is compromised by its debt burden and the high cost of HIV/AIDS treatment drugs. One of the most outspoken figures on the relationship between disease and debt on the continent has been Botswana's president, Festus Mogae. In a direct appeal to the wealthier nations, he observed:

> Your wealth in recent years increased by trillions and therefore what we owe is peanuts. It will not affect anybody, not the balance sheets of banks or anybody. It's just a matter of principle. You are insisting on repayment as a matter of principle, but it has no financial consequences for anybody else except the debtor. For him it's a lot of money.... Pharmaceutical companies have come forward and offered us discounts. Some of these discounts are very generous but are still more than our faint means can allow us to afford, and therefore we are still not able to take full advantage of the offer.... We are saying the rest of the world, including and especially the United States and the rest of the G-7, at the governmental level should do something to make it possible for us to access these treatments that are currently available.

During the 70 or so years of European colonial rule over most of Africa, its nations' economies were shaped to the advantage of the imperialists. Cash crops such as cocoa, coffee, and rubber began to be grown for the European market. Some African farmers benefited from these crops, but the cash-crop economy also involved large foreign-run plantations. It also encouraged the trends toward use of migrant labor and the decline in food production. Many people became dependent for their livelihood on the forces of the world market, which, like the weather, were beyond their immediate control.

Mining also increased during colonial times, again for the benefit of the colonial rulers. The ores were extracted from African soil by European companies. African labor was employed, but the machinery came from abroad. The copper, diamonds, gold, iron ore, and uranium were shipped overseas to be processed and marketed in the Western economies. Upon independence, African governments received a varying percentage of the take through taxation and consortium agreements. But mining remained an enclave industry, sometimes described as a "state within a state" because such industries were run by outsiders who established communities that used imported machinery and technicians and exported the products to industrialized countries.

Inflationary conditions in other parts of the world have had adverse effects on Africa. The raw materials that Africans export today often receive low prices on the world market, while the manufactured goods that African countries import are expensive. Local African industries lack spare parts and ma-

chinery, and farmers frequently cannot afford to transport crops to market. As a result, the whole economy slows down. Thus, Africa, because of the policies of former colonial powers and current independent governments, is tied into the world economy in ways that do not always serve its peoples' best interests.

THE PROBLEMS OF GOVERNANCE

Outside forces are not the only cause of Africa's current crises. Too often, Africa has been a misgoverned continent. After independence, the idealism that characterized various nationalist movements, with their promises of popular self-determination, gave way in most states to cynical authoritarian regimes. By 1989, only Botswana, Mauritius, soon-to-be-independent Namibia, and, arguably, The Gambia and Senegal could reasonably claim that their governments were elected in genuinely free and fair elections.

During the 1980s, the government of Robert Mugabe in Zimbabwe, in Southern Africa, undoubtedly enjoyed majority support, but political life in the country was already seriously marred by violence and intimidation aimed at the Mugabe regime's potential opposition. Past multiparty contests in the North African nations of Egypt, Morocco, and Tunisia, as well as in the West African state of Liberia, had been manipulated to assure that the ruling establishments remained unchallenged. Elsewhere, the continent was divided between military and/or one-party regimes, which often combined the seemingly contradictory characteristics of weakness and absolutism at the top. While a few of the one-party states, most notably Tanzania, then offered people genuine, if limited, choices of leadership, most were, to a greater or lesser degree, simply vehicles of personal rule.

But since 1990 there has been a democratic reawakening in Africa, which has toppled the political status quo in some areas and threatened its survival throughout the continent. Whereas in 1989 some 35 nations were governed as single-party states, by 1994 there were none, though Swaziland and Uganda were experimenting with no-party systems. In a number of countries— Benin, Cape Verde, Central African Republic, Congo, Madagascar, Mali, Malawi, Niger, São Tomé and Príncipe, Senegal, South Africa, and Zambia—ruling parties were decisively rejected in multiparty elections, while elections in other areas led to a greater sharing of power between the old regimes and their formerly suppressed oppositions.

In many countries, the democratic transformation is still ongoing and remains fragile. There have been accusations of manipulation and voting fraud in a growing number of countries in the past decade; while in Algeria, The Gambia, Niger, Nigeria, and Sierra Leone, the seeming will of the electorates has been overridden by military coups.

A very fragile democracy has since been restored to Nigeria, Africa's most populous state, while military rule in Sierra Leone has given way to an ongoing attempt to restore a democratic consensus through UN–sanctioned intervention by international peacekeeping forces. In The Gambia, where three decades of multiparty democracy were ended through a military coup, there have also been elections. But their legitimacy has been questioned.

Events in Benin have most closely paralleled the recent changes of Central/Eastern Europe. Benin's military-based, Marxist-Leninist regime of Mathieu Kérékou was pressured into relinquishing power to a transitional civilian government made up of technocrats and former dissidents. (Television broadcasts of this "civilian coup" enjoyed large audiences in neighboring countries.) In several other countries, such as Equatorial Guinea, Gabon, and Togo, mounting opposition has resulted in the semblance without the substance of free elections by long-ruling military autocrats. In the Democratic Republic of the Congo, attempts to establish a framework for reform through a multiparty consultative conference were overshadowed by the almost complete collapse of state structures. A victory by externally backed rebels in 1997 was followed by renewed civil war and foreign intervention. Continued conflict in the country has contributed to the further destabilization of neighboring states. Many people fear that these countries may soon experience turmoil similar to that which has engulfed Ethiopia, Liberia, Rwanda, Somalia, and Uganda, where military autocrats have been overthrown by armed rebels.

Why did most postcolonial African governments, until recently, take on autocratic forms? And why are these forms now being so widely challenged? There are no definitive answers to either of these questions. One common explanation for authoritarianism in Africa has been the weakness of the states themselves. Most African governments have faced the difficult task of maintaining national unity with diverse, ethnically divided citizenries. Although the states of Africa may overlay and overlap historic kingdoms, most are products of colonialism. Their boundaries were fashioned during the late-nineteenth-century European partition of the continent, which divided and joined ethnic groups by lines drawn in Europe. The successful leaders of African independence movements worked within the colonial boundaries; and when they joined together in the Organization of African Unity (OAU), they agreed to respect the territorial status quo.

While the need to stem interethnic and regional conflict has been one justification for placing limits on popular self-determination, another explanation can be found in the administrative systems that the nationalist leaderships inherited. All the European colonies in Africa functioned essentially as police states. Not only were various forms of opposition curtailed, but intrusive security establishments were created to watch over and control the indigenous populations. Although headed by Europeans, most colonial security services employed local staff members who were prepared to assume leadership roles at independence. A wave of military coups swept across West Africa during the 1960s; elsewhere, aspiring dictators like "Life President" Ngwazi Hastings Banda of Malawi were quick to appreciate the value of the inherited instruments of control.

Africa's economic difficulties have also frequently been cited as contributing to its political underdevelopment. On one hand, Nigeria's last civilian government, for example, was certainly undermined in part by the economic crisis that engulfed it due to falling oil revenues. On the other hand, in a pattern reminiscent of recent changes in Latin America, economic difficulties resulting in high rates of inflation and indebtedness seem to be tempting some African militaries, such as Benin's, to return to the barracks and allow civilian politicians to assume responsibility for the implementation of inevitably harsh austerity programs.

External powers have long sustained African dictatorships through their grants of military and economic aid—and, on occasion, direct intervention. For example, a local attempt in 1964 to restore constitutional rule in Gabon was thwarted by French paratroopers, while Joseph Desiré Mobutu's kleptocratic hold over Zaire (now the D.R.C.) relied from the very beginning on overt and covert assistance from the United States and other Western states. The former Soviet bloc and China also helped in the past to support their share of unsavory African allies, in places like Ethiopia, Equatorial Guinea, and Burundi. But the end of the Cold War has led to a reduced desire on the part of outside powers to prop up their unpopular African allies. At the same time, the major international lending agencies have increasingly concerned themselves with the perceived need to adjust the political as well as economic structures of debtor nations. This new emphasis is justified in part by the alleged linkage between political unaccountability and economic corruption and mismanagement.

The ongoing decline of socialism on the continent is also having a significant political effect. Some regimes have professed a Marxist orientation, while others have felt that a special African socialism could be built on the communal and cooperative traditions of their societies. In countries such as Guinea-Bissau and Mozambique, a revolutionary socialist orientation was introduced at the grassroots level during the struggles for independence, within areas liberated from colonialism. The various socialist governments have not been free of personality cults, nor from corruption and oppressive measures. And many governments that have eschewed the socialist label have, nonetheless, developed public corporations and central-planning methods similar to those governments that openly profess Marxism. In recent years, virtually all of Africa's governments, partly in line with IMF and World Bank requirements but also because of the inefficiency and losses of many of their public corporations, have placed greater emphasis on private-sector development.

REASONS FOR OPTIMISM

Although the problems facing African countries have grown since independence, so have the continent's collective achievements. The number of people who can read and write in local languages as well as in English, French, or Portuguese has increased enormously. More people can peruse newspapers, follow instructions for fertilizers, and read the labels on medicine bottles. Professionals trained in modern technology who, for example, plan electrification schemes, organize large office staffs, or develop medical facilities are more available because of the large number of African universities that have developed since the end of colonialism. Health care has also expanded and improved in most areas. Outside of the areas that have been ravaged by war, life expectancy has generally increased and infant mortality rates have declined.

The problems besetting Africa have caused deep pessimism in some quarters, with a few observers going so far as to question whether the postcolonial division of the continent into multiethnic states is viable. But the states themselves have proved to be surprisingly resilient. Central authority has reemerged in such traumatized, once seemingly ungovernable countries as Uganda, Mozambique, and most recently Sierra Leone. Despite the terrible wars that are still being waged in a few nations, mostly in the form of civil wars, postwar African governments have been notably successful in avoiding armed conflict with one another (although this fact has been severely tested by external interventions in the Democratic Republic of the Congo). Of special significance is South Africa's recent transformation into a nonracial democracy, which has been accompanied by its emergence as the leading member of the Southern African Development Community and has brought an end to its previous policy of regional destabilization.

Another positive development is the increasing attention that African governments and intra-African agencies are giving to women, as was exemplified in a global population summit held in Cairo, Egypt, in 1994. The pivotal role of women in agriculture and other activities is increasingly being recognized and supported. In many countries, prenatal and hospital care for mothers and their babies have increased, conditions for women workers in factories have improved, and new cooperatives for women's activities have been developed. Women are also playing a more prominent role in the political life of many African countries.

The advances that have been made in Africa are important ones, but they could be undercut by continued economic decline. Africa needs debt relief and outside aid just to maintain the gains that have been made. Yet as an African proverb observes, "Someone else's legs will do you no good in traveling." Africa, as the individual country reports in this volume observe, is a continent of many and varied resources. There are mineral riches and a vast agricultural potential. However, the continent's people, the youths who make up more than half the population and the elders whose wisdom is revered, are its greatest resource. The rest of the world, which has benefited from the continent's material resources, can also learn from the social strengths of African families and communities.

Central Africa

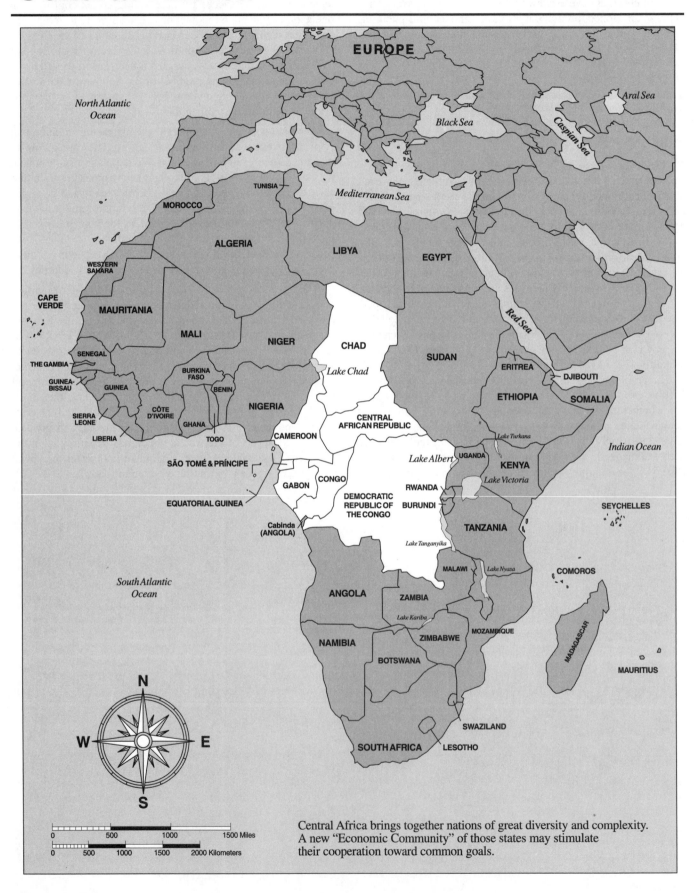

Central Africa brings together nations of great diversity and complexity. A new "Economic Community" of those states may stimulate their cooperation toward common goals.

Central Africa: Possibilities for Cooperation

The Central African region, as defined in this book, brings together countries that have not shared a common past; nor do they necessarily seem destined for a common future. Cameroon, Chad, Central African Republic, Congo (Congo-Brazzaville), the Democratic Republic of the Congo (the D.R.C., or Congo-Kinshasa, formerly known as Zaire), Equatorial Guinea, Gabon, and São Tomé and Príncipe are not always grouped together as one region. Indeed, users of this volume who are familiar with the continent may also associate the label "Central Africa" with such states as Angola and Zambia rather than with some of the states mentioned here. Geographically, Chad is more closely associated with the Sahelian nations of West Africa than with the heavily forested regions of Central Africa to its south. Similarly, the southern part of Democratic Republic of Congo has longstanding cultural and economic links with Angola and Zambia, which in this text are associated with the states of Southern Africa, largely because of their political involvements.

Yet the eight countries that are designated here as belonging to Central Africa have much in common. French is a predominant language in all the states except Equatorial Guinea and São Tomé and Príncipe. All except São Tomé and Príncipe and the Democratic Republic of the Congo share a common currency, the CFA franc. And while Chad's current economic prospects appear to be exceptionally poor, the natural wealth found throughout the rest of Central Africa makes the region as a whole one of enormous potential. Finally, in the postcolonial era, all the Central African governments have made some progress in realizing their developmental possibilities through greater regional cooperation.

The countries of Central Africa incorporate a variety of peoples and cultures, resources, environments, systems of government, and national goals. Most of the modern nations overlay both societies that were village-based and localized, and societies that were once part of extensive state formations. Islam has had little influence in the region, except in Chad and northern Cameroon. In most areas, Christianity coexists with indigenous systems of belief. Sophisticated wooden sculptures are one of the cultural achievements associated with most Central African societies. To many people, the carvings are only material manifestations of the spiritual potential of complex local cosmologies. However, the art forms are myriad and distinctive, and their diversity is as striking as the common features that they share.

The postcolonial governments of Central Africa have ranged from outwardly conservative regimes (in Gabon and the Democratic Republic of Congo) to self-proclaimed revolutionary Marxist-Leninist orders (in Congo and São Tomé and Príncipe). More fundamentally, all of the states in the region have in the past fallen under the control of unelected autocracies, whose continued existence has been dependent on the coercive capacities of military forces—sometimes external ones. But the authoritarian status quo has been challenged in recent years. In Central African Republic, Congo, and São Tomé and Príncipe, democratic openings have resulted in the peaceful election of

(United Nations photo)

In Africa, cooperative work groups such as the one pictured above often take on jobs that would be done by machinery in industrialized countries.

new governments. However, the elected government in Congo has since been overthrown by forces loyal to its former dictator. Elsewhere in the region, opposition parties have been legalized but otherwise have made limited progress.

GEOGRAPHIC DISTINCTIVENESS

All the states of the Central African region except Chad encompass equatorial rain forests. Citizens who live in these regions must cope with a climate that is hot and moist while facing the challenges of utilizing (and in some cases, unfortunately, clearing) the resources of the great forests. The problems of living in these areas account, in part, for the relatively low, albeit growing, population densities of most of the states. The difficulty

21

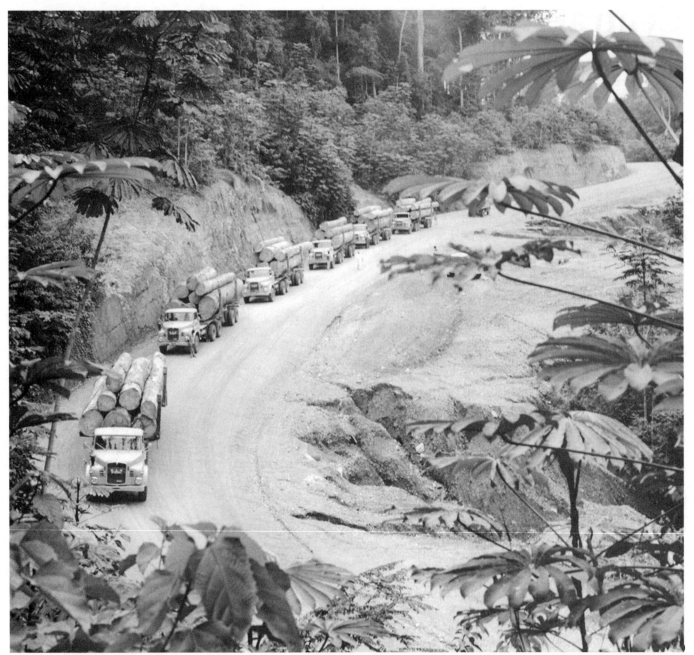

(United Nations photo)

Regional cooperation will be necessary to utilize the natural resources of Central African countries without damaging their precious but fragile environment. Environmentalists warn that the Central African forests are rapidly being destroyed.

of establishing roads and railroads impedes communication and thus economic development. The peoples of the rain-forest areas tend to cluster along riverbanks and existing rail lines. In modern times, largely because of the extensive development of minerals, many inhabitants have moved to the cities, accounting for a comparatively high urban population in all the states.

Central Africa's rivers have long been its lifelines. The watershed in Cameroon between the Niger and Congo (or Zaire) Rivers provides a natural divide between the West and Central African regions. The Congo River is the largest in the region, but the Oubangi, Chari, Ogooue, and other rivers are important also for the communication and trading opportunities they offer. The rivers flow to the Atlantic Ocean, a fact that has en-

couraged the orientation of Central Africa's external trade toward Europe and the Americas.

Many of the countries of the region have similar sources of wealth. The rivers are capable of generating enormous amounts of hydropower. The rain forests are also rich in lumber, which is a major export of most of the countries. Other forest products, such as rubber and palm oil, are widely marketed.

Extensive lumbering and clearing activities for agriculture, have created worldwide concern about the depletion of the rain forests. As a result, in recent years there have been some organized boycotts in Europe of the region's hardwood exports, although far more trees are felled to process plywood.

(United Nations photo)

Refugees from the conflict in the Republic of the Congo are seen here in the make-shift living quarters that they erected in the area allotted to them outside Elizabethville.

As one might expect, Central Africa as a whole is one of the areas least affected by the drought conditions that periodically plague Africa. Nevertheless, serious drought is a well-known visitor in Chad, Central African Republic, and the northern regions of Cameroon, where it contributes to local food shortages. Savanna lands are found in some parts lying to the north and the south of the forests. Whereas rain forests have often inhibited travel, the savannas have long been transitional areas, great avenues of migration linking the regions of Africa, while providing agricultural and pastoral opportunities for their residents.

The Central African countries share other resources besides the products of the rain forest. Cameroon, Congo, and Gabon derive considerable revenues from their petroleum reserves. Other important minerals include diamonds, copper, gold, manganese, and uranium. The processes involved in the exploitation of these commodities, as well as the demand for them in the world market, are issues of common concern among the producing nations. Many of the states also share an interest in exported cash crops such as coffee, cocoa, and cotton, whose international prices are subject to sharp fluctuations. The similarity of their environments and products provides an economic incentive for Central African cooperation.

LINKS TO FRANCE

Many of the different ethnic groups in Central Africa overlap national boundaries. Examples include the Fang, who are found in Cameroon, Equatorial Guinea, and Gabon; the Bateke of Congo and Gabon; and the BaKongo, who are concentrated in Angola as well as in Congo and the Democratic Republic of the Congo. Such cross-border ethnic ties are less important as sources of regional unity than the European colonial systems that the countries inherited. While Equatorial Guinea was controlled by Spain, São Tomé and Príncipe by Portugal, and the D.R.C. by Belgium, the predominant external power in the region remains France. Central African Republic, Chad, Congo, and Gabon were all once part of French Equatorial Africa. Most of Cameroon was also governed by the French, who were awarded the bulk of the former German colony of the Kamerun as a "trust territory" in the aftermath of World War I. French administration provided the five states with similar colonial experiences.

Early colonial development in the former French colonies and the Democratic Republic of the Congo was affected by European "concessions" companies, institutions that were sold extensive rights (often 99-year leases granting political as well as

23

(United Nations photo)

Here, in the Democratic Republic of the Congo, a bulldozer levels an access road to the Kamnyola Bridge over the Livumvi River.

economic powers) to exploit such local products as ivory and rubber. At the beginning of the twentieth century, just 41 companies controlled 70 percent of all of the territory of contemporary Central African Republic, Congo, and Gabon. Mining operations as well as large plantations were established that often relied on forced labor. Individual production by Africans was also encouraged, often through coercion rather than economic incentives. While the colonial companies encouraged production and trade, they did little to aid the growth of infrastructure or long-term development. Only in the D.R.C. was industry promoted to any great extent.

In general, French colonial rule, along with that of the Belgians, Portuguese, and Spanish and the activities of the companies, offered few opportunities for Africans to gain training and education. There was also little encouragement of local entrepreneurship. An important exception to this pattern was the policies pursued by Felix Eboue, a black man from French Guiana (in South America) who served as a senior administrator in the Free French administration of French Equatorial Africa during the 1940s. Eboue increased opportunities for the urban elite in Central African Republic, Congo, and Gabon. He also played an important role in the Brazzaville Conference of 1944, which, recognizing the part that the people of the French colonies had played in World War II, abolished forced labor and granted citizenship to all. Yet political progress toward self-government was uneven. Because of the lack of local labor development, there were too few people at independence who were qualified to shoulder the bureaucratic and administrative tasks of the re-

gimes that took power. People who could handle the economic institutions for the countries' benefit were equally scarce. And in any case, the nations' economies remained for the most part securely in outside—largely French—hands.

The Spanish on the Equatorial Guinea island of Fernando Po, and the Portuguese of São Tomé and Príncipe, also profited from their exploitation of forced labor. Political opportunities in these territories were even more limited than on the African mainland. Neither country gained independence until fairly recently: Equatorial Guinea in 1968, São Tomé and Príncipe in 1975.

In the years since independence, most of the countries of Central Africa have been influenced, pressured, and supported by France and the other former colonial powers. French firms in Central African Republic, Congo, and Gabon continue to dominate the exploitation of local resources. Most of these companies are only slightly encumbered by the regulations of the independent states in which they operate, and all are geared toward European markets and needs. Financial institutions are generally branches of French institutions, and all the former French colonies as well as Equatorial Guinea are members of the Central African Franc (CFA) Zone. French expatriates occupy senior positions in local civil-service establishments and in companies; many more of them are resident in the region today than was true 30 years ago. In addition, French troops are stationed in Central African Republic, Chad, and Gabon, regimes that owe their very existence to past French military interventions. Besides being a major trading partner, France has contributed significantly to the budgets of its former possessions, especially the poorer states of Central African Republic and Chad.

Despite having been under Belgian rule, the Democratic Republic of the Congo is an active member of the Francophonic bloc in Africa. In 1977, French troops put down a rebellion in the southeastern part of Zaire (as the D.R.C. was then known). Zaire, in turn, sent its troops to serve beside those of France in Chad and Togo. Since playing a role in the 1979 coup that brought the current regime to power, France has also had a predominant influence in Equatorial Guinea.

REGIONAL COOPERATION AND CONFLICT

Although many Africans in Central Africa recognize that closer links among their countries would be beneficial, there have been fewer initiatives toward political unity or economic integration in this region than in East, West, or Southern Africa. In the years before independence, Barthelemy Boganda, of what is now Central African Republic, espoused and publicized the idea of a "United States of Latin Africa," which was to include Angola and Zaire as well as the territories of French Equatorial Africa, but he was frustrated by Paris as well as by local politicians. When France offered independence to its colonies in 1960, soon after Boganda's death, the possibility of forming a federation was discussed. But Gabon, which was wealthier than the other countries, declined to participate. Central African Republic, Chad, and Congo drafted an agreement that would have created a federal legislature and executive branch governing all three countries, but local jealousies defeated this plan.

There have been some formal efforts at economic integration among the former French states. The Customs and Economic Union of the Central African States (UDEAC) was established in 1964, but its membership has been unstable. Chad and Central African Republic withdrew to join Zaire in an alternate organization. (Central African Republic later returned, bringing the number of members to six.) The East and Central African states together planned an "Economic Community" in 1967, but it never materialized.

In the 1980s there were efforts to make greater progress toward economic cooperation. Urged on by the United Nations Economic Commission for Africa, and with the stimulus of the Lagos Plan of Action, representatives of Central African states met in 1982 to prepare for a new economic grouping. In 1983,

all the Central African states as well as Rwanda and Burundi in East Africa signed a treaty establishing the Economic Community of Central African States (ECCA). ECCA's goals were broader than those of UDEAC. Members hoped that the union would stimulate industrial activity, increase markets, and reduce the dependence on France and other countries for trade and capital. But with dues often unpaid and meetings postponed, ECCA has so far failed to meet its potential.

Hopes that the Central African states could work collectively toward a brighter future have been further compromised by the spread in recent years of extreme political instability and brutal and far-reaching armed conflicts both within and between states in the region. This is particularly true of the ongoing civil war in the region's biggest state, the Democratic Republic of the Congo; this war has not only divided the region but has involved states as far afield as Libya, Nigeria, and South Africa.

In the mid-1990s, there was a growing optimism that democratization might enable the region to become a center of a continental renaissance. This view was greatly boosted by the fall of the corrupt, authoritarian regime of Mobutu Sese Seko, which led to the nominal transformation of Zaire into the Democratic Republic of the Congo. But such hopes suffered a serious setback with the resurgence of dictatorship and kleptocracy under Mobutu's successor, Laurent Kabila, who remained in power through external backers and through implicit appeals for genocide against ethnic groups perceived as his enemies. With such continued divide-and-misrule, the chances of Boganda's vision becoming a reality appeared more remote. But since the assassination of Laurent Kabila and the rise of his son Joseph Kabila in the D.R.C., prospects for peace in Central Africa's largest state seem realistic. After years of autocracy, there is an opportunity for a long-term cease-fire and cessation of hostilities among the warring factions within the region. Most notably, by the end of 2002, foreign state military forces had largely pulled out of the country. Regional initiatives may have failed in the past, but this is not to say that there will not be more effective cooperation in the future. The key element for the creation of new regional organizations rests in the ability of the Central African states to sit at a negotiating table and recognize the existence of their common interest. In this respect, peace in the D.R.C. represents the best hope for stability and development in Central Africa.

Cameroon (Republic of Cameroon)

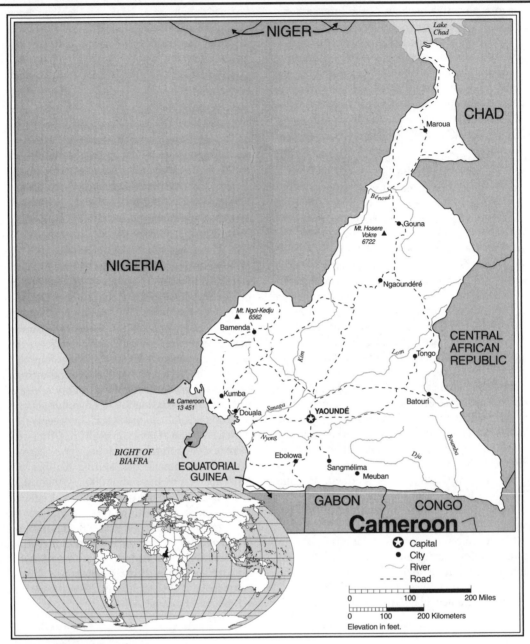

Cameroon Statistics

GEOGRAPHY

Area in Square Miles (Kilometers): 183,568
(475,400) (about the size of California)
Capital (Population): Yaoundé
(1,119,000)
Environmental Concerns: deforestation;
overgrazing; desertification; poaching;
overfishing; water-borne disease
Geographical Features: diverse, with
coastal plain in southwest, dissected
plain in center, mountains in west, and
plains in north

Climate: from tropical to semiarid

PEOPLE

Population

Total: 16,185,000
Annual Growth Rate: 2.36%
Rural/Urban Population Ratio: 54/46
Major Languages: English; French; Fulde;
Ewondo; Duala; Bamelke; Bassa; Bali;
others

Ethnic Makeup: 31% Cameroonian
Highlander; 19% Equatorial Bantu; 11%
Kirdi; 10% Fulani; 29% others
Religions: 40% indigenous beliefs; 40%
Christian; 20% Muslim

Health

Life Expectancy at Birth: 54 years (male);
55 years (female)
Infant Mortality: 68.8/1,000 live births
Physicians Available: 1/11,848 people
HIV/AIDS Rate in Adults: 7.73%

Education

Adult Literacy Rate: 63.4%
Compulsory (Ages): 6–12; free

COMMUNICATION

Telephones: 95,000 main lines
Televisions: 72/1,000 people
Internet Users: 20,000 (2000)

TRANSPORTATION

Highways in Miles (Kilometers): 20,580 (34,300)
Railroads in Miles (Kilometers): 693 (1,111)
Usable Airfields: 49
Motor Vehicles in Use: 153,000

GOVERNMENT

Type: unitary republic
Independence Date: January 1, 1960 (from UN trusteeship under French administration)

Head of State/Government: President Paul Biya; Prime Minister Peter Mafany Musonge
Political Parties: Democratic Rally of the Cameroon People; National Union for Democracy and Progress; Social Democratic Front; Cameroonian Democratic Union; Union of Cameroonian Populations; others
Suffrage: universal at 20

MILITARY

Military Expenditures (% of GDP): 1.4%
Current Disputes: various border conflicts, especially with Nigeria

ECONOMY

Currency ($ U.S. equivalent): 529.43 CFA francs =$1
Per Capita Income/GDP: $1,700/$26.4 billion
GDP Growth Rate: 4.9%
Inflation Rate: 2%

Unemployment Rate: 30%
Labor Force by Occupation: 70% agriculture; 13% industry and commerce; 17% other
Population Below Poverty Line: 48%
Natural Resources: petroleum; timber; bauxite; iron ore; hydropower
Agriculture: coffee; cocoa; cotton; rubber; bananas; oilseed; grain; roots; livestock; timber
Industry: petroleum production and refining; food processing; light consumer goods; textiles; lumber
Exports: $2.1 billion (primary partners Italy, France, Netherlands)
Imports: $1.5 billion (primary partners France, Germany, United States, Japan)

SUGGESTED WEB SITES

http://www.sas.upenn.edu/
  African_Studies/
  Country_Specific/Cameroon.html
http://www.cameroon.net
http://www.telp.com/cameroon/
  home.htm

Cameroon Country Report

Over the past decade, Cameroonians have been united by at least two things: support for their world-class national soccer team, the Indomitable Lions, and condemnation of neighboring Nigeria about a long-simmering border dispute over the Bakassi Peninsula. The latter conflict was referred by both countries to the International Court of Justice for resolution. But when the Court ruled in Cameroon's favor in October 2002, the government of Nigeria reneged on its previous agreement to accept the verdict.

DEVELOPMENT

The Cameroon Development Corporation coordinates more than half of the agricultural exports and, after the government, employs the most people. Cocoa and coffee comprise more than 50% of Cameroon's exports. Lower prices for these commodities in recent years have reduced the country's income.

Politically, Cameroonians have remained deeply divided, notwithstanding incumbent President Paul Biya's claim of victory with 92 percent of the vote in the last (October 1998) elections. The poll, which was boycotted by the largest opposition parties, was a followup to the controversial elections of March 1992, which

ended a quarter-century of one-party rule by Biya's Cameroon People's Democratic Party (CPDM). Although the CPDM relinquished its monopoly of power, it retained control of the government. With a plurality of 88 out of 180 seats, the CPDM was able to form a coalition government with the Movement for the Defense of Democracy, which won six seats. Two other parties, the National Union for Democracy and Progress (UNDP) and the Union of Cameroonian Populations (UPC), divided most of the remaining seats.

Like the 1998 elections, the legitimacy of the 1992 poll was compromised by the boycott of some opposition parties—most notably the Social Democratic Front (SDF), the Democratic Union (CDU), and a faction of the UPC—allowing the CPDM to win numerous constituencies by default. The situation was later aggravated, in October 1992, when the CPDM's Paul Biya was declared the victor in a snap election accompanied by opposition allegations of vote-rigging.

Cameroon's fractious politics is partially a reflection of its diversity. In geographical terms, the land is divided between the tropical forests in the south, the drier savanna of the north-central region, and the mountainous country along its western border, which forms a natural division between West and Central Africa. In terms of religion, the country has many

Christians, Muslims, and followers of indigenous belief systems. More than a dozen major languages, with numerous dialects, are spoken. The languages of southern Cameroon are linguistically classified as Bantu. The "Bantu line" that runs across the country, roughly following the course of the Sanaga River, forms a boundary between the Bantu languages of Central, East, and Southern Africa and the non-Bantu tongues of North and West Africa. Many scholars believe that the roots of the Bantu language tree are buried in Cameroonian soil. Cameroon is also unique among the continental African states in sharing two European languages, English and French, as its official mediums. Relations between Anglophone and Francophone Cameroon have been troubled in recent years. In October 2001, violence flared between government security forces and protesters favoring the separation of English-speaking Cameroon.

Cameroon's use of both English and French is a product of its unique colonial heritage. Three European powers have ruled over Cameroon. The Germans were the first. From 1884 to 1916, they laid the foundation of much of the country's communications infrastructure and, primarily through the establishment of European-run plantations, export agriculture. During World War I, the area was divided between the British and French, who subsequently

(United Nations photo by Shaw McCutcheon)

Cameroon has experienced political unrest in recent years as various factions have moved to establish a stable form of government. At the heart of the political turmoil is the need to raise the living standards of the population through an increase in agricultural production. These farmers with their cattle herds are one part of this movement.

ruled their respective zones as League of Nations (later the United Nations) mandates. French "Cameroun" included the eastern four fifths of the former German colony, while British "Cameroon" consisted of two narrow strips of territory that were administered as part of its Nigerian territory.

FREEDOM

While Cameroon's human-rights record has improved since its return to multipartyism, political detentions and harassment continue. Amnesty International has drawn attention to the alleged starvation of detainees at the notorious Tchollire prison. Furthermore, the nation's vibrant free press has become a prime target of repression, with several editors arrested in 1998.

In the 1950s, Cameroonians in both the British and French zones began to agitate for unity and independence. At the core of their nationalist vision was the "Kamerun Idea," a belief that the period of German rule had given rise to a pan-Cameroonian identity. The largest and most radical of the nationalist movements in the French zone was the Union of the Cameroonian People,

which turned to armed struggle. Between 1955 and 1963, when most of the UPC guerrillas were defeated, some 10,000 to 15,000 people were killed. Most of the victims belonged to the Bamileke and Bassa ethnic groups of southwestern Cameroon, which continues to be the core area of UPC support. (Some sources refer to the UPC uprising as the Bamileke Rebellion.)

To counter the UPC revolt, the French adopted a dual policy of repression against the guerrillas' supporters and the devolution of political power to local non-UPC politicians. Most of these "moderate" leaders, who enjoyed core followings in both the heavily Christianized southeast and the Muslim north, coalesced as the Cameroonian Union, whose leader was Ahmadou Ahidjo, a northerner. In pre-independence elections, Ahidjo's party won just 51 out of the 100 seats. Ahidjo thus led a divided, war-torn state to independence in 1960.

In 1961, the southern section of British Cameroon voted to join Ahidjo's republic. The northern section opted to remain part of Nigeria. The principal party in the south was the Kamerun National Democratic Party, whose leader, John Foncha, became the vice president of the Cameroon republic, while Ahidjo served as president. The

former British and French zones initially maintained their separate local parliaments, but the increasingly authoritarian Ahidjo pushed for a unified form of government. In 1966, all of Cameroon's legal political groups were dissolved into Ahidjo's new Cameroon National Union (CNU), creating a de facto one-party state. Trade unions and other mass organizations were also brought under CNU control. In 1972, Ahidjo proposed the abolition of the federation and the creation of a constitution for a unified Cameroon. This was approved by a suspiciously lopsided vote of 3,217,058 to 158.

In 1982, Ahidjo, believing that his health was graver than was actually the case, suddenly resigned. His handpicked successor was Paul Biya. To the surprise of many, the heretofore self-effacing Biya quickly proved to be his own man. He brought young technocrats into the ministries and initially called for a more open and democratic society. But as he pressed forward, Biya came into increasing conflict with Ahidjo, who tried to reassert his authority as CNU chairman. The ensuing power struggle took on overtones of an ethnic conflict between Biya's largely southern Christian supporters and Ahidjo's core following of northern Muslims. In 1983, Ahidjo lost and went into exile. The next year, he was tried and convicted, in absentia, for allegedly plotting Biya's overthrow.

HEALTH/WELFARE

The overall literacy rate in Cameroon, about 63%, is among the highest in Africa. There exists, however, great disparity in regional figures as well as between males and females. In addition to public schools, the government devotes a large proportion of its budget to subsidizing private schools.

In April 1984, only two months after the conviction, Ahidjo's supporters in the Presidential Guard attempted to overthrow Biya. The revolt was put down, but up to 1,000 people were killed. In the coup's aftermath, Biya combined repression with attempts to restructure the ruling apparatus. In 1985, the CNU was overhauled as the Cameroon People's Democratic Movement. However, President Biya became increasingly reliant on the support of his own Beti group.

An upsurge of prodemocracy agitation began in 1990. In March, the Social Democratic Front was formed in Bamenda, the main town of the Anglophonic west, over government objections. In May, as many as 40,000 people from the vicinity of Bamenda, out of a total population of about 100,000, attended an SDF rally. Government troops opened fire on school children returning from the demonstration. This action led to a wave of unrest, which spread to the capital city of Yaoundé. The government media tried to portray the SDF as a subversive movement of "English speak-

ers," but it attracted significant support in Francophonic areas. Dozens of additional opposition groups, including the UNDP (which is loyal to the now-deceased Ahidjo's legacy) and the long-underground UPC, joined forces with the SDF in calling for a transition government, a new constitution, and multiparty elections.

Throughout much of 1991, Cameroon's already depressed economy was further crippled by opposition mass action, dubbed the "Ghost Town Campaign." A series of concessions by Biya culminated in a November agreement between Biya and most of the opposition (the SDF being among the holdouts) to formulate a new constitution and prepare for elections.

ACHIEVEMENTS

The strong showing by Cameroon's national soccer team, the Indomitable Lions, in the 1990 and 1994 World Cup competitions is a source of pride for sports fans throughout Africa. Their success, along with the record numbers of medals won by African athletes in the 1988 and 1992 Olympics, is symbolic of the continent's coming of age in international sports competitions.

One unrealized hope has been that democratic reform would help move Cameroon away from its consistent Transparency International rating as one of the world's most corrupt countries. Endemic corruption has become associated with environmental degradation. In recent years conservationists have been especially concerned about the construction of an oil pipeline, funded by the World Bank, without an environmental-impact study, and

the allocation of about 80 percent of the country's forest for logging.

Timeline: PAST

1884
The establishment of the German Kamerun Protectorate

1916
The partition of Cameroon; separate British and French mandates are established under the League of Nations

1955
The UPC (formed in 1948) is outlawed for launching revolts in the cities

1960
The Independent Cameroon Republic is established, with Ahmadou Ahidjo as the first president

1961
The Cameroon Federal Republic reunites French Cameroon with British Cameroon after a UN-supervised referendum

1972
The new Constitution creates a unitary state

1980s
Ahidjo resigns and is replaced by Paul Biya; Lake Nyos releases lethal volcanic gases, killing an estimated 2,000 people

1990s
Nationwide agitation for a restoration of multiparty democracy; Biya retains the presidency in disputed elections

PRESENT

2000s
New clashes over the Bakassi Peninsula

Central African Republic

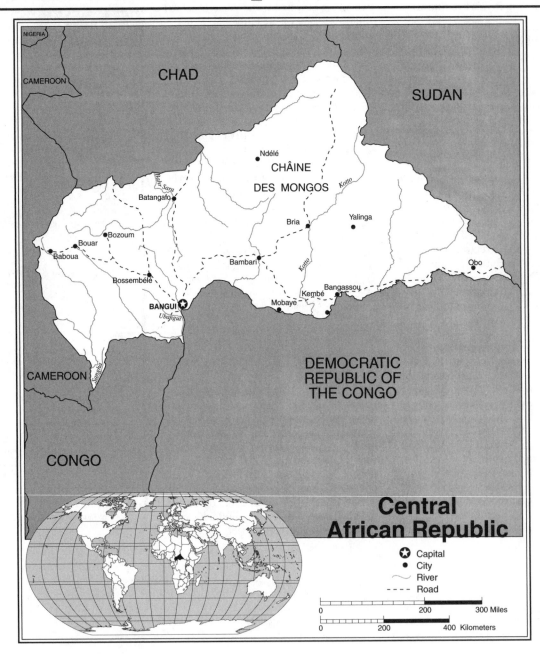

Central African Republic Statistics

GEOGRAPHY

Area in Square Miles (Kilometers): 240,324
(662,436) (about the size of Texas)

Capital (Population): Bangui (666,000)

Environmental Concerns: poaching;
desertification; deforestation; potable
water

Geographical Features: vast, flat to
rolling, monotonous plateau; scattered
hills in the northeast and southwest;
landlocked

Climate: tropical

PEOPLE

Population

Total: 3,643,000

Annual Growth Rate: 1.8%

Rural/Urban Population Ratio: 60/40

Major Languages: French; Songo; Arabic;
Hunsa; Swahili

Ethnic Makeup: 34% Baya; 27% Banda;
21% Mandja; 10% Sara; 8% others

Religions: 35% indigenous beliefs; 25%
Protestant; 25% Roman Catholic; 15%
Muslim

Health

Life Expectancy at Birth: 42 years (male);
45 years (female)

Infant Mortality: 103.8/1,000 live births
Physicians Available: 1/18,660 people
HIV/AIDS Rate in Adults: 13.84%

Education

Adult Literacy Rate: 60%
Compulsory (Ages): 6–14

COMMUNICATION

Telephones: 10,000 main lines
Televisions: 5/1,000 people
Internet Users: 1,500 (2001)

TRANSPORTATION

Highways in Miles (Kilometers): 14,286
  (23,810)
Railroads in Miles (Kilometers): none
Usable Airfields: 51
Motor Vehicles in Use: 20,000

GOVERNMENT

Type: republic
Independence Date: August 13, 1960
  (from France)
Head of State/Government: President
  Ange-Félix Patassé; Prime Minister
  Martin Ziguele
Political Parties: Movement for the
  Liberation of the Central African People;
  Central African Democratic Assembly;
  Movement for Democracy and
  Development; others
Suffrage: universal at 21

MILITARY

Military Expenditures (% of GDP): 2.2%
Current Disputes: internal strife

ECONOMY

Currency ($U.S. Equivalent): 529.43 CFA
  francs = $1

Per Capita Income/GDP: $1,300/$4.6
  billion
GDP Growth Rate: 1.8%
Inflation Rate: 3.6%
Unemployment Rate: 8%
Natural Resources: diamonds; uranium;
  timber; gold; petroleum; hydropower
Agriculture: cotton; coffee; tobacco;
  manioc; millet; corn; bananas; timber
Industry: diamond mining; sawmills;
  breweries; textiles; footwear; assembly
  of bicycles and motorcycles
Exports: $166 million (primary partners
  Benelux, Côte d'Ivoire, Spain)
Imports: $154 million (primary partners
  France, Cameroon, Benelux)

SUGGESTED WEB SITES

http://www.sas.upenn.edu/
  African_Studies/
  Country_Specific/Cent-A-R.html
http://www.emulateme.com/car.htm

Central African Republic Country Report

In March 2003, the government of the Central African Republic (CAR) was overthrown when rebels loyal to former Army chief-of-staff General François Bozize seized control of the capital, Bangui. The country's president, Ange-Félix Patassé, was out of the country at the time. For many years Patassé, elected in 1993 and re-elected in 1998, had struggled to bring political and financial stability to the poor, landlocked country, following three decades of military dictatorship.

In 1996, the French military rescued the government from an army mutiny. The intervention was resented by many as a sign of France's continuing control. A national-reconciliation agreement was signed in March 1998 allowing UN peacekeepers to oversee elections in 1999 and the training of a new army.

DEVELOPMENT

C.A.R.'s timber industry has suffered from corruption and environmentally destructive forms of exploitation. However, the nation has considerable forestry potential, with dozens of commercially viable and renewable species of trees.

But in 2000, further unrest was sparked by government's failure to pay civil servants their back wages, resulting in a general strike organized by 15 opposition groups. In May 2001, Patassé survived an attempted coup, with the help of Libyan, Chadian, and Congolese rebel forces. Former president André Kolingba and Army chief-of-staff Bozize led the coup. Thereafter a curfew was instituted, which was lifted a year later. The lifting of the curfew was meant to signal the return of "security and peace." But in October 2002, rebels loyal to Bozize seized control of much of Bangui before being driven out by Libyan and progovernment forces.

Since gaining independence in 1960, the political, economic, and military presence of France has remained pervasive in C.A.R. At the same time, the country's natural resources, as well as French largess, have been dissipated. Yet with diamonds, timber, and a resilient peasantry, the country is better endowed than many of its neighbors. In recent years, Libya has taken a special interest in the country's mostly still untapped natural wealth.

C.A.R.'s population has traditionally been divided between the so-called river peoples and savanna peoples, but most are united by the Songo language. What the country has lacked is a leadership committed to national development rather than to internationally sanctioned waste.

BOGANDA'S VISION

In 1959, as Central African Republic moved toward independence, Barthelemy Boganda, a former priest and the leader of the territory's nationalist movement, did not share the euphoria exhibited by many of his colleagues. To him, the French path to independence was a trap. Where there once had been a united French Equatorial Africa (A.E.F.), there were now five separate states, each struggling toward its own nationhood. Boganda had led the struggle to transform the territory into a true Central African Republic. But in 1958, French president Charles de Gaulle overruled all objections in forcing the breakup of the A.E.F. Boganda believed that, thus balkanized, the Central African states would each be too weak to achieve true independence, but he still hoped that A.E.F. reunification might prove possible after independence.

FREEDOM

C.A.R.'s human-rights record remains poor. Its security forces are linked to summary executions and torture. Other human-rights abuses include harsh prison conditions, arbitrary arrest, detention without trial, and restrictions on freedom of assembly. President Patassé granted amnesty to former senior officials of the Kolingba regime and mutineers.

In 1941, Boganda had founded the Popular Movement for the Social Evolution of Black Africa (MESAN). While Boganda was a pragmatist willing to use moderate means in his struggle, his vision was radical, for he hoped to unite French, Belgian, and Portuguese territories into an independent republic. His movement succeeded in gaining a local following among the peasantry as well as intellectuals. In 1958, Boganda led the territory to self-government, but he died in a mysterious plane crash just before independence.

Boganda's successors have failed to live up to his stature. At independence, David Dacko, a nephew of Boganda's who succeeded to the leadership of MESAN but also cultivated the political support of local French settlers who had seen Boganda as an agitator, led the country. Dacko's MESAN became the vehicle of the wealthy elite.

HEALTH/WELFARE

The literacy rate is low in Central African Republic—63%. Teacher training is currently being emphasized, especially for primary-school teachers. Poaching has diminished C.A.R.'s reputation as one of the world's last great wildlife refuges.

A general strike in December 1965 was followed by a military coup on New Year's Eve, which put Dacko's cousin, Army Commander Jean-Bedel Bokassa, in power. Dacko's overthrow was justified by the need to launch political and economic reforms. But more likely motives for the coup were French concern about Dacko's growing ties with China and Bokassa's own budding megalomania.

The country suffered greatly under Bokassa's eccentric rule. During the 1970s, he was often portrayed, alongside Idi Amin of Uganda, as an archetype of African leadership at its worst. It was more the sensational nature of his brutality—such as public torture and dismemberment of prisoners—rather than its scale that captured headlines. In 1972, he made himself "president-for-life." Unsatisfied with this position, in 1976 he proclaimed himself emperor, in the image of his hero Napoleon Bonaparte. The French government underwrote the $22 million spent on his coronation ceremony, which attracted widespread coverage in the global media.

ACHIEVEMENTS

Despite recurrent drought, a poor infrastructure, and inefficient official marketing, the farmers of Central African Republic have generally been able to meet most of the nation's basic food needs.

In 1979, reports surfaced that Bokassa himself had participated in the beating deaths of schoolchildren who had protested his decree that they purchase new uniforms bearing his portrait. The French government finally decided that its ally had become a liability. While Bokassa was away on a state visit to Libya, French paratroopers returned Dacko to power. In 1981, Dacko was once more toppled, in a coup that installed Prime Minister (General) André Kolingba. In 1985, Kolingba's provisional military regime was transformed into a one-party state. But in 1991, under a combination of local and French pressure, he agreed to the legalization of opposition parties.

Pressure for multiparty politics had increased as the government sank deeper into debt, despite financial intervention on the part of France, the World Bank, and the International Monetary Fund. Landlocked C.A.R.'s economy has long been constrained by high transport costs. But a perhaps greater burden has been the smuggling of its diamonds and other resources, including poached ivory, by officials who are high up in the government.

Chad (Republic of Chad)

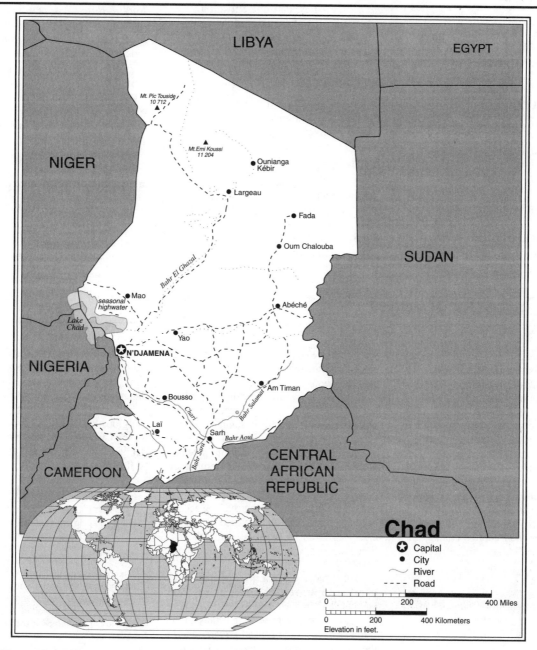

Chad Statistics

GEOGRAPHY

Area in Square Miles (Kilometers): 496,000 (1,284,634) (about 3 times the size of California)

Capital (Population): N'Djamena (826,000)

Environmental Concerns: soil and water pollution; desertification; insufficient potable water; waste disposal

Geographical Features: broad, arid plains in the center; desert in the north; mountains in the northwest; lowlands in the south; landlocked

Climate: tropical in the south; desert in the north

PEOPLE

Population

Total: 8,998,000

Annual Growth Rate: 3.27%

Rural/Urban Population Ratio: 77/23

Major Languages: French; Arabic; Sara; Sango; others

Ethnic Makeup: 200 distinct groups

Religions: 51% Muslim; 35% Christian; 7% animist; 7% others

Health

Life Expectancy at Birth: 49 years (male); 53 years (female)

Infant Mortality: 93.4/1,000 live births

Physicians Available: 1/27,765 people

HIV/AIDS Rate in Adults: 5%–7%

33

Education

Adult Literacy Rate: 40%
Compulsory (Ages): 6–14

COMMUNICATION

Telephones: 10,300 main lines
Televisions: 8/1,000 people
Internet Users: 1,000 (2000)

TRANSPORTATION

Highways in Miles (Kilometers): 19,620
    (32,700)
Railroads in Miles (Kilometers): none
Usable Airfields: 49
Motor Vehicles in Use: 24,000

GOVERNMENT

Type: republic
Independence Date: August 11, 1960
    (from France)

Head of State/Government: President
    Idriss Déby; Prime Minister Nagoum
    Yamassoum
Political Parties: Patriotic Salvation
    Movement; National Union for
    Development and Renewal; many others
Suffrage: universal at 18

MILITARY

Military Expenditures (% of GDP): 1.9%
Current Disputes: civil war; border
    conflicts over Lake Chad area

ECONOMY

Currency ($ U.S. equivalent): 617 CFA
    francs = $1
Per Capita Income/GDP: $1,030/$8.9
    billion
GDP Growth Rate: 8%
Inflation Rate: 3%

Labor Force by Occupation: 80%+
    agriculture
Population Below Poverty Line: 80%
Natural Resources: petroleum; uranium;
    natron; kaolin; fish (Lake Chad)
Agriculture: subsistence crops; cotton;
    peanuts; fish; livestock
Industry: livestock products; breweries;
    natron; soap; textiles; cigarettes;
    construction materials
Exports: $172 million (primary partners
    Portugal, Germany, Thailand)
Imports: $223 million (primary partners
    France, Nigeria, Cameroon)

SUGGESTED WEB SITES

http://www.chadembassy.org/site/
    index.cfm
http://www.sas.upenn.edu/
    African_Studies/
    Country_Specific/Chad.html

Chad Country Report

After decades of civil war between northern- and southern-based armed movements, in 1997 Chad completed its transition to civilian rule, under the firm guidance of its president, Idriss Déby. A former northern warlord who seized power in 1990, in 1996 Déby achieved a second-round victory in the country's first genuinely contested presidential elections since its independence in 1960. The result was seen as an endorsement of his government's gradual progress in rebuilding Chad's state structures, which had all but collapsed by the 1980s, when the fractured country was referred to as the "Lebanon of Africa."

DEVELOPMENT

Chad has potential petroleum and mineral wealth that would greatly help the economy if stable central government can be created. Deposits of chromium, tungsten, titanium, gold, uranium, and tin as well as oil are known to exist. Roads are in poor condition and are dangerous.

In June 2001, Chad's highest court confirmed Déby's reelection, after a controversial poll in which the results of about one quarter of the polling stations were cancelled due to alleged irregularities. Six unsuccessful candidates were briefly picked up for questioning by police after the poll. While Déby's success—through both the ballot and bullet—in defeating, marginalizing, and/or reconciling rival factions has restored a semblance of statehood to Chad over the past five years, he contin-

ues to preside over a bankrupt government whose control over much of the countryside is tenuous.

In 1998, a new armed insurgency broke out in the north, led by former defense chief Youssouf Togimi. In January 2002, a Libyan-brokered peace deal was agreed to by the government and Togoimi's rebels (known as the Movement for Democracy and Justice in Chad, or MDJT). But further clashes have put implementation of the peace agreement into jeopardy. The accord provides for a ceasefire, release of prisoners, integration of the rebels into the national army, and government positions for MDJT leaders.

Déby also faces continuing challenges to his control over the south, where there have been calls for Chad to become a federal, rather than unitary, state. Many in the south still resent the central government in the capital city of N'Djamena, believing it to represent predominantly northern interests. On-going fighting within the Central African Republic, Chad's southern neighbor, could further destabilize the situation.

CIVIL WAR

Chad's conflicts are partially rooted in the country's ethnic and religious divisions. It has been common for outsiders to portray the struggle as being between Arab-oriented Muslim northerners and black Christian southerners, but Chad's regional and ethnic allegiances are much more complex. Geographically, the country is better divided into three zones: the northern Sahara,

a middle Sahel region, and the southern savanna. Within each of these ecological areas live peoples who speak different languages and engage in a variety of economic activities. Wider ethno-regional and religious loyalties have emerged as a result of the Civil War, but such aggregates have tended to be fragile and their allegiances shifting.

FREEDOM

Despite some modest improvement, Chad's human-rights record remains poor. Its security forces are linked to torture, extra-judicial killings, beatings, disappearances, and rape. A recent Amnesty International report on Chad was entitled "Hope Betrayed." Antigovernment rebel forces are also accused of atrocities. The judiciary is not independent.

At Chad's independence, France turned over power to François Tombalbaye, a Sara-speaking Christian southerner. Tombalbaye ruled with a combination of repression, ethnic favoritism, and incompetence, which quickly alienated his regime from broad sectors of the population. A northern-based coalition of armed groups, the National Liberation Front, or Frolinat, launched an increasingly successful insurgency. The intervention of French troops on Tombalbaye's behalf failed to stem the rebellion. In 1975, the army, tired of the war and upset by the president's increasingly conspicuous brutality, overthrew

Tombalbaye and established a military regime, headed by Felix Malloum.

Malloum's government was also unable to defeat Frolinat; so, in 1978, it agreed to share power with the largest of the Frolinat groups, the Armed Forces of the North (FAN), led by Hissène Habré. This agreement broke down in 1979, resulting in fighting in N'Djamena. FAN came out ahead, while Malloum's men withdrew to the south. The triumph of the "northerners" immediately led to further fighting among various factions—some allied to Habré, others loyal to his main rival within the Frolinat, Goukkouni Oueddie. Earlier Habré had split from Oueddie, whom he accused of indifference toward Libya's unilateral annexation in 1976 of the Aouzou Strip, along Chad's northern frontier. At the time, Libya was the principal foreign backer of Frolinat.

In 1992, there were reports of catastrophic famine in the countryside. Limited human services were provided by external aid agencies. Medicines are in short supply or completely unavailable.

In 1980, shortly after the last French forces withdrew from Chad, the Libyan Army invaded the country, at the invitation of Oueddie. Oueddie was then proclaimed the leader in a "Transitional Government of National Unity" (GUNT), which was established in N'Djamena. Nigeria and other neighboring states, joined by France and the United States, pressed for the withdrawal of the Libyan forces. This pressure grew in 1981 after Libyan leader Muammar al-Qadhafi announced the merger of Chad and Libya. Following a period of intense multinational negotiations, the Libyan military presence was reduced at Oueddie's request.

The removal of the Libyan forces from most of Chad was accompanied by revived fighting between GUNT and FAN, with the latter receiving substantial U.S. support, via Egypt and Sudan. A peacekeeping force assembled by the Organization of African Unity proved ineffectual. The collapse of GUNT in 1982 led to a second major Libyan invasion. The Libyan offensive was countered by the return of French forces, assisted by Zairian troops and by smaller contingents from several other Francophonic African countries. Between 1983 and 1987, the country was virtually partitioned along the 16th Parallel, with Habré's French-backed, FAN–led coalition in the south and the Libyan-backed remnants of GUNT in the north.

A political and military breakthrough occurred in 1987. Habré's efforts to unite the country led to a reconciliation with Malloum's followers and with elements within GUNT. Oueddie himself was apparently placed under house arrest in Libya. Emboldened, Habré launched a major offensive north of the 16th Parallel that rolled back the better-equipped Libyan forces, who by now included a substantial number of Lebanese mercenaries. A factor in the Libyan defeat was U.S.–supplied Stinger missiles, which allowed Habré's forces to neutralize Libya's powerful air force (Habré's government lacked significant air power of its own). A cease-fire was declared after the Libyans had been driven out of all of northern Chad with the exception of a portion of the disputed Aouzou Strip.

In 1988, Qadhafi announced that he would recognize the Habré government and pay compensation to Chad. The announcement was welcomed—with some skepticism—by Chadian and other African leaders, although no mention was made of the conflicting claims to the Aouzou Strip.

In precolonial times, the town of Kanem was a leading regional center of commerce and culture. Since independence in 1960, perhaps Chad's major achievement has been the resiliency of its people under the harshest of circumstances. The holding of truly contested elections is also a significant accomplishment.

The long-running struggle for Chad took another turn in November 1990, with the sudden collapse of Habré's regime in the face of a three-week offensive by guerrillas loyal to his former army commander, Idriss Déby. Despite substantial Libyan (and Sudanese) backing for his seizure of power, Déby had the support of France, Nigeria, and the United States (Habré had supported Iraq's annexation of Kuwait). A 1,200-man French force began assisting Déby against rebels loyal to Habré and other faction leaders.

Between January and April 1993, Déby's hand was strengthened by the successful holding of a "National Convention," in which a number of formerly hostile groups agreed to cooperate with the government in drawing up a new constitution. In April 1994, his government was further boosted by Qadhafi's unexpected decision to withdraw his troops from the Aouzou Strip, leaving Chad in undisputed control of the territory. The move followed an International Court of Justice ruling in Chad's favor.

A BETTER FUTURE?

The long, drawn-out conflict in Chad has led to immense suffering. Up to a half a million people—the equivalent of 10 percent of the total population—have been killed in the fighting.

Even if peace could be restored, the overall prospects for national development are bleak. The country has potential mineral wealth, but its geographic isolation and current world prices are disincentives to investors. Local food self-sufficiency should be obtainable despite the possibility of recurrent drought, but geography limits the potential of export crops. Chad thus appears to be an extreme case of the more general African need for a radical transformation of prevailing regional and global economic interrelationships. Had outside powers devoted half the resources to Chad's development over the past decades as they have provided to its civil conflicts, perhaps the country's future would appear brighter.

Timeline: PAST

1960
Independence is achieved under President François Tombalbaye

1965–1966
Revolt breaks out among peasant groups; FROLINAT is formed

1978
Establishment of a Transitional Government of National Unity (GUNT) with Hissène Habré and Goukkouni Oueddie

1980s
Habré seizes power and reunites the country in a U.S.–supported war against Libya

1990s
Habré is overthrown by Idriss Déby; Déby promises to create a multiparty democracy, but conditions remain anarchic

PRESENT

2000s
Chad's northern provinces bordering Libya remain heavily landmined

Persistent armed insurgency in the north

Déby is confirmed as reelected, after a controversial poll

Congo (Republic of the Congo; Congo-Brazzaville)

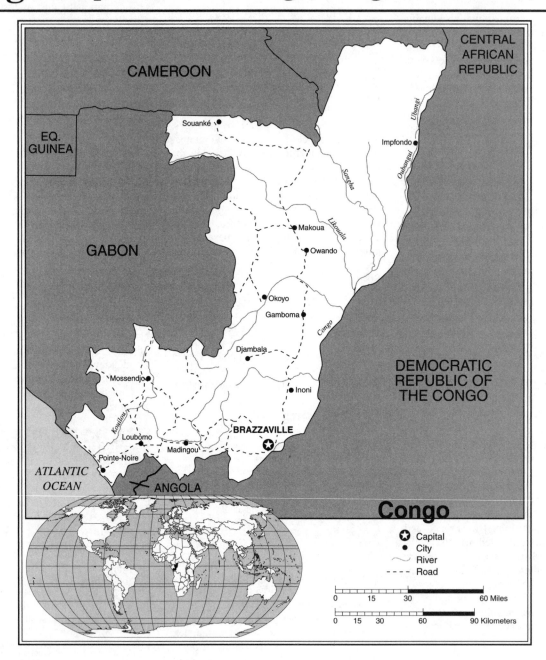

Congo Statistics

GEOGRAPHY

Area in Square Miles (Kilometers): 132,000 (342,000) (about the size of Montana)

Capital (Population): Brazzaville (1,360,000)

Environmental Concerns: air and water pollution; deforestation

Geographical Features: coastal plain; southern basin; central plateau; northern basin

Climate: tropical; particularly enervating climate astride the equator

PEOPLE

Population

Total: 2,959,000

Annual Growth Rate: 2.18%

Rural/Urban Population Ratio: 38/62

Major Languages: French; Lingala; Kikongo; Teke; Sangha; M'Bochi; others

Ethnic Makeup: 48% BaKongo; 20% Sangha; 17% Teke; 15% others

Religions: 50% Christian; 48% indigenous beliefs; 2% Muslim

Health

Life Expectancy at Birth: 44 years (male); 51 years (female)

Infant Mortality: 97/1,000 live births
Physicians Available: 1/3,873 people
HIV/AIDS Rate in Adults: 6.43%

Education

Adult Literacy Rate: 75%
Compulsory (Ages): 6–16

COMMUNICATION

Telephones: 22,000 main lines
Televisions: 17/1,000 people
Internet Users: 500 (2000)

TRANSPORTATION

Highways in Miles (Kilometers): 7,680 (12,800)
Railroads in Miles (Kilometers): 494 (797)
Usable Airfields: 33
Motor Vehicles in Use: 47,000

GOVERNMENT

Type: republic
Independence Date: August 15, 1960 (from France)
Head of State/Government: President Denis Sassou-Nguesso is both head of state and head of government
Political Parties: Democratic and Patriotic Forces; Congolese Movement for Democracy and Integral Development; many others
Suffrage: universal at 18

MILITARY

Military Expenditures (% of GDP): 2.8%
Current Disputes: civil conflicts; boundary issue with Democratic Republic of Congo

ECONOMY

Currency ($ U.S. Equivalent): 529.43 CFA francs = $1

Per Capita Income/GDP: $900/$2.5 billion
GDP Growth Rate: 4.2%
Inflation Rate: 3%
Natural Resources: timber; potash; lead; zinc; uranium; petroleum; natural gas; copper; phosphates
Agriculture: cassava; cocoa; coffee; sugarcane; rice; peanuts; vegetables; forest products
Industry: processing of agricultural and forestry goods; cement; brewing; petroleum
Exports: $2.6 billion (primary partners United States, South Korea, China)
Imports: $725 million (primary partners France, United States, Italy)

SUGGESTED WEB SITES

http://www.sas.upenn.edu/
African_Studies/
Country_Specific/Congo.html

Congo Country Report

Once considered one of Africa's most promising economies, over the past decade the Republic of the Congo—not to be confused with its larger neighbor the Democratic Republic of the Congo (D.R.C., the former Zaire)—has been afflicted by civil strife that has killed thousands while displacing up to one third of the population. In March 2002, the current president, General Denis Sassou-Nguesso, claimed 89 percent of the vote in presidential elections in which his two main political rivals, former president Pascal Lissouba and prime minister Bernard Kolelas, were barred from the contest. The poll was supposed to be the culmination of a two-year peace process that returned the country to democracy under a new Constitution. But in the same month, renewed fighting broke out between government forces and "Ninja" rebels loyal to Kolelas. In June, battles between government troops and the Ninja spread to Brazzaville, killing about 100 people.

The current round of political conflict in Congo began in 1997, when forces loyal to Sassou-Nguesso, who had previously ruled as a virtual dictator between 1979 and 1992, launched a rebellion. They seized control of Brazzaville in October 1997. The return of Sassou-Nguesso, who had garnered only 17 percent of the vote while losing the presidency to Lissouba just two years earlier, would not have been possible without the intervention of Angolan government forces, whose overt violation of Congolese sovereignty attracted little international comment. The Angolans were apparently motivated by allegations that Lissouba was supportive of UNITA rebels in their own country.

DEVELOPMENT

Congo's Niari Valley has become the nation's leading agricultural area, due to its rich alluvial soils. The government has been encouraging food-processing plants to locate in the region.

The fall of the capital did not end the fighting, which resulted in the Congolese Army as well as the country as a whole becoming largely split along north–south regional lines. Opposition to Sassou-Nguesso has been concentrated among the BaKongo people of the south, who make up about half of the total population. With continued Angolan backing, Sassou-Nguesso's forces were able to drive back those of his rivals.

At the end of 1999, many of the rebels agreed to a cease-fire accord. This was followed by a peace agreement that provided for a national dialogue; demilitarization of political parties; and the reorganization of the army, including the re-admission of rebel units into the security forces. By September 2001, some 15,000 rebels had been disarmed in a cash-for-arms scheme. This was rewarded by the International Mone-

tary Fund, which cancelled some $4 billion in debt. But other rebels have not as yet been bought off.

In December 2001, Lissouba was convicted in absentia by the High Court of treason and corruption charges, and sentenced to 30 years hard labor. Although Sassou-Nguesso's forces have maintained the upper hand, political stability and economic growth are unlikely to return to Congo in the absence of a truly inclusive process of reconciliation.

FREEDOM

Until 1990, political opposition groups, along with Jehovah's Witnesses and certain other religious sects, were vigorously suppressed. The new Constitution provides for basic freedoms of association, belief, and speech.

The overthrow of Lissouba was all the more unfortunate in that his government had seemed to have been making progress in negotiating an end to the violence among the political factions that has plagued the country for more than a decade. A 1995 agreement was supposed to lead to the disarmament of party militias. But instead, efforts to assure the disarmament of Sassou-Nguesso's men set off the revolt in June 1997. Previously, in November 1993, large sections of Brazzaville had been a battleground between troops loyal

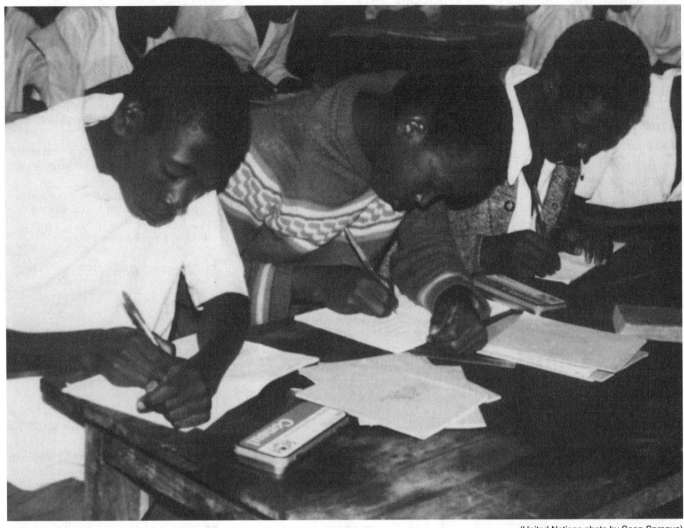

(United Nations photo by Sean Sprague)

Since Congo achieved its independence in 1960, it has made enormous educational strides. Almost all children in the country now attend school, which has had a tremendous effect on helping to realize the potential of the country's natural resources.

to the elected Pan-African Union of Social Democracy (UPADS) government of President Lissouba and the Ninjas—armed supporters of opposition leader Bernard Kolelas's Union for the Renewal of Democracy (URD). After several weeks of fighting, peace was finally restored with the intervention of an Organization of African Unity mediator. The crisis underscored the continuing fragility of Congo's difficult political transition into a multiparty democracy.

Sassou-Nguesso had previously ruled Congo as the head of a self-proclaimed, Marxist-Leninist one-party state. But in 1990, the ruling Congolese Workers Party (PCT) agreed to abandon both its past ideology and its monopoly of power. In 1991, a four-month-long "National Conference" met to pave the way for a new constitutional order. An interim government headed by Andre Milongo was appointed, pending elections, while Sassou-Nguesso

was stripped of all but ceremonial power. In the face of coup attempts by elements of the old order and a deteriorating economy, legislative and executive elections were finally held in the second half of 1992, resulting in Lissouba's election and a divided National Assembly.

A new National Assembly election in 1993 resulted in a decisive UPADS victory over a URD–PCT alliance, but the losers rejected the results for five months. With the economy experiencing a prolonged depression, political tensions remained high.

CONGO REPUBLIC

The Republic of the Congo takes its name from the river that forms its southeastern border with the Democratic Republic of Congo. Because Zaire prior to 1971 also called itself the Congo, the two countries are sometimes confused. Close historical and ethnic ties do in fact exist between the

nations. The BaKongo are the largest ethnolinguistic group in Congo and western former Zaire as well as in northern Angola. During the fifteenth and sixteenth centuries, this group was united under the powerful Kingdom of the Kongo, which ruled over much of Central Africa while establishing commercial and diplomatic ties with Europe. But the kingdom had virtually disappeared by the late nineteenth century, when the territory along the northwest bank of the Congo River—the modern republic—was annexed by France, while the southeast bank—Zaire—was placed under the brutal rule of King Leopold of Belgium.

Despite the establishment of this political division, cultural ties between Congo and the former Zaire, the former French and Belgian Congos, remained strong. Brazzaville sits across the river from the D.R.C. capital of Kinshasha. The metropolitan region formed by these two centers has, through such figures as the late Congolese artist Franco, given rise to *soukous*, a musical style that is now popular in such places as Tokyo and Paris as well as throughout much of Africa.

ECONOMIC DEVELOPMENT

Brazzaville, which today houses well over a third of Congo's population, was established during the colonial era as the administrative headquarters of French Equatorial Africa, a vast territory that included the present-day states of Chad, Central African Republic, Gabon, and Congo Republic. As a result, the city expanded, and the area around it developed as an imperial crossroads. The Congolese paid a heavy price for this growth. Thousands died while working under brutal conditions to build the Congo-Ocean Railroad, which linked Brazzaville with Pointe-Noire on the coast. Many more suffered as forced laborers for foreign concessionaires during the early decades of the twentieth century.

While the economies of many African states stagnated or declined during the 1970s and 1980s, Congo generally experienced growth, a result of its oil wealth. Hydrocarbons account for 90 percent of the total value of the nation's exports. But the danger of this dependence has been apparent since 1986, when falling oil prices led to a sharp decline in gross domestic product. An even greater threat to the nation's economic health is its mounting debt. As a result of heavy borrowing during the oil-boom years, by 1989 the total debt was estimated to be 50 percent greater than the value of the country's annual economic output. The annual cost of servicing the

debt was almost equal to domestic expenditure.

The debt led to International Monetary Fund pressure on Congo's rulers to introduce austerity measures as part of a Structural Adjustment Program (SAP). The PCT regime and its interim successor were willing to move away from the country's emphasis on central planning toward a greater reliance on market economics. But after an initial round of severe budgetary cutbacks, both administrations found it difficult to reduce their spending further on such things as food subsidies and state-sector employment.

With nearly two thirds of Congo's population now urbanized, there has been deep concern about the social and political consequences of introducing harsher austerity measures. Many urban-dwellers are already either unemployed or underemployed; even those with steady formal-sector jobs have been squeezed by wages that fail to keep up with inflation. The country's powerful trade unions, which are hostile to SAP, have been in the forefront of the democratization process.

ACHIEVEMENTS

There are a number of Congolese poets and novelists who combine their creative efforts with teaching and public service. Tchicaya U'Tam'si, who died in 1988, wrote poetry and novels and worked for many years for UNESCO.

Although most Congolese are facing tough times in the short run, the economy's long-term prospects remain promising. Besides oil, the country is endowed with a wide variety of mineral reserves. Timber has long been a major industry. And after years of neglect, the agricultural sector is growing. The goal of a return to food self-sufficiency appears achievable. Cocoa, coffee, tobacco, and sugarcane are major cash crops, while palm-oil estates are being rehabilitated.

The small but well-established Congolese manufacturing sector also has much

potential. Congo's urbanized population is relatively skilled, thanks to the enormous educational strides that have been made since independence. Almost all children in Congo now attend school. Prior to the devastation that occurred during the 1997 revolt, the infrastructure serving Brazzaville and Pointe-Noire, coupled with the previous government's emphasis on private-sector growth, was potentially attractive to outside investors.

Timeline: PAST

1910
Middle Congo becomes part of French Equatorial Africa

1944
Conference establishes French Union; Felix Eboue establishes positive policies for African advancement

1960
Independence is achieved, with Abbe Fulbert Youlou as the first president

1963
A general strike brings the army and a more radical government (National Revolutionary Movement) to power

1968–1969
A new military government under Marien Ngouabi takes over; the Congolese Workers' Party is formed

1977
Ngouabi is assassinated; Colonel Yhombi-Opango rules

1979–1992
Denis Sassou-Nguesso is president

1990s
Pascal Lissouba is elected president; former dictator Sassou-Nguesso again seizes power

PRESENT

2000s
Congo tries to recover from the civil conflict of the late 1990s

Security problems remain despite the peace process

Democratic Republic of the Congo
(Congo-Kinshasa; formerly Zaire)

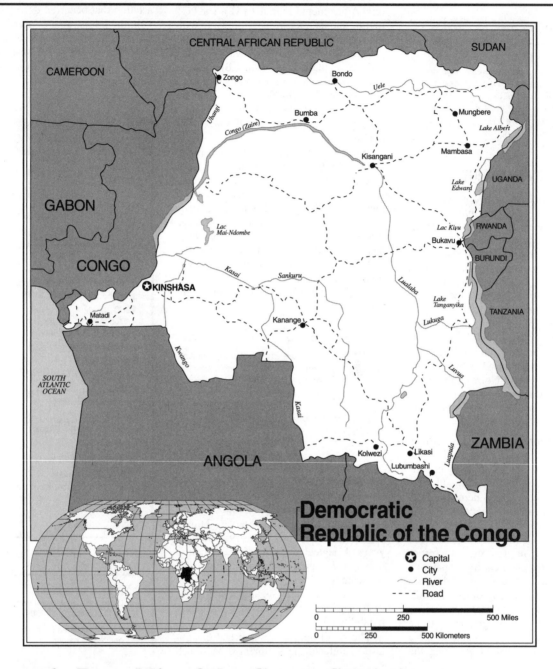

Democratic
Republic of the Congo

★ Capital
● City
〜 River
--- Road

0 — 250 — 500 Miles
0 — 250 — 500 Kilometers

Democratic Republic of the Congo Statistics

GEOGRAPHY

Area in Square Miles (Kilometers): 905,063 (2,300,000) (1/4 the size of the United States)

Capital (Population): Kinshasa (5,064,000)

Environmental Concerns: poaching; water pollution; deforestation; soil erosion; mining

Geographical Features: the vast central basin is a low-lying plateau; mountains in the east; dense tropical rain forest

Climate: tropical equatorial

PEOPLE

Population

Total: 55,226,000
Annual Growth Rate: 2.79%
Rural/Urban Population Ratio: 70/30
Major Languages: French; Lingala; Kingwana; others

Ethnic Makeup: Bantu majority; more than 200 African groups

Religions: 70% Christian; 20% indigenous beliefs; 10% Muslim

Health

Life Expectancy at Birth: 47 years (male); 51 years (female)

Infant Mortality: 98/1,000 live births

Physicians Available: 1/15,584 people

HIV/AIDS Rate in Adults: 5%

Education

Adult Literacy Rate: 77.3%

Compulsory (Ages): 6–12

COMMUNICATION

Telephones: 21,000 main lines

Daily Newspaper Circulation: 3/1,000 people

Internet Users: 1,500 (1999)

TRANSPORTATION

Highways in Miles (Kilometers): 97,340 (157,000)

Railroads in Miles (Kilometers): 3,206 (5,138)

Usable Airfields: 232

Motor Vehicles in Use: 530,000

GOVERNMENT

Type: dictatorship; presumably in transition to representative government

Independence Date: June 30, 1960 (from Belgium)

Head of State/Government: President Joseph Kabila is both head of state and head of government

Political Parties: Popular Movement of the Revolution; Democratic Social Christian Party; others

Suffrage: universal and compulsory at 18

MILITARY

Military Expenditures (% of GDP): 4.6%

Current Disputes: civil war; boundary issues with surrounding states; military conflicts with neighboring states

ECONOMY

Currency ($ U.S. equivalent): 3,275 new Zaires = $1

Per Capita Income/GDP: $590/$32 billion

GDP Growth Rate: –4%

Labor Force by Occupation: 65% agriculture; 19% services; 16% industry

Natural Resources: cobalt; copper; cadmium; petroleum; zinc; diamonds; manganese; tin; gold; silver; bauxite; iron ore; coal; hydropower; timber; others

Agriculture: coffee; palm oil; rubber; tea; manioc; root crops; corn; fruits; sugarcane; wood products

Industry: mineral mining and processing; consumer products; cement; diamonds

Exports: $750 million (primary partners Belgium, United States, South Africa)

Imports: $1.02 billion (primary partners South Africa, Benelux, Nigeria)

SUGGESTED WEB SITES

```
http://www.sas.upenn.edu/
  African_Studies/
  Country_Specific/DR_Congo.html
```

Democratic Republic of the Congo Country Report

Due to its geographical location and staggering mineral resources, the Democratic Republic of Congo (D.R.C.), also formerly known as Zaire, is perhaps the most strategically important nation in Africa, with the potential to become one of its richest. Unfortunately, after years of almost unbelievable levels of corruption, external interference, and greed, it suffers from the lack of a strong central government with credible structures. At times in its history, it has been difficult to imagine a nation in more distress than the D.R.C. After being ruled from 1965 to 1997 by the notoriously corrupt dictator Joseph Mobutu, in recent years the D.R.C. has all but fallen apart as armed internal and external forces have divided the country.

A series of agreements in 2002 led to the withdrawal of most external-state forces from Congolese soil and the promise of a "government of national unity." With an on-again, off-again peace process now bearing some fruit, there may be some grounds for cautious optimism.

The D.R.C.'s latest conflict began in August 1998, when rebels of the Congolese Assembly for Democracy, heavily backed by the Rwandan and Ugandan militaries, advanced rapidly from the east. This occurred shortly after the 15-month-old D.R.C. regime headed by Laurent Kabila, who had replaced Mobutu, called for the withdrawal of all foreign forces from Congolese soil. Early rebel hopes of a quick victory were frustrated by the intervention of Angolan, Namibian, and Zimbabwean forces. The fighting quickly became deadlocked, leading to mediation efforts under the chairmanship of Botswana's former president Sir Ketumile Masire, and backed by the United Nations and the Organization of African Unity. These negotiations appeared to be getting nowhere until January 2001, when Laurent Kabila was assassinated and replaced as president by his son Joseph. Meanwhile, the rebel forces had become divided, with their Rwandan and Ugandan mentors at times turning on each other. By May 2001, the International Rescue Committee, a New York–based refugee agency, had estimated that the war had directly and indirectly killed some 2.5 million people.

Laurent Kabila had come to power as the leader of the Alliance of Democratic Forces for the Liberation of Congo-Zaire, a rebel movement that was itself heavily dependent on external support from Rwanda, Uganda, and Zimbabwe. In the final months of 1996, Kabila's forces seized control of eastern Zaire, and they arrived in Kinshasa in May 1997. Once in power, Kabila changed the name of the country from Zaire, an identity that had been imposed by Mobutu in 1971, to its former title: Democratic Republic of the Congo.

DEVELOPMENT

Western aid and development assistance were drastically reduced in 1992, but new aid was pledged in 1994 as a reward for Mobutu's cooperation in dealing with the Rwandan conflict. An agreement was signed with Egypt for the long-term development of Zaire's hydroelectric power.

Hopes that the new government would bring an end to the chronic corruption and mismanagement that have long plagued the country, or lead to an opening for democracy and improved human rights, soon proved misplaced. A leading Mobutu opponent, Etienne Tshisekedi wa Mulumba, was assaulted by Kabila's men within days of their victory. Tshisekedi remains the head of a nonviolent opposition coalition, the Union for Democracy and Social Progress, that is calling for a transition to democracy through a negotiated "government of national unity." So far, this plea has been all but ignored by outside powers as well as the warring factions.

Meanwhile, the country's already decayed social and economic infrastructure continues to disintegrate. The informal sector now dominates the local economy. In the process, national unity is being challenged by the reemergence of secessionist tendencies in the mineral-rich provinces of Katanga (Shaba) and Kasai. Western and eastern former Zaire has been inundated by an influx of some 2 million refugees from the killing fields of Rwanda and Burundi. With its vast size and wealth, D.R.C.'s fate will ultimately affect the future of neighboring states as well as its own citizens. Since independence, its instability has been a destabilizing influence for its neighbors in Central, Eastern, and Southern Africa. At peace, it could become the hub of an African renaissance.

Geographically, the country is located at Africa's center. It encompasses the entire Congo River Basin, whose waters are the potential source of 13 percent of the world's hydroelectric power. This immense area, about one quarter the size of the United States, encompasses a variety of land forms. It contains good agricultural possibilities and a wide range of natural resources, some of which have been intensively exploited for decades.

The D.R.C. links Africa from west to east. Its very narrow coastline faces the Atlantic. Forces from the East African coast have long influenced the landlocked eastern D.R.C. In the mid-nineteenth century, Swahili, Arab, and Nyamwezi traders from Tanzania established their hegemony over much of southeastern Zaire, pillaging the countryside for ivory and slaves. While the slave trade has left bitter memories, the Swahili language has spread to become a lingua franca throughout the eastern third of the country.

The 55 million people of the Democratic Republic of the Congo belong to more than 200 different ethnic groups, speak some 700 languages and dialects, and have varied lifestyles. Boundaries established in the late nineteenth century hemmed in portions of the Azande, Konga, Chokwe, and Songye peoples, yet they maintain contact with their kin in other countries.

Many important precolonial states were centered here, including the Luba, Kuba, and Lunda kingdoms, the latter of which, in earlier centuries, exploited the salt and copper of southeastern Zaire. The Kingdom of the Kongo, located at the mouth of the Congo River, flourished during the fifteenth and sixteenth centuries, establishing important diplomatic and commercial relations with Portugal. The elaborate political systems of these kingdoms are an important heritage for the Democratic Republic of the Congo.

LEOPOLD'S GENOCIDE

The European impact, like the Swahili and Arab influences from the east, had deeply destructive results. The Congo Basin was explored and exploited by private individuals before it came under Belgian domination. As a private citizen, King Leopold of Belgium sponsored H. M. Stanley's expeditions to explore the basin. In 1879, Leopold used Stanley's "treaties" as a justification for setting up the "Congo Independent State" over the whole region. This state was actually a private proprietary colony. To turn a profit on his vast enterprise, Leopold acted under the assumption that the people and resources in the territory were his personal property. His commercial agents and various concessionaires, to whom he leased portions of his colony, began to brutally coerce the local African population into providing ivory, wild rubber, and other commodities. The armed militias sent out to collect quotas of rubber and other goods committed numerous atrocities against the people, including destroying whole villages.

FREEDOM

The regimes in the Democratic Republic of the Congo have shown little respect for human rights. One Amnesty International report concluded that all political prisoners were tortured, and death squads were active.

No one knows for sure how many Africans perished in Congo Independent State as a result of the brutalities of Leopold's agents. Some critics estimate that the territory's population was reduced by 10 *million* people over a period of 20 years. Many were starved to death as forced laborers. Others were massacred in order to induce survivors to produce more rubber. Women and children were suffocated in "hostage houses" while their men did their masters' bidding. Thousands fled to neighboring territories.

For years the Congo regime was able to keep information of its crimes from leaking overseas, but eventually reports from missionaries and others did emerge. Public outrage was stirred by accounts such as E. D. Morel's *Red Rubber* and Mark Twain's caustic *King Leopold's Soliloquy,* as well as gruesome pictures of men, women, and children whose hands had been severed by troops (who were expected to produce the hands for their officers as evidence of their diligence). Joseph Conrad's fictionalized account of his experiences, *The Heart of Darkness,* became a popular literary classic. Finally even the European imperialists, during an era when their racial arrogance was at its height, could no longer stomach Leopold, called by some "the king with ten million murders on his soul."

During the years of Belgian rule, 1908 to 1960, foreign domination was less genocidal, but a tradition of abuse had nevertheless been established. The colonial authorities still used armed forces for "pacification" campaigns, tax collection, and labor recruitment. Local collaborators were turned into chiefs and given arbitrary powers that they would not have had under indigenous political systems. Concessionary companies continued to use force to recruit labor for their plantations and mines. The colonial regime encouraged the work of Catholic missionaries. Health facilities as well as a paternalistic system of education were developed. A strong elementary-school system was part of the colonial program, but the Belgians never instituted a major secondary-school system, and there was no institution of higher learning. By independence, only 16 Congolese had been able to earn university degrees, all but two in non-Belgian institutions. A small group of high-school–educated Congolese, known as *évolués* ("evolved ones"), served the needs of an administration that never intended nor planned for Congo's independence.

In the 1950s, the independence movements that were emerging throughout Africa affected the Congolese, especially townspeople. The Belgians began to recognize the need to prepare for a different future. Small initiatives were allowed; in 1955, nationalist associations were first permitted, and a 30-year timetable for independence was proposed. This sparked heated debate. Some évolués agreed with the Belgians' proposal. Others, including the members of the Alliance of the Ba-Kongo (ABAKO), an ethnic association in Kinshasa, and the National Congolese Movement (MNC), led by Prime Minister Patrice Lumumba, rejected it.

A serious clash at an ABAKO demonstration in 1959 resulted in some 50 deaths. In the face of mounting unrest, further encouraged by the imminent independence of the French Congo (Republic of the Congo), the Belgians conceded a rapid transition to independence. A constitutional conference in January 1960 established a federal-government system for the future independent state. But there was no real preparation for this far-reaching political change.

THE CONGO CRISIS

Democratic Republic of the Congo became independent on June 30, 1960, under the leadership of President Joseph Kasavubu and Prime Minister Patrice Lumumba. Within a week, an army mutiny had stimulated widespread disorder. The scars of Congo's uniquely bitter colonial experience showed. Unlike in Africa's other postcolonial states, hatred of the white former masters turned to violence in Congo, resulting in the hurried flight of the majority of its large European community. Ethnic and regional bloodshed took a much greater toll among the African population. The wealthy Katanga Province (now Shaba) and South Kasai seceded.

Lumumba called upon the United Nations for assistance, and troops came from a variety of countries to serve in the UN force. Later, as a result of a dispute with President Kasavubu, Lumumba sought Soviet aid. Congo could have become a Cold War battlefield, but the army, under Lumumba's former confidant, Joseph Desiré Mobutu, intervened. Lumumba was arrested and turned over to the Katanga rebels; he was later assassinated. Western interests and, in particular, the U.S. Central Intelligence Agency (CIA) played a substantial if not fully revealed role in the downfall of the idealistic Lumumba and the rise of his cynical successor, Mobutu. Rebellions by Lumumbists in the northeast and Katanga secessionists, supported by foreign mercenaries, continued through 1967.

MOBUTUISM

Mobutu seized full power in 1965, ousting Kasavubu in a military coup. With ruthless energy, he eliminated the rival political factions within the central government and crushed the regional rebellions. Mobutu banned party politics. In 1971, he established the Second Republic as a one-party state in which all power was centralized around the "Founding President." Every citizen, at birth, was legally expected to be a disciplined member of Mobutu's Popular Revolutionary Movement (MPR). With the exception of some religious organizations, virtually all social institutions were to function as MPR organs. The official ideology of the MPR republic became "Mobutuism"—the words, deeds, and decrees of "the Guide" Mobutu. All citizens were required to sing his praises daily at the workplace, at schools, and at social gatherings. In hymns and prayers, the name Mobutu was often substituted for that of Jesus. A principal slogan of Mobutuism was "authenticity." Supposedly this meant a rejection of European values and norms for African ones.

But it was Mobutu alone who defined what was authentic. He added to his own name the title *Sese Seko* ("the All Powerful") while declaring all European personal names illegal. He also established a national dress code; ties were outlawed, men were expected to wear his abacost suit, and women were obliged to wear the *paigne,* or wrapper. (The former Zaire was perhaps the only place in the world where the necktie was a symbol of political resistance.) The name of the country was changed from Congo to *Zaire,* a word derived from the sixteenth-century Portuguese mispronunciation of the (Ki)Kongo word for "river."

Outside of Zaire, some took Mobutu's protestations of authenticity at face value, while a few other African dictators, such as Togo's Gnassingbé Eyadéma, emulated aspects of his fascist methodology. But the majority of Zairians grew to loathe his "cultural revolution."

Authenticity was briefly accompanied by a program of nationalization. Union Minière and other corporations were placed under government control. In 1973 and 1974, plantations, commercial institutions, and other businesses were also taken over, in what was called a "radicalizing of the Zairian Revolution."

But the expropriated businesses simply enriched a small elite. In many cases, Mobutu gave them away to his cronies, who often sold off the assets. Consequently, the economy suffered. Industries and businesses were mismanaged or ravaged. Some individuals became extraordinarily wealthy, while the population as a whole became progressively poorer with each passing year. Mobutu allegedly became the wealthiest person in all of Africa, with a fortune estimated in excess of $5 billion (about equal to Zaire's national debt), most of which was invested and spent outside of Africa. He and his relatives owned mansions all over the world.

Until his last year in power, no opposition to Mobutu was allowed. Those critical of the regime faced imprisonment, torture, or death. The Roman Catholic Church and the Kimbanguist Church of Jesus Christ Upon This Earth were the only institutions able to speak out. Strikes were not allowed. In 1977 and 1978, new revolts in the Shaba Province were crushed by U.S.–backed Moroccan, French, and Belgian military interventions. Thus in 1997, rebels under Laurent Kabila ousted the ailing Mobutu and renamed the country the Democratic Republic of the Congo.

ECONOMIC DISASTER

The country's economic potential was developed by and for the Belgians, but by 1960, that development had gone further than in most other African colonial territories. It started with a good economic base, but the chaos of the early 1960s brought development to a standstill, and the Mobutu years were marked by regression. Development projects were initiated, but often without careful planning. World economic conditions, including falling copper and cobalt prices, contributed to Zaire's difficulties.

But the main obstacle to any sort of economic progress was the rampant corruption of Mobutu and those around him. The governing system in Zaire was a kleptocracy (rule by thieves). A well-organized system of graft transferred wealth from ordinary citizens to officials and other elites. With Mobutu stealing billions and those closest to him stealing millions, the entire society operated on an invisible tax system; for example, citizens had to bribe nurses for medical care, bureaucrats for documents, principals for school admission, and police to stay out of jail. For most civil servants, who were paid little or nothing, accepting bribes was a necessary activity. This fundamental fact also applied to most soldiers, who thus survived by living off the civilian population.

Ordinary people have suffered. By 1990, real wages of urban workers in the country were only 2 percent of what they were in 1960. Rural incomes had also deteriorated. The official 1990 price paid to coffee farmers, for example, was only one fifth of what it was in 1954 under the hugely exploitive Belgian regime. The situation has worsened since, due to periods of hyperinflation.

Much of the state's coffee and other cash crops have long been smuggled, more often than not through the connivance of senior government officials. Thus, although the country's agriculture has great economic potential, the returns from this sector continue to shrink. Despite its immense size and plentiful rainfall, it must import about 60 percent of its food requirements. Rural people move to the city or, for lack of employment, move back to the country and take up subsistence agriculture, rather than cash-crop farming, in order to ensure their own survival. The deterioration of roads and bridges has led to the decline of all trade.

In 1983, the government adopted International Monetary Fund austerity measures, but this only cut public expenditures. It had no effect on the endemic corruption, nor did it increase taxes on the rich. Under Mobutu's regime, more than 30 percent of former Zaire's budget went for debt servicing.

In June 1997, Kabila announced short-term economic priorities, including job creation, road and hospital rebuilding, and a national fuel-supply pipeline. But it was unclear where the money would come from to implement these plans.

U.S. SUPPORT FOR MOBUTU

Mobutu's regime was able to sidestep its financial crises and maintain power through the support of foreign powers, especially Belgium, France, Germany, and the United States. A U.S. intelligence report prepared in the mid-1950s concluded that the then–Belgian Congo was indeed the hub of Africa and thus vital to America's strategic interests. U.S. policy was thus the first to promote and then to perpetuate Mobutu as a pro-Western source of stability in the region. Mobutu himself skillfully cultivated this image.

Mobutu collaborated with the United States in opposing the Marxist-oriented Popular Movement for the Liberation of Angola. By so doing, he not only set himself up as an important Cold War ally but also was able to pursue regional objectives of his own. The National Front for the Liberation of Angola, long championed by the CIA as a counterforce to the MPLA, was led by an in-law of Mobutu, Holden Roberto. Mobutu also long coveted Angola's oil-rich enclave of Cabinda and thus sought CIA and South African assistance for the "independence" movement there. In recent years, millions of U.S. dollars were spent upgrading the airstrip at Kamina in Shaba Province, used by the CIA to supply the guerrillas of the National Union for the Total Independence of Angola, another faction opposed to the MPLA government. In 1989, Mobutu attempted to set himself up as a mediator between the government and the UNITA rebels, but even the latter grew to distrust him.

The United States had long known of Mobutu's human-rights violations and of the oppression and corruption that characterized his regime; high-level defectors as well as victims had publicized its abuses. Since 1987, Mobutu responded with heavily financed public-relations efforts aimed at lobbying U.S. legislators. U.S. support for Mobutu continued, but the eventual collapse of his authority led Washington belatedly to search for alternatives.

Mobutu also allied himself with other conservative forces in Africa and the Middle East. Moroccan troops came to his aid during the revolts in Shaba Province in 1977 and 1978. For his part, Mobutu was a leading African supporter of Morocco's stand with regard to the Western Sahara dispute. Under his rule the country was also an active member of the Francophonic African bloc. In 1983, Mobutu dispatched 2,000 Zairian troops to Chad in support of the government of Hissène Habré, then under attack from Libya, while in 1986, his men again joined French forces in propping up the Eyadéma regime in Togo. He also maintained and strengthened his ties with South Africa (today, the former Zaire imports almost half its food from that state). In 1982, he renewed the diplomatic ties with Israel that had been broken after the Arab–Israeli War of 1973. Israelis subsequently joined French and Belgians as senior advisers and trainers working within the former Zairian Army. In 1990, the outbreak of violent unrest in Kinshasha once more led to the intervention of French and Belgian troops.

Despite Mobutu's cultivation of foreign assistance to prop up his dictatorship, internal opposition grew. In 1990, he tried to head off his critics both at home and abroad by promising to set up a new Third Republic, based on multiparty democracy. Despite this step, repression intensified.

In October 1996, while Mobutu was in Europe recovering from cancer surgery, rebel troops under the leadership of Laurent Kabila seized their first major town, Uvira. Thousands of Rwandan Hutu refugees were forced to flee back to Rwanda. When Mobutu returned home in April 1997, he declared a nationwide state of emergency. Kabila's supporters then closed down Kinshasa as part of the campaign to oust Mobutu. Following negotiations with South Africa's Mandela, Mobutu left Kinshasa and went into exile.

Timeline: PAST

1879
Leopold sets up the Congo Independent State as his private kingdom

1906
Congo becomes a Belgian colony

1960
Congo gains independence; civil war begins; a UN force is involved; Patrice Lumumba is murdered

1965
Joseph Desiré Mobutu takes command in a bloodless coup

1971
The name of the state is changed to Zaire

1990s
Central authority crumbles; millions of Rwandan and Burundian refugees flood into Zaire; Mobutu is overthrown

PRESENT

2000s
Civil war continues, but hopes of a permanent cease-fire have emerged

Laurent Kabila is assassinated and replaced by his son, Joseph Kabila

Equatorial Guinea (Republic of Equatorial Guinea)

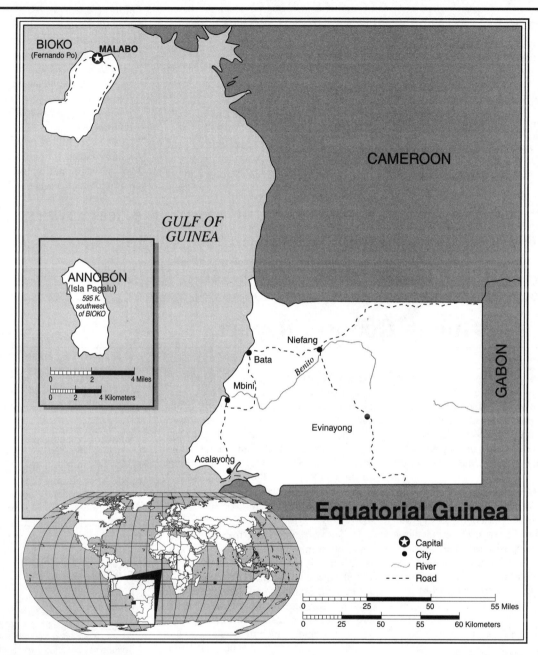

BIOKO
(Fernando Po)

MALABO

CAMEROON

GULF OF GUINEA

ANNOBÓN
(Isla Pagalu)
595 K.
southwest
of BIOKO

0   2   4 Miles
0   2   4 Kilometers

Niefang

Bata

Benito

Mbini

GABON

Evinayong

Acalayong

Equatorial Guinea

⊛ Capital
● City
〜 River
- - - Road

0   25   50   55 Miles
0   25   50   55   60 Kilometers

Equatorial Guinea Statistics

GEOGRAPHY

Area in Square Miles (Kilometers): 10,820
  (28,023) (about the size of Maryland)
Capital (Population): Malabo (58,000)
Environmental Concerns: desertification;
  lack of potable water
Geographical Features: coastal plains rise
  to interior hills; volcanic islands;
  mainland portion and 5 inhabited islands
Climate: tropical

PEOPLE

Population

Total: 499,000
Annual Growth Rate: 2.47%
Rural/Urban Population Ratio: 57/43
Major Languages: Spanish; French; Bubi;
  Fang; Ibo
Ethnic Makeup: primarily Bioko, Rio
  Muni; fewer than 1,000 Europeans

Religions: nominally Christian,
  predominantly Roman Catholic;
  indigenous beliefs

Health

Life Expectancy at Birth: 52 years (male);
  56 years (female)
Infant Mortality: 91/1,000 live births
Physicians Available: 1/3,532 people
HIV/AIDS Rate in Adults: 0.51%

Education

Adult Literacy Rate: 78.5%
Compulsory (Ages): 6–11; free

COMMUNICATION

Telephones: 4,000 main lines
Televisions: 88/1,000 people
Internet Users: 600 (2000)

TRANSPORTATION

Highways in Miles (Kilometers): 1,786
  (2,880)
Railroads in Miles (Kilometers): none
Usable Airfields: 3
Motor Vehicles in Use: 7,600

GOVERNMENT

Type: republic
Independence Date: October 12, 1968
  (from Spain)

Head of State/Government: President
  (Brigadier General) Teodoro Obiang
  Nguema Mbasogo; Prime Minister
  Candido Muatetema Rivas
Political Parties: Democratic Party for
  Equatorial Guinea; Progressive
  Democratic Alliance; Popular Action of
  Equatorial Guinea; Convergence Party
  for Social Democracy; others
Suffrage: universal at 18

MILITARY

Military Expenditures (% of GDP): 2.5%
Current Disputes: maritime boundary
  disputes with Cameroon, Gabon, and
  Nigeria

ECONOMY

Currency ($ U.S. Equivalent): 529.43
  CFA francs = $1

Per Capita Income/GDP: $2,100/$1.04
  billion
GDP Growth Rate: 6%
Inflation Rate: 6%
Unemployment Rate: 30%
Natural Resources: timber; petroleum;
  gold; manganese; uranium
Agriculture: cocoa; coffee; timber; rice;
  yams; cassava; bananas; palm oil;
  livestock
Industry: fishing; sawmilling; petroleum;
  natural gas
Exports: $2.1 billion (primary partners
  China, Japan, United States)
Imports: $736 million (primary partners
  United States, France, Spain)

SUGGESTED WEB SITES

http://www.sas.upenn.edu/
  African_Studies/E-Guinea.html

Equatorial Guinea Country Report

Few countries have been more consistently misruled than Equatorial Guinea. Having been traumatized during its first decade of independence by the sadistic Macias Nguema (1968–1979), the country continues to decay under his nephew and former security chief, Obiang Nguema Mbasogo.

In June 2002, the country's High Court handed 68 people prison sentences of up to 20 years for an alleged coup plot against President Obiang Nguema. Those jailed included a main opposition leader, Placido Mico Abogo. The European Union noted with concern that alleged "confessions" from among the accused seemed to have been obtained under duress.

DEVELOPMENT

The exploitation of oil and gas by U.S., French, and Spanish companies should greatly increase government revenues. The U.S. company Walter International recently finished work on a gas-separation plant.

Many observers look upon the trial as another betrayal of repeated promises of political reform. Since the last (1999) elections, which were characterized by widespread intimidation, including arrest and torture against the opposition, Obiang has sought to encourage exiled opponents to return to Equatorial Guinea. A few have re-

turned to register their parties for elections, scheduled for 2003, and eight of the principal opposition groups have formed a coalition to try to remove Obiang from power. In response, the president appears to be trying to distance himself from his own government. In 2001, he replaced his cabinet for failing to "respect the majority opinion of the people," immediately prior to the formation of the opposition's coalition of eight parties.

Obiang officially transformed his regime into a multiparty democracy back in 1992. But this gesture is now dismissed as a thinly disguised sham for the benefit of the French, Spanish, and Americans who provide assistance to his regime. Outside interest in the country is focused on its newfound oil wealth, which in recent years has fueled economic growth that has so far been of little benefit to ordinary people.

As a result of the government's failure to honor its commitments, all the significant opposition groups boycotted the 1993 elections, describing them as a farce. Subsequent opposition attempts to come to an accommodation with the government were set back in 1995, when a prominent opposition leader was arrested. In 1996, Obiang claimed 97.85 percent of the vote in a new presidential poll.

Equatorial Guinea's current suffering contrasts with the mood of optimism that characterized the country when it gained its independence from Spain in 1968. Confi-

dence was then buoyed by a strong and growing gross domestic product, potential mineral riches, and exceptionally good soil.

The republic is comprised of two small islands, Fernando Po (now officially known as Bioko) and Annobón, and the larger and more populous coastal enclave of Rio Muni. Before the two islands and the enclave were united, during the 1800s, as Spain's only colony in sub-Saharan Africa, all three areas were victimized by their intense involvement in the slave trade.

FREEDOM

In September 1998, Amnesty International cautiously welcomed a decree by President Obiang Nguema Mbasogo commuting the death sentences of 15 political opponents, including 4 exiles judged in absentia, who had been convicted in a summary trial the previous June. The reprieves were considered a vindication of those arguing for the continued need to put international pressure on the regime.

Spain's major colonial concern was the prosperity of the large cocoa and coffee plantations that were established on the islands, particularly on Fernando Po. Because of resistance from the local Bubi, labor for these estates was imported from elsewhere in West Africa. Coercive recruitment and poor working conditions led to frequent charges of slavery.

HEALTH/WELFARE

At independence, Equatorial Guinea had one of the best doctor-to-population ratios in Africa, but Macias's rule left it with one of the lowest. Health care is gradually reviving, however, with major assistance coming from public and private sources.

Despite early evidence of its potential riches, Rio Muni was largely neglected by the Spanish, who did not occupy its interior until 1926. In the 1930s and 1940s, much of the enclave was under the political control of the Elar-ayong, a nationalist movement that sought to unite the Fang, Rio Muni's principal ethnic group, against both the Spanish and the French rulers in neighboring Cameroon and Gabon. The territory has remained one of the world's least developed areas.

In 1968, then–Fascist-ruled Spain entrusted local power to Macias Nguema, who had risen through the ranks of the security service. Under his increasingly deranged misrule, virtually all public and private enterprise collapsed; indeed, between 1974 and 1979, the country had no budget. One third of the nation's population went into exile; tens of thousands of others were either murdered or allowed to die of disease and starvation. Many of the survivors were put to forced labor, and the rest were left to subsist off the land. Killings were carried out by boys conscripted between the ages of seven and 14.

Although no community in Equatorial Guinea was left unscarred by Macias's tyranny, the greatest disruption occurred on the islands. By 1976, the entire resident-alien population had left, along with most surviving members of the educated class. On Annobón, the government blocked all international efforts to stem a severe cholera epidemic in 1973. The near-total depopulation of the island was completed in 1976, when all able-bodied men on Annobón, along with another 20,000 from Rio Muni, were drafted for forced labor on Fernando Po.

ACHIEVEMENTS

At independence, 90% of all children attended school, but the schools were closed under Macias. Since 1979, primary education has revived and now incorporates most children.

If Equatorial Guinea's first decade of independence was hell, the years since have at best been purgatory. No sector of the economy is free of corruption. Uncontrolled—and in theory illegal—logging is destroying Rio Muni's environment, while in Malabo, the police routinely engage in theft. Food is imported and malnutrition commonplace. It has been reported that the remaining population of Annobón is being systematically starved while Obiang Nguema Mbasogo collects huge payments from international companies that use the island as a toxic-waste dump.

At least one fifth of the Equato-Guinean population continue to live in exile, mostly in Cameroon and Gabon. This community has fostered a number of opposition groups. The government relies financially on French and Spanish aid. But Madrid's commitment has been strained by criticism from the Spanish press, which has been virtually alone in publicizing the continued suffering of Equatorial Guinea's people.

Timeline: PAST

1500s
Europeans explore modern Equatorial Guinea

1641
The Dutch establish slave-trading stations

1778
Spain claims the area of Equatorial Guinea; de facto control is not completed until 1926

1930
The League of Nations investigates charges of slavery on Fernando Po

1958
The murder of nationalist leader Acacio Mane leads to the founding of political parties

1963
Local autonomy is granted

1968
Independence; Macias Nguema begins his reign

1979
A coup ends the dictatorial regime of Macias Nguema; Obiang Nguema Mbasogo becomes the new ruler

1990s
A shift to multipartyism is accompanied by wave of political detentions; Obiang Nguema Mbasogo claims electoral victory

PRESENT

2000s
The exploitation of large oil reserves boosts the economy, but ordinary people still suffer

Gabon (Gabonese Republic)

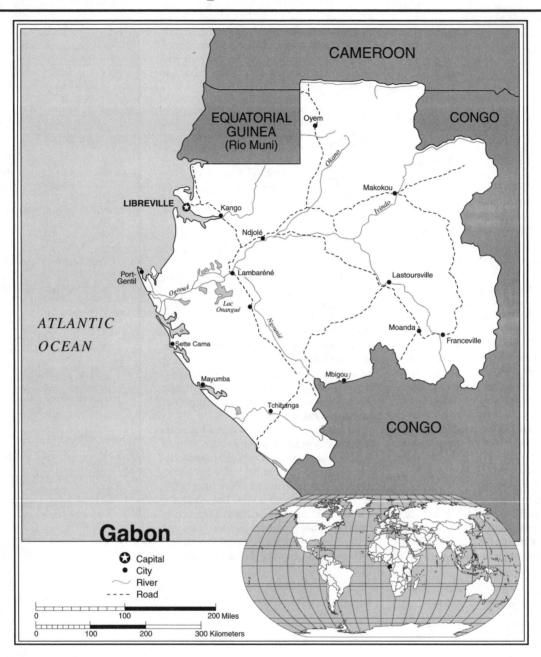

Gabon

Capital
City
River
Road

0   100   200 Miles
0   100   200   300 Kilometers

Gabon Statistics

GEOGRAPHY

Area in Square Miles (Kilometers): 102,317
  (264,180) (about the size of Colorado)
Capital (Population): Libreville (573,000)
Environmental Concerns: deforestation;
  poaching
Geographical Features: narrow coastal
  plain; hilly interior; savanna in the east
  and south
Climate: tropical

PEOPLE

Population

Total: 1,234,000
Annual Growth Rate: 0.97%
Rural/Urban Population Ratio: 19/81
Major Languages: French; Fang; Myene;
  Eshira; Bopounou; Bateke; Bandjabi
Ethnic Makeup: about 95% African,
  including Eshira, Fang, Bapounou, and
  Bateke; 5% European

Religions: 55%–75% Christian; less than
  1% Muslim; remainder indigenous
  beliefs

Health

Life Expectancy at Birth: 48 years (male);
  50 years (female)
Infant Mortality: 93.5/1,000 live births
Physicians Available: 1/2,337 people
HIV/AIDS Rate in Adults: 9%

48

Education

Adult Literacy Rate: 63.2%
Compulsory (Ages): 6–16

COMMUNICATION

Telephones: 39,000 main lines
Televisions: 35/1,000 people
Internet Users: 15,000 (2001)

TRANSPORTATION

Highways in Miles (Kilometers): 4,650 (7,500)
Railroads in Miles (Kilometers): 402 (649)
Usable Airfields: 59
Motor Vehicles in Use: 33,000

GOVERNMENT

Type: republic; multiparty presidential regime
Independence Date: August 17, 1960 (from France)

Head of State/Government: President El Hadj Omar Bongo; Prime Minister Jean-Francois Ntoutoume-Emane
Political Parties: Gabonese Democratic Party; Gabonese Party for Progress; National Woodcutters Rally; others
Suffrage: universal at 21

MILITARY

Military Expenditures (% of GDP): 2%
Current Disputes: maritime boundary dispute with Equatorial Guinea

ECONOMY

Currency ($ U.S. equivalent): 529.43 CFA francs = $1
Per Capita Income/GDP: $5,500/$6.7 billion
GDP Growth Rate: 2.5%
Inflation Rate: 1.5%
Unemployment Rate: 21%

Labor Force by Occupation: 60% agriculture; 25% services and government; 15% industry and commerce
Natural Resources: petroleum; iron ore; manganese; uranium; gold; timber; hydropower
Agriculture: cocoa; coffee; palm oil
Industry: petroleum; lumber; mining; chemicals; ship repair; food processing; cement; textiles
Exports: 2.5 billion (primary partners United States, France, China)
Imports: $921 million (primary partners France, Côte d'Ivoire, United States)

SUGGESTED WEB SITES

```
http://www.gabonnews.com
http://www.sas.upenn.edu/
    African_Studies/
    Country_Specific/Gabon.html
http://www.presidence-gabon.com/
    index-a.html
```

Gabon Country Report

Since independence, Gabon has achieved one of the highest per capita gross domestic products in Africa, due to exploitation of the country's natural riches, especially its oil. But there is a wide gap between such statistical wealth and the real poverty that still shapes the lives of most Gabonese. Disparities in income have helped fuel crime. In a controversial response, in July 2002 the government razed four village suburbs of the capital city Libreville, saying that the areas had become havens for foreign criminal gangs.

DEVELOPMENT

The Trans-Gabonais Railway is one of the largest construction projects in Africa. Work began in 1974 and, after some delays, most of the line is now complete. The railway has opened up much of Gabon's interior to commercial development.

At the top of the local governing elite is President Omar Bongo, whose main palace, built a decade ago at a reported cost of $300 million, symbolizes his penchant for grandeur. Shortly after taking office, in 1967, Bongo institutionalized his personal rule as the head of a one-party state. Until recently, his Democratic Party of Gabon (PDG) held a legal monopoly of power. But, although the PDG's Constitution restricted the presidency to the "Founder President," for many years it was Gabon's

former colonial master, France, not the ruling party's by-laws, that upheld the Bongo regime.

The French colonial presence in Gabon dates back to 1843. Between 1898 and 1930, many Gabonese were subject to long periods of forced labor, cutting timber for French concessions companies. World War II coincided with a period of political liberalization in the territory under the Free French government of Felix Emboue, a black man born in French Guiana. Educated Gabonese were promoted for the first time to important positions in the local administration. In the 1950s, two major political parties emerged to compete in local politics: the Social Democratic Union of Gabon (UDSG), led by Jean-Hilaire Aubame; and the Gabonese Democratic Bloc (BDG) of Indjenjet Gondjout and Leon M'ba.

In the 1957 elections, the UDSG received 60 percent of the popular vote but gained only 19 seats in the 40-seat Assembly. Leon M'ba, who had the support of French logging interests, was elected leader by 21 BDG and independent deputies. As a result, it was M'ba who was at the helm when Gabon gained its independence, in 1960. This birth coincided with M'ba's declaration giving himself emergency powers, provoking a period of prolonged constitutional crisis.

In January 1964, M'ba dissolved the Assembly over its members' continued re-

fusal to accept a one-party state under his leadership. In February, the president himself was forced to resign by a group of army officers. Power was transferred to a civilian "Provisional Government," headed by Aubame, which also included BDG politicians such as Gondjout and several prominent, unaffiliated citizens. However, no sooner had the Provisional Government been installed than Gabon was invaded by French troops. Local military units were massacred in the surprise attack, which returned M'ba to office. Upon his death, M'ba was succeeded by his hand-picked successor, Omar Bongo.

FREEDOM

Since 1967 Bongo has maintained power through a combination of repression and the deft use of patronage. The current transition to a multiparty process, however, has led to an improvement in human rights.

It has been suggested that France's 1964 invasion was motivated primarily by a desire to maintain absolute control over Gabon's uranium deposits, which were then vital to France's nuclear-weapons program. Many Gabonese have since believed that their country has remained a de facto French possession. France has maintained its military presence, and the Presidential Guard, mainly officered by Moroccan and

French mercenaries, outguns the Gabonese Army. France dominates Gabon's resource-rich economy.

HEALTH/WELFARE

The government claims to have instituted universal, compulsory education for Gabonese up to age 16. Independent observers doubt the government's claim but concur that major progress has been made in education. Health services have also expanded greatly.

In recent decades, Gabon's status quo has been challenged by its increasingly urbanized population. Although Bongo was able to co-opt or exile many of the figures who had once opposed M'ba, a new generation of opposition has emerged both at home and in exile. During the 1980s, the underground Movement for National Recovery (MORENA) emerged as the leading opposition group. In 1989, Bongo began talks with some elements within MORENA, which led to a division within its ranks. But the breakup of MORENA failed to stem the emergence of new groups calling for a return to multiparty democracy.

Demonstrations and strikes at the beginning of 1990 led to the legalization of opposition parties. But the murder of a prominent opposition leader in May led to serious rioting at Port-Gentil, Gabon's second city. In response, France sent troops to the area. Multiparty elections for the National Assembly, in September–October 1990, resulted in a narrow victory for the PDG, amid allegations of widespread fraud. In 1992, most opposition groups united as the Coordination of Democratic Opposition. Bongo's victory claim in the December 1993 presidential elections was widely disbelieved. In September 1994, he agreed to the formation of a coalition "Transitional Government" and the drafting of a new Constitution, which was approved by 96 percent of the voters in July 1995.

ACHIEVEMENTS

Gabon will soon have a second private television station, funded by a French cable station. Profits will be used to fund films that will be shown on other African stations. Gabon's first private station is funded by Swiss and Gabonese capital.

In 1996, the PDG won a sweeping, but controversial, victory in parliamentary elections. This was followed up in 1998 by a landslide reelection victory for Bongo. The PDG's continuing hold on power was reconfirmed in the 2002 parliamentary elections. Following the poll, the National Woodcutters Rally, a party that has strong support in Libreville, agreed to take up Bongo's offer to serve as junior partners with the PDG in Gabon's first coalition government. Although opposition politicians claimed that the results had been rigged, independent observers have credited the PDG with success in retaining support of the rural base while co-opting opponents into its fold.

Timeline: PAST

1849
Libreville is founded by the French as a settlement for freed slaves

1910
Gabon becomes a colony within French Equatorial Africa

1940
The Free French in Brazzaville seize Gabon from the pro-Vichy government

1960
Independence is gained; Leon M'ba becomes president

1967
Omar Bongo becomes Gabon's second president after M'ba's death

1968
The Gabonese Democratic Party (PDG) becomes the only party of the state

1990s
Bongo agrees to multiparty elections but seeks to put limits on the opposition; riots in Port-Gentil

PRESENT

2000s
The PDG retains power

São Tomé and Príncipe (Democratic Republic of São Tomé and Príncipe)

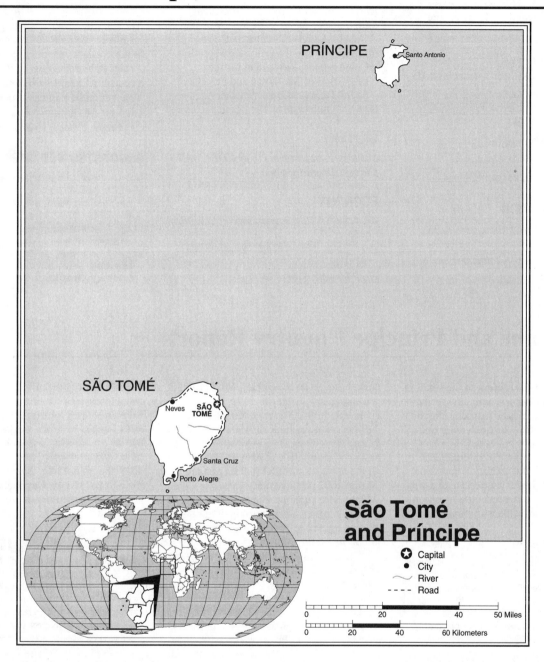

São Tomé and Príncipe Statistics

GEOGRAPHY

Area in Square Miles (Kilometers): 387 (1,001) (about 5 times the size of Washington, D.C.)

Capital (Population): São Tomé (67,000)

Environmental Concerns: deforestation; soil erosion; soil exhaustion

Geographical Features: volcanic; mountainous; islands in the Gulf of Guinea; the smallest country in Africa

Climate: tropical

PEOPLE

Population

Total: 170,100

Annual Growth Rate: 3.1%

Rural/Urban Population Ratio: 54/46

Major Languages: Portuguese; Fang; Kriolu

Ethnic Makeup: Portuguese-African mixture; African minority

Religions: 80% Christian; 20% others

Health

Life Expectancy at Birth: 64 years (male);
67 years (female)
Infant Mortality: 47.5/1,000 live births
Physicians Available: 1/1,881 people

Education

Adult Literacy Rate: 73%
Compulsory (Ages): for 4 years between
ages 7–14

COMMUNICATION

Telephones: 3,000 main lines
Televisions: 154/1,000 people
Internet Users: 6,500 (2001)

TRANSPORTATION

Highways in Miles (Kilometers): 198
(320)
Railroads in Miles (Kilometers): none
Usable Airfields: 2

GOVERNMENT

Type: republic
Independence Date: July 12, 1975 (from
Portugal)
Head of State/Government: President
Fradique de Menezes; Prime Minister
Gabriel Arcanjo Ferreira da Costa
Political Parties: Party for Democratic
Convergence; Movement for the
Liberation of São Tomé and Príncipe–
Social Democratic Party; others
Suffrage: universal at 18

MILITARY

Military Expenditures (% of GDP): 0.8%
Current Disputes: none

ECONOMY

Currency ($ U.S. equivalent): 130 dobras
= $1
Per Capita Income/GDP: $1,200/$189
million
GDP Growth Rate: 4%

Inflation Rate: 7%
Labor Force by Occupation: mainly
engaged in subsistence agriculture and
fishing
Natural Resources: fish; hydropower
Agriculture: cacao; coconut palms; coffee;
bananas; palm kernels; copra
Industry: light construction; textiles; soap;
beer; fish processing; timber
Exports: $4.1 million (primary partners
Netherlands, Portugal, Spain)
Imports: $40 million (primary partners
Portugal, France, United Kingdom)

SUGGESTED WEB SITES

```
http://www.sao-tome.com
http://www.state.gov/
  www.background_notes/
  sao_tome_0397_bgn.html
http://www.emulateme.com/
  saotome.htm
http://www.sas.upenn.edu/
  African_Studies/
  Country_Specific/Sao_Tome.html
```

São Tomé and Príncipe Country Report

In August 1995, soldiers in the small is-land-nation of São Tomé and Príncipe briefly deposed Miguel Trovoada, the country's first democratically elected president. The coup quickly collapsed, however, in the face of domestic and international opposition. While the country's new democracy survived, it was vulnerable to a weak economy, which showed little prospect of significant improvement anytime soon.

DEVELOPMENT

Local food production has been significantly boosted by a French-funded plan. Japan is assisting in fishery development. There is concern that tourist fishermen may adversely affect the local fishing industry.

The islands held their first multiparty elections in January 1991. The elections resulted in the defeat of the former ruling party, the Liberation Movement of São Tomé and Príncipe–Social Democratic Party (MLSTP–PSD), by Trovoada's Party for Democratic Convergence–Group of Reflection (PDC–GR). Subsequent elections in December 1992, however, reversed the PDC–GR advantage in Parliament, leading to an uneasy division of power. This division was reinforced with Trovoada's reelection in 1996, fol-

lowed by an even greater MLSTP–PSD parliamentary victory in 1998.

In the July 2001 presidential elections businessman Fradique de Menezes was declared the winner. He was sworn into office in early September. However, the victory of the opposition MLSTP–PSD party in the March 2002 parliamentary elections led de Menezes to appoint Gabriel da Costa as prime minister, and both main political parties agreed to share power and form a broad-based government.

The government was confronted with a massive civil-servants strike in 2001 to press for higher pay. Officials said the country's external debt in 1998 amounted to U.S. $270 million, far more than the country's annual gross domestic product.

São Tomé and Príncipe gained its independence in 1975, after a half-millennium of Portuguese rule. During the colonial era, economic life centered around the interests of a few thousand Portuguese settlers, particularly a handful of large-plantation owners who controlled more than 80 percent of the land. After independence, most of the Portuguese fled, taking their skills and capital and leaving the economy in disarray. But production on the plantations has since been revived.

The Portuguese began the first permanent settlement of São Tomé and Príncipe in the late 1400s. Through slave labor, the islands developed rapidly as one of the

world's leading exporters of sugar. Only a small fraction of the profits from this boom were consumed locally; and high mortality rates, caused by brutal working conditions, led to an almost insatiable demand for more slaves. Profits from sugar declined after the mid-1500s due to competition from Brazil and the Caribbean. A period of prolonged depression set in.

FREEDOM

Before 1987, human rights were circumscribed in São Tomé and Príncipe. Gradual liberalization has now given way to a commitment to political pluralism. The current government has a good record of respect for human rights. Major problems are an inefficient judicial system, harsh prison conditions, and acts of police brutality. Outdated labor practices on the plantations limit worker rights.

In the early 1800s, a second economic boom swept the islands, when they became leading exporters of coffee, and, more important, cocoa. (São Tomé and Príncipe's position in the world market has since declined, yet these two cash crops, along with copra, have continued to be economic mainstays.) Although slavery was officially abolished during the nineteenth century, forced labor was maintained by the Portuguese into modern times. Involuntary contract workers, known as *servicais,* were imported to labor on the islands' plantations, which had notoriously high mortality rates. Sporadic labor unrest and occasional incidents of international outrage led to some improvement in working conditions, but fundamental reforms came about only after independence. A historical turning point for the islands was the Batepa Massacre in 1953, when several hundred African laborers were killed following local resistance to labor conditions.

HEALTH/WELFARE

Since independence, the government has had enormous progress in expanding health care and education. The Sãotoméan infant mortality rate is now among the lowest in Africa, and average life expectancy is among the highest. About 65% of the population between 6 and 19 years of age now attend school.

Between 1975 and 1991, São Tomé and Príncipe was ruled by the MLSTP–PSD, which had emerged in exile as the island's leading anticolonial movement, as a one-party state initially committed to Marxist-Leninism. But in 1990, a new policy of *abertura,* or political and economic "opening," resulted in the legalization of opposition parties and the introduction of direct elections with secret balloting. Press restrictions were also lifted, and the nation's security police were purged. The democratization process was welcomed by previously exiled opposition groups, most of which united as the PDC–GR. The changed political climate was also reflected in the establishment of an independent labor movement. Previously, strikes were forbidden.

ACHIEVEMENTS

São Tomé and Príncipe shares in a rich Luso-African artistic tradition. The country is particularly renowned for poets such as Jose de Almeida and Francisco Tenreiro, who were among the first to express in the Portuguese language the experiences and pride of Africans.

The move toward multiparty politics was accompanied by an evolution to a market economy. Since 1985, a "Free Trade Zone" was established, state farms were privatized, and private capital was attracted to build up a tourist industry. These moves were accompanied by a major expansion of Western loans and assistance to the islands—an inflow of capital that now accounts for nearly half of the gross domestic product.

The government also focused its development efforts on fishing. In 1978, a 200-mile maritime zone was declared over the tuna-rich waters around the islands. The state-owned fishing company, Empesca, began upgrading the local fleet, which still consists mostly of canoes using old-fashioned nets. The influx of aid and investment has resulted in several years of sustained economic growth.

The current inhabitants of São Tomé and Príncipe are primarily of mixed African and European descent. During the colonial period, the society was stratified along racial lines. At the top were the Europeans —mostly Portuguese. Just below them were the mesticos or *filhos da terra*, the mixed-blood descendants of slaves. Descendants of slaves who arrived later were known as *forros*. Contract workers were labeled as *servicais*, while their chil-

dren became known as *tongas*. Still another group was the *angolares*, who reportedly were the descendants of shipwrecked slaves. All of these colonial categories were used to divide and rule the local population; the distinctions have begun to diminish, however, as an important sociological factor on the islands.

Timeline: PAST

1500s
The Portuguese settle São Tomé and Príncipe

1876
Slavery is abolished, but forced labor continues

1953
The Portuguese massacre hundreds of islanders

1972
Factions within the liberation movement unite to form the MLSTP in Gabon

1975
Independence

1979
Manuel Pinto da Costa deposes and exiles Miguel Trovoada, the premier and former number-two man in the MLSTP

1990s
Economic and political liberalization; multiparty elections

PRESENT

2000s
Fradique de Menezes wins the presidency

East Africa

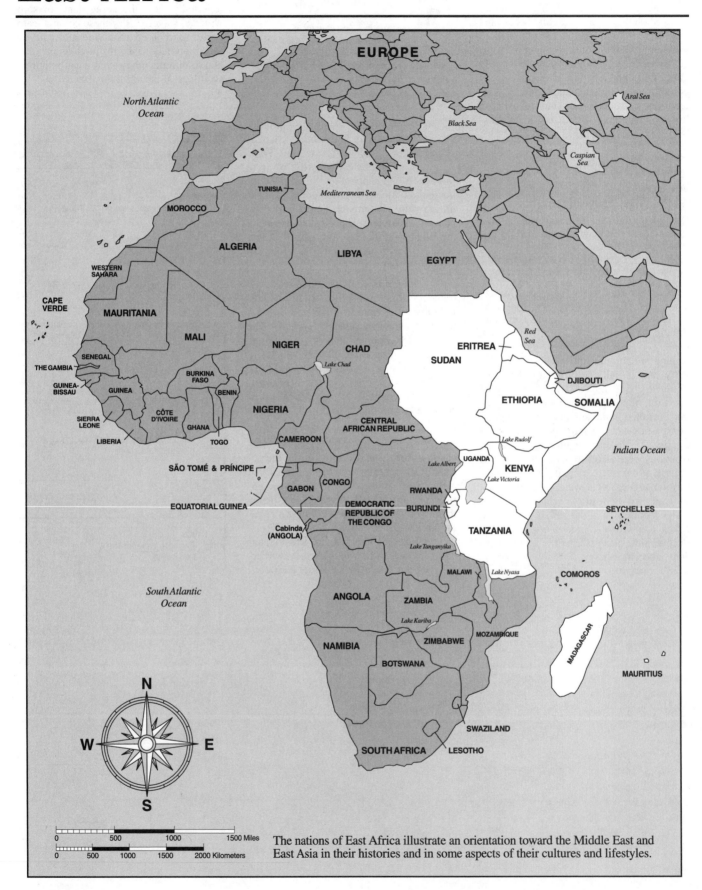

The nations of East Africa illustrate an orientation toward the Middle East and East Asia in their histories and in some aspects of their cultures and lifestyles.

East Africa: A Mixed Inheritance

The vast East African region, ranging from Sudan in the north to Tanzania and the Indian Ocean islands in the south, is an area of great diversity. Although the islands are the homes of distinctive civilizations with ties to Asia, their interactions with the African mainland give their inclusion here validity. Ecological features such as the Great Rift Valley, the prevalence of cattle-herding lifestyles, and long-standing participation in the Indian Ocean trading networks are some of the region's unifying aspects.

CATTLE-HERDING SOCIETIES

A long-horned cow would be an appropriate symbol for East Africa. Most of the region's rural inhabitants, who make up the majority of people from the Horn, to Lake Malawi, to Madagascar, value cattle for their social as well as economic importance. The Nuer of Sudan, the Somalis near the Red Sea (who, like many other peoples of the Horn, herd camels as well as cattle, goats, and sheep), and the Maasai of Tanzania and Kenya are among the pastoral peoples whose herds provide their livelihoods. Farming communities such as the Kikuyu of Kenya, the Baganda of Uganda, and the Malagasy of Madagascar also prize cattle.

Much of the East African landmass is well suited for herding. Whereas the rain forests of West and Central Africa are generally infested with tsetse flies, whose bite is fatal to livestock, most of East Africa is made up of belts of tropical and temperate savanna, which are ideal for grazing. Thus pastoralism has long been predominant in the savanna zones of West and Southern, as well as East, Africa. Tropical rain forests are found in East Africa only on the east coast of Madagascar and scattered along the mainland's coast. Much of the East African interior is dominated by the Great Rift Valley, which stretches from the Red Sea as far south as Malawi. This geological formation is characterized by mountains as well as valleys, and it features the region's great lakes, such as Lake Albert, Lake Tanganyika, and Lake Malawi.

People have been moving into and through the East African region since the existence of humankind; indeed, most of the earliest human fossils have been unearthed in this region. Today, almost all the mainland inhabitants speak languages that belong to either the Bantu or Nilotic linguistic families. There has been much historical speculation about the past migration of these peoples, but recent research indicates that both linguistic groups have probably been established in the area for a long time, although oral traditions and other forms of historical evidence indicate locally important shifts in settlement patterns into the contemporary period. Iron working and, in at least a few cases, small-scale steel production have been a part of the regional economy for more than 2,000 years. Long-distance trade and the production of various crafts have also existed since ancient times.

The inhabitants of the region have had to deal with insufficient and unreliable rainfall. Drought and famine in the Horn and in areas of Kenya and Tanzania have in recent years changed lifestyles and dislocated many people.

ISLAMIC INFLUENCE

Many of the areas of East Africa have been influenced—since at least as far back as Roman times and perhaps much further— by the Middle East and other parts of Asia. Over the past thousand years, most parts of East Africa, including the Christian highlands of Ethiopia and the inland interlake states such as Buganda, Burundi, and Rwanda, became familiar to the Muslim Arab traders of the Swahili and Red Sea coasts and the Sudanese interior. Somalia, Djibouti, and Sudan, which border the Red Sea and are close to the Arabian Peninsula, have been the countries most influenced by Arab Islamic culture. Mogadishu, the capital of Somalia, began as an Islamic trading post in the tenth century A.D. The Islamic faith, its various sects or brotherhoods, the Koran, and the Shari'a (the Islamic legal code) are predominant throughout the Horn, except in the Ethiopian and Eritrean highlands and southern Sudan. In recent years, many Somalis, Sudanese, and others have migrated to the oil-rich states of Arabia to work.

Farther south, in the communities and cultures on the perimeters of the east coast, Arabs and local Bantu-speaking Africans combined, from as early as the ninth century but especially during the 1200s to 1400s, to form the culture and the language that we now call Swahili. In the first half of the nineteenth century, Seyyid Said, the sultan of Oman, transferred his capital to Zanzibar, in recognition of the outpost's economic importance. Motivated by the rapid expansion of trade in ivory and slaves, many Arab–Swahili traders began to establish themselves and build settlements as far inland as the forests of eastern Democratic Republic of the Congo. As a result, some of the non-coastal peoples also adopted Islam, while Swahili developed into a regional lingua franca.

The whole region from the Horn to Tanzania continued to be affected by the slave trade through much of the nineteenth century. Slaves were sent north from Uganda and southern Sudan to Egypt and the Middle East, and from Ethiopia across the Red Sea. Others were taken to the coast by Arab, Swahili, or African traders, either to work on the plantations in Zanzibar or to be transported to the Persian Gulf and the Indian Ocean islands.

In the late 1800s and early 1900s, South Asian laborers from what was then British India were brought in by the British to build the East African railroad. South Asian traders already resided in Zanzibar; others now came and settled in Kenya and Tanzania, becoming shopkeepers and bankers in inland centers, such as Kampala and Nairobi, as well as on the coast, in Mombasa and Dar es Salaam or in smaller stops along the railroad. South Asian laborers were also sent in large numbers to work on the sugar plantations of Mauritius; their descendants there now make up about two thirds of that island's population.

The subregions of East Africa include the following: the countries of the Horn, East Africa proper, and the islands. The Horn includes Djibouti, Ethiopia, Eritrea, Somalia, and Sudan, which are associated here with one another not so much because of a common heritage or on account of any compatibility of

(United Nations photo by Ray Witlin)

In the drought-affected areas of East Africa, people must devote considerable time and energy to the search for water.

their governments (indeed, they are often hostile to one another), but because of the movements of peoples across borders in recent times. *East Africa proper* is comprised of Kenya, Tanzania, and Uganda, which do have underlying cultural ties and a history of economic relations, in which Rwanda and Burundi have also shared. The *Indian Ocean islands* include the Comoros, Madagascar, Mauritius, and Seychelles, which, notwithstanding the expanses of ocean that separate them, have certain cultural aspects and current interests in common.

THE HORN

Ethiopia traditionally has had a distinct, semi-isolated history that has separated the nation from its neighbors. This early Christian civilization, which was periodically united by a strong dynasty but at other times was disunited, was centered in the highlands of the interior, surrounded by often hostile lowland peoples. Before the nineteenth century, it was in infrequent contact with other Christian societies. In the 1800s, however, a series of strong rulers reunified the highlands and went on to conquer surrounding peoples such as the Afar, Oromo, and Somali. In the process, the state expanded to its current boundaries. While the empire's expansion helped it to preserve its independence during Africa's colonial partition, sectarian and ethnic divisions—a legacy of the imperial state-building process—now threaten to tear the polity apart.

Ethiopia and the rest of the Horn have been influenced by outside powers, whose interests in the region have been primarily rooted in its strategic location. In the nineteenth century, both Britain and France became interested in the Horn, because the Red Sea was the link between their countries and the markets of Asia. This was especially true after the completion of the Suez Canal in 1869. Both of the imperial powers occupied ports on the Red Sea at the time. They then began to compete over the upper Nile in modern Sudan. In the 1890s, French forces, led by Captain Jean Baptiste Marchand, literally raced from the present-day area of Congo to reach the center of Sudan before the arrival of a larger British expeditionary force, which had invaded the region from Egypt. Ultimately, the British were able to consolidate their control over the entire Sudan.

Italian ambitions in the Horn were initially encouraged by the British, in order to counter the French. Italy's defeat by the Ethiopians at the Battle of Adowa in 1896 did not deter its efforts to dominate the coastal areas of Eritrea and southeastern Somalia. Later, under Benito Mussolini, Italy briefly (1936–1942) occupied Ethiopia itself.

During the Cold War, great-power competition for control of the Red Sea and the Gulf of Aden, with their strategic locations near the oil fields of the Middle East as well as along the Suez shipping routes, continued between the United States and the Soviet Union. Local events sometimes led to shifts in alignments. Before 1977, for instance, the United States was closely allied with Ethiopia, and the Soviet Union with Somalia. However, in 1977–1978, Ethiopia, having come under a self-proclaimed Marxist-Leninist government, allied itself with the

Soviet Union, receiving in return the support of Cuban troops and billions of dollars' worth of Socialist-bloc military aid, on loan, for use in its battles against Eritrean and Somali rebels. The latter group, living in Ethiopia's Ogaden region, were seeking to become part of a greater Somalia. In this irredentist adventure, they had the direct support of invading Somalia troops. Although the United States refused to counter the Soviets by in turn backing the irredentists, it subsequently established relations with the Somali government at a level that allowed it virtually to take over the former Soviet military facility at Berbera.

Discord and Drought

The countries of the Horn, unlike the other states in the region, are politically alienated from one another. There is thus little prospect of an effective regional community emerging among them in the foreseeable future. Although the end of the Cold War has reduced the interest of external powers, local animosities continue to wreak havoc in the region. The Horn continues to be bound together and torn apart by millions of refugees fleeing armed conflicts in all of the states. Ethiopia, Somalia, and Sudan have suffered under especially vicious authoritarian regimes that resorted to the mass murder of dissident segments of their populations. Although the old regimes have been overthrown in Ethiopia and Somalia, peace has yet to come to either society. Having gained its independence only in 1993, Eritrea, Africa's newest nation, has struggled to overcome the devastating legacy of its 30-year liberation struggle against Ethiopia. Eritrea's well-being has been further compromised by reverses in a border war with Ethiopia. Recent battlefield victories against the Eritreans have revived the passions of some Ethiopians who have never fully accepted Eritrea's succession. The stability of neighboring Djibouti, once a regional enclave of calm, has also been compromised in recent years by sometimes violent internal political conflicts.

The horrible effects of these wars have been magnified by recurrent droughts and famines. Hundreds of thousands of people have starved to death in the past decade, while many more have survived only because of international aid efforts.

Ethiopians leave their homes for Djibouti, Somalia, and Sudan for relief from war and famine. Sudanese and Somalis flee to Ethiopia for the same reasons. Today, every country harbors not only refugees but also dissidents from neighboring lands and has a citizenry related to those who live in adjoining countries. Peoples such as the Afar minority in Djibouti often seek support from their kin in Eritrea and Ethiopia. Many Somali guerrilla groups have used Ethiopia as a base, while Somali factions have continued to give aid and comfort to Ethiopia's rebellious Ogaden population. Ethiopian factions allegedly continue to assist southern rebels against the government of Sudan, which had long supported the Tigray and Eritrean rebel movements of northern Ethiopia.

At times, the states of the region have reached agreements among themselves to curb their interference in one another's affairs. But they have made almost no progress in the more fundamental task of establishing internal peace, thus assuring that the region's violent downward spiral continues.

THE SOUTHERN STATES OF EAST AFRICA

The peoples of Kenya, Tanzania, and Uganda as well as Burundi and Rwanda have underlying connections rooted in the past. The kingdoms of the Lakes Region of Uganda, Rwanda, and Burundi, though they have been politically superseded in the post-colonial era, have left their legacies. For example, myths about a heroic dynasty of rulers, the Chwezi, who ruled over an early Ugandan-based kingdom, are widespread. Archaeological evidence attests to the actual existence of the Chwezi, probably in the sixteenth century. Peoples in western Kenya and Tanzania, who have lived under less centralized systems of governance but nonetheless have rituals similar to those of the Ugandan kingdoms, also share the traditions of the Chwezi dynasty, which have become associated with a spirit cult.

The precolonial kingdoms of Rwanda and Burundi, both of which came under German and, later, Belgian control during the colonial era, were socially divided between a ruling warrior class, the Tutsis, and a much larger peasant class, the Hutus. Although both states are now independent republics, their societies remain bitterly divided along these ethnoclass lines. In Rwanda, the feudal hegemony of the Tutsis was overthrown in a bloody civil conflict in 1959, which led to the flight of many Tutsis. But in 1994, the sons of these Tutsi exiles came to power, after elements in the former Hutu-dominated regime organized a genocidal campaign against all Tutsis. In the belief that the Tutsis were back on top, millions of Hutus then fled the country. In Burundi, Tutsi rule was maintained for decades through a repressive police state, which in 1972 and 1988 resorted to the mass murder of Hutus. Elections in 1993 resulted in the country's first Hutu president at the head of a government that included members of both groups, but he was murdered by the predominantly Tutsi army. Since then, the country has been teetering on the brink of yet another catastrophe, as some of its politicians try to promote reconciliation.

Kenya and Uganda were taken over by the British in the late nineteenth century, while Tanzania, originally conquered by Germany, became a British colony after World War I. In Kenya, the British encouraged the growth of a settler community. Although never much more than 1 percent of the colony's resident population, the British settlers were given the best agricultural lands in the rich highlands region around Nairobi; and throughout most of the colonial era, they were allowed to exert a political and economic hegemony over the local Africans. The settler populations in Tanzania and Uganda were smaller and less powerful. While the settler presence in Kenya led to land alienation and consequent immiseration for many Africans, it also fostered a fair amount of colonial investment in infrastructure. As a result, Kenya had a relatively sophisticated economy at the time of its independence, a fact that was to complicate proposals for its economic integration with Tanzania and Uganda.

In the 1950s, the British established the East African Common Services Organization to promote greater economic cooperation among its Kenyan, Tanganyikan (Tanzanian), and Ugandan territories. By the early 1960s, the links among the states were so close that President Julius Nyerere of Tanzania proposed that his country delay its independence until Kenya also gained its freedom, in hopes that the two countries would then join together. This did not occur.

In 1967, the Common Services Organization was transformed by its three (now independent) members into a full-fledged "common market," known as the East African Community (EAC). The EAC collectively managed the railway system, development of harbors, and international air, postal, and telecommunication facilities. It also maintained a common currency, development bank, and other economic, cultural, and scientific services. Peoples moved freely across the borders of the three EAC states. However, the EAC soon began to unravel, as conflicts over its operations grew. It finally collapsed in 1977. The countries disputed the benefits of the association, which seemed to have been garnered primarily by Kenya. The ideologies and personalities of its leaders at the time—Nyerere, Jomo Kenyatta of Kenya, and Idi Amin of Uganda—differed greatly. Relations between Kenya and Tanzania deteriorated to the point that the border between them was closed for several years.

In 1984, Kenya, Tanzania, and Uganda signed an "East African Mediation Agreement," which allowed for the division of the EAC's assets and liabilities, along with the reopening of the Kenya–Tanzania border. This final chapter of the old Community laid the groundwork for renewed cooperation, which ultimately, in late 2001, led to the EAC's reestablishment in a lavish ceremony at Arusha, Tanzania.

By the end of the 1980s, the value of the Community to the three economies has become clear. But political factors continued to complicate the quest for integraiton. In 1986, Kenya and Tanzania, along with Rwanda, Burundi, Sudan, and then Zaire (today the Democratic Republic of the Congo) pledged to prevent their territories from being used by exiles seeking to destabilize their neighbors. While this broader agreement went unenforced, political relations among Kenya, Tanzania, and Uganda (which in 1986 came under the control of Yoweri Museveni's National Resistance Movement, after years of suffering under the brutal regimes of Amin and Milton Obote) began to improve.

From 1981, the three states were also linked in a loose nineteen-member state "Preferential Trade Area" for southern and eastern Africa. This body laid the basis for further cooperation in the areas of security, trade, and joint hydroelectric projects. Although members of the Economic Community of Central African States, Rwanda and Burundi were also linked with Kenya, Tanzania, and Uganda as a subregion of the UN Commission for Africa.

In 1993, the three states established a "Permanent Tripartite Commission" to look into reviving the East African Community. By then the leaders of all three countries—Daniel Mkapa of Tanzania, Daniel Arap Moi of Kenya, and Museveni—had been implementing confidence-building measures. The 1993 agreement had the goal of establishing a common market and currency zone for the region. But both Tanzania and Uganda were reluctant to move forward due to Kenya's continued industrial advantages. The 2001 treaty has allayed these concerns by dropping a strict time frame for the removal of trade restrictions.

While full economic and political union for the EAC members (who are likely to be expanded to include Rwanda, Burundi, and perhaps Ethiopia) remains a long-term goal, some important structures have already been put in place: the East African Court of Justice, the East African Legislative Assembly, and the Secretariat. In 1998, a common East African passport was introduced that allows citizens of the three nations to cross one another's borders freely. Progress is also reportedly being made toward free currency convertibility, reduced tariffs, and in the areas of defense and foreign policy. All of these steps have generally been greeted with popular support. As the Tanzanian statesman Salim Salim noted: "You can choose a friend but you cannot choose a brother.... In this case Kenyans and Ugandans are our brothers."

THE ISLANDS

The Comoros, Madagascar, Mauritius, and Seychelles each have their own unique characteristics. They all have some important traits in common. All four island nations have been strongly influenced historically by contacts with Asia as well as with mainland Africa and Europe. Madagascar and the Comoros have populations that originated in Indonesia and the Middle East as well as in Africa; the Malagasy language is related to Indonesian Malay. The citizens of Mauritius and Seychelles are of European as well as African and Asian origin.

All four island groups have also been influenced by France. Mauritius and Seychelles were not permanently inhabited until the 1770s, when French settlers arrived with their African slaves. The British subsequently took control of these two island groups and, during the 1830s, abolished slavery. Thereafter the British encouraged migration from South Asia and, to a lesser extent, from China to make up for labor shortages on the islands' plantations. Local French-based creoles remain the major languages on the islands.

In 1978, all the islands, along with opposition groups from the French possession of Réunion, formed the Indian Ocean Commission. Originally a body with a socialist orientation, the commission campaigned for the independence of Réunion and the return of the island of Diego Garcia by Britain to Mauritius, as well as the dismantling of the U.S. naval base located there. By the end of the 1980s, however, the export-oriented growth of Mauritius and the continuing prosperity of Seychelles' tourist-based economy were helping to push all nations toward a greater emphasis on market economics in their multilateral, as well as internal, policy initiatives. Madagascar and the Comoros have recently offered investment incentives for Mauritius-based private firms. Mauritians have also played prominent roles in the development of tourism in the Comoros.

In addition to their growing economic ties, the Comoros and Mauritius, and to a somewhat lesser extent, Madagascar and Seychelles, have created linkages with South Africa. In 1995, Mauritius followed South Africa's lead to become the 12th member of the Southern African Development Community (SADC).

Burundi (Republic of Burundi)

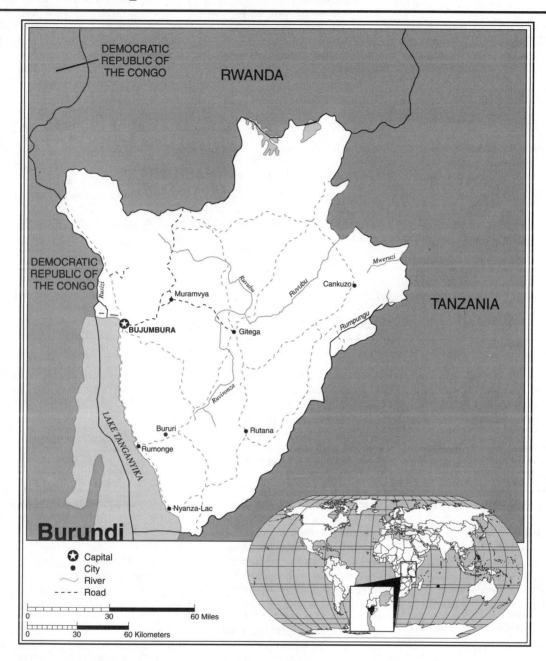

Burundi Statistics

GEOGRAPHY

Area in Square Miles (Kilometers): 10,759
(27,834) (about the size of Maryland)

Capital (Population): Bujumbura
(346,000)

Environmental Concerns: soil erosion;
deforestation; habitat loss

Geographical Features: hilly and
mountainous, dropping to a plateau in the
east; some plains; landlocked

Climate: tropical to temperate;
temperature varies with altitude

PEOPLE

Population

Total: 6,373,000
Annual Growth Rate: 2.36%
Rural/Urban Population Ratio: 92/8
Major Languages: Kirundi; French;
Swahili; others

Ethnic Makeup: 85% Hutu; 14% Tutsi; 1%
Twa and others
Religions: 67% Christian; 23% indigenous
beliefs; 10% Muslim

Health

Life Expectancy at Birth: 45 years (male);
47 years (female)
Infant Mortality: 69.7/1,000 live births
Physicians Available: 1/31,777 people
HIV/AIDS Rate in Adults: 11.32%

59

Education

Adult Literacy Rate: 35.3%
Compulsory (Ages): 7–13; free

COMMUNICATION

Telephones: 20,000 main lines
Televisions: 7/1,000 people
Internet Users: 6,000 (2002)

TRANSPORTATION

Highways in Miles (Kilometers): 8,688 (14,480)
Railroads in Miles (Kilometers): none
Usable Airfields: 7
Motor Vehicles in Use: 20,000

GOVERNMENT

Type: republic

Independence Date: July 1, 1962 (from UN trusteeship under Belgian administration)
Head of State/Government: President (General) Pierre Buyoyo is both head of state and head of government
Political Parties: Union for National Progress; Burundi Democratic Front; others
Suffrage: universal for adults

MILITARY

Military Expenditures (% of GDP): 5.3%
Current Disputes: severe interethnic conflict, transcending national borders

ECONOMY

Currency ($ U.S. equivalent): 1,072 francs = $1
Per Capita Income/GDP: $600/$3.7 billion

GDP Growth Rate: 1.4%
Inflation Rate: 14%
Population Below Poverty Line: 70%
Natural Resources: nickel; uranium; rare earth oxides; peat; cobalt; copper; platinum; vanadium; arable land; hydropower
Agriculture: coffee; cotton; tea; corn; sorghum; sweet potatoes; bananas; manioc; livestock
Industry: light consumer goods; assembly of imported components; public-works construction; food processing
Exports: $24 million (primary partners European Union, United States, Kenya)
Imports: $125 million (primary partners European Union, Tanzania, Zambia)

SUGGESTED WEB SITES

http://www.sas.upenn.edu/
African_Studies/
Country_Specific/Burundi.html

Burundi Country Report

Notwithstanding the tireless diplomatic efforts of South African mediators, Burundi has in recent years remained divided between the forces loyal to its transitional government and the predominantly ethnic-Hutu rebels of the Forces for the Defense of Democracy (FDD) and the National Liberation Forces (FNL). Talks among the groupings throughout 2002 failed to achieve national reconciliation.

DEVELOPMENT

Burundi's sources of wealth are limited. There is no active development of mineral resources, although nickel has been located and may be mined soon. There is little industry, and the coffee industry, which contributes 75% to 90% of export earnings, has declined.

Hutu rebels have been fighting with Burundi's government since 1993, when Tutsi paratroopers assassinated the small Central African country's first democratically elected president, who was a Hutu. Despite being in the minority, Tutsis have effectively controlled the nation of 6 million people for all but a few months since independence in 1961. The current transitional government took office in November 2001 to implement a power-sharing agreement, mediated in August 2000 by former South African president Nelson Mandela, between Hutu and Tutsi political parties. But the rebels maintain that true power remains in the hands of the Tutsi-dominated military.

Burundi's current divisions have deeper roots. In 1972 and again in 1988, tens of thousands of people, mostly Hutus, perished in genocidal attacks. In more recent years, the situation has been further complicated by the escalation of conflict between Tutsis and Hutus in the neighboring states of Rwanda and the Democratic Republic of the Congo (D.R.C.).

In the past the violence was initiated by members of the Tutsi governing elite seeking to maintain their privileged status through brutal military control. Today, the army's hold on the countryside is increasingly being challenged by an armed movement of the FDD and FNL. This movement is spreading counterterror in the name of the country's Hutu majority. What has remained the same over the years is the general indifference of the outside world to Burundi's horrific record of ethnic conflict.

In 1993, the country seemed poised to enter a bright new era when, in their first democratic elections, Burundians chose their first Hutu head of state, Melchior Ndadaye, and a Parliament dominated by the Hutu Front for Democracy in Burundi (FRODEBU) party. But within months Ndadaye was assassinated, setting the scene for subsequent Hutu–Tutsi violence, in which at least 200,000 people have been killed. In early 1994, Parliament elected another Hutu, Cyprien Ntaryamira, as president. However, he was killed when a plane he was traveling in was sabotaged in April—the same incident that killed the reformist president of neighboring Rwanda.

FREEDOM

Beset by on-going genocide, there is currently no genuine freedom in Burundi, for either its ethnic majority or minority populations. People continue to flee the country by the tens of thousands.

After talks among the main parties, another Hutu, Sylvestre Ntibantunganya, was appointed president in October 1994. But within months the mainly Tutsi Union for National Progress (UPRONA) party withdrew from the government and Parliament, sparking off a new wave of ethnic violence. In the process the capital city, Bujumbura, was largely emptied of its Hutu majority, while many ordinary Tutsis fled from much of the countryside. In July 1996, the army overthrew Ntibantunganya, bring back to power the Tutsi general Pierre Buyoyo, who had ruled the country from 1987 to 1993. Subsequent talks among the Burundian political parties, mediated first by former Tanzanian president Julius Nyerere and then by Mandela, failed to reach agreement on crucial issues. These included the role of the Burundian Army and the dismantling of "regroupment camps," which are said to hold more than 800,000 Hutu civilians.

A DIVIDED SOCIETY

Burundi's population is ethnosocially divided into three distinctive groups. At the bottom of the social hierarchy are the Twa, commonly stereotyped as "pygmies." Believed to be the earliest inhabitants of the country, today the Twa account for only about 1 percent of the population. The largest group, constituting 85 percent of the population, are the Hutus, most of whom subsist as farmers. The dominant group are the Tutsis, 14 percent of the population.

Among the Tutsis, who are subdivided into clans, status has long been associated with cattle-keeping. Leading Tutsis continue to form an aristocratic ruling class over the whole of Burundi society. Until 1966, the leader of Burundi's Tutsi aristocracy was the *Mwami,* or king.

The Burundi kingdom goes back at least as far as the sixteenth century. By the late 1800s, when the kingdom was incorporated into German East Africa, the Tutsis had subordinated the Hutus, who became clients of local Tutsi aristocrats, herding their cattle and rendering other services. The Germans and subsequently the Belgians, who assumed paramount authority over the kingdom after World War I, were content to rule through Burundi's established social hierarchy. But many Hutus as well as Tutsis were educated by Christian missionaries.

HEALTH/WELFARE

Much of the educational system has been in private hands, especially the Roman Catholic Church. Burundi lost many educated and trained people during the Hutu massacres in the 1970s and 1980s.

In the late 1950s, Prince Louis Rwagazore, a Tutsi, tried to accommodate Hutu as well as Tutsi aspirations by establishing the nationalist reform movement known as UPRONA. Rwagazore was assassinated before independence, but UPRONA led the country to independence in 1962, with King Mwambutsa IV retaining considerable power as head of state. The Tutsi elite remained dominant, but the UPRONA cabinets contained representation from the two major groups. This attempt to balance the interests of the Tutsi and Hutu broke down in 1965, when Hutu politicians within both UPRONA and the rival People's Party won 80 percent of the vote and the majority of the seats in both houses of the bicameral Legislature. In response, the king abolished the Legislature before it could convene. A group of Hutu army officers then attempted to overthrow the government. Mwambutsa fled the country, but

Tutsi officers, led by Michel Micombero, crushed the revolt in a countercoup.

In the aftermath of the uprising, Micombero took power amid a campaign of reprisals in which, it is believed, some 5,000 Hutus were killed. He deposed Mwambutsa's son, Ntare V, from the kingship and set up a "Government of Public Safety," which set about purging Hutu members from the government and the army. The political struggle involved inter-clan competition among the Tutsis as well as the maintenance of their hegemony over the Hutus.

Under Micombero, Burundi continued to be afflicted with interethnic violence, occasional coup attempts, and pro-monarchist agitation. A major purge of influential Hutus was carried out in 1969. In 1972, Ntare V was lured to Uganda by Idi Amin, who turned him over to Micombero. Ntare was placed under arrest upon his arrival and was subsequently murdered by his guards.

ACHIEVEMENTS

Burundians were briefly united in July 1996 by the victory of their countryman Venuste Niyongabo in the men's 5,000-meter race at the Atlanta Summer Olympic Games. He dedicated his gold medal (the first for a Burundi citizen) to the hope of national reconciliation.

A declaration of martial law then set off another explosion of violence. In response to an alleged uprising involving the deaths of up to 2,000 Tutsis, government supporters began to massacre large numbers of Hutus. Educated Hutus were especially targeted in a two-month campaign of selective genocide, which is generally estimated to have claimed 200,000 victims (estimates range from 80,000 to 500,000 deaths for the entire period, with additional atrocities being reported through 1973). More than 100,000 Hutus fled to Uganda, Rwanda, Zaire (present-day D.R.C.), and Tanzania. Among the governments of the world, only Tanzania and Rwanda showed any deep concern for the course of events. China, France, and Libya used the crisis to significantly upgrade their military aid to the Burundi regime.

In 1974, Micombero formally transformed Burundi into a single-party state under UPRONA. Although Micombero was replaced two years later in a military coup by Colonel Jean-Baptiste Bagaza, power remained effectively in the hands of members of the Tutsi elite who controlled UPRONA, the civil service, and the army. In 1985, Bagaza widened existing state persecution of Seventh Day Adventists and

Jehovah's Witnesses to include the Roman Catholic Church, to which a majority of Burundi's population belong, suspecting it of fostering seditious—that is, pro-Hutu—sympathies. (The overthrow of Bagaza by Pierre Buyoyo, in a 1987 military coup, led to a lifting of the anti-Catholic campaign.)

Timeline: PAST

1795
Mwami Ntare Rugaamba expands the boundaries of the Nkoma kingdom

1919
The area is mandated to Belgium by the League of Nations after the Germans lose World War I

1958–1961
Prince Louis Rwagazore leads a nationalist movement and founds UPRONA

1961
Rwagazore is assassinated; independence is achieved

1965–1966
A failed coup results in purges of Hutus in the government and army; Michel Micombero seizes power

1972
Government forces massacre 200,000 Hutu

1976
Jean-Baptiste Bagaza comes to power in a military coup

1987
Bagaza is overthrown in a military coup led by Pierre Buyoyo

1990s
Buyoyo loses in multiparty elections; Melchior Ndadye becomes Burundi's first Hutu president; Buyoyo regains power in a military coup

PRESENT

2000s
Unrest in neighboring states complicates the ethnic conflict in Burundi

Nelson Mandela mediates for national reconciliation

Ethnic violence erupted again in 1988. Apparently some Tutsis were killed by Hutus in northern Burundi, in response to rumors of another massacre of Hutus. In retaliation, the army massacred between 5,000 and 25,000 Hutus. Another 60,000 Hutus took temporary refuge in Rwanda, while more than 100,000 were left homeless. In 1991, the revolutionary Party for the Liberation of the Hutu People, or Palipehutu, launched its own attacks on Tutsi soldiers and civilians, leading to further killing on all sides.

LAND ISSUES

Burundi is one of the poorest countries in the world, despite its rich volcanic soils and generous international development assistance (it has been one of the highest per capita aid recipients on the African continent). In addition to the dislocations caused by cycles of interethnic violence, the nation's development prospects are seriously compromised by geographic isolation and population pressure on the land. About 25 percent of Burundi's land is under cultivation—generally by individual farmers trying to subsist on plots of no more than three acres. Another 60 percent of the country is devoted to pasture for mostly Tutsi livestock. Hutu farmers continue to be tied by patron–client relationships to Tutsi overlords.

In the 1980s, the government tied its rural development efforts to an unpopular villagization scheme. This issue has complicated on-going attempts to reach some kind of accommodation between the Tutsi elite and Hutu masses. Having cautiously increased Hutu participation in his government while reserving ultimate power in the hands of the all-Tutsi Military Committee of National Salvation, Buyoyo agreed to the restoration of multiparty politics in 1991. A new Constitution was approved in March 1992; it allowed competition between approved, ethnically balanced, parties. In the resulting July 1993 elections, Buyoyo's UPRONA was defeated by the Front for Democracy in Burundi. FRODEBU's leader, Ndadaye, was sworn in as the head of a joint FRODEBU–UPRONA government. His subsequent assassination by Tutsi hard-liners in the military set off a new wave of interethnic killings. The firm stand against the coup by Buyoyo and the Tutsi/UPRONA prime minister, Sylvie Kinigi, helped to calm the situation, but attempts to make a fresh start collapsed in 1994 when a plane carrying Ntaryamira, Burundi's newly elected head of state, and his Rwandan counterpart was shot down over Rwanda. The latest coup followed UPRONA's withdrawal from the government following the massacre of more than 300 Tutsis by FDD, who by September 1996 were attempting to besiege the capital.

During a four-week period from late October to November 1996, the Tutsi-led Burundian military massacred at least 1,000 civilians. The government forces fought with Hutu rebels, as some 50,000 Hutus returned from camps that had been closed in Zaire. The Tutsi-dominated military set up more than a dozen "protection zones" for Hutu civilians while soldiers continued battling Hutu rebels. Strife continued as an estimated 200,000 Burundians were living in refugee camps in Tanzania.

Comoros (Union of Comoros)

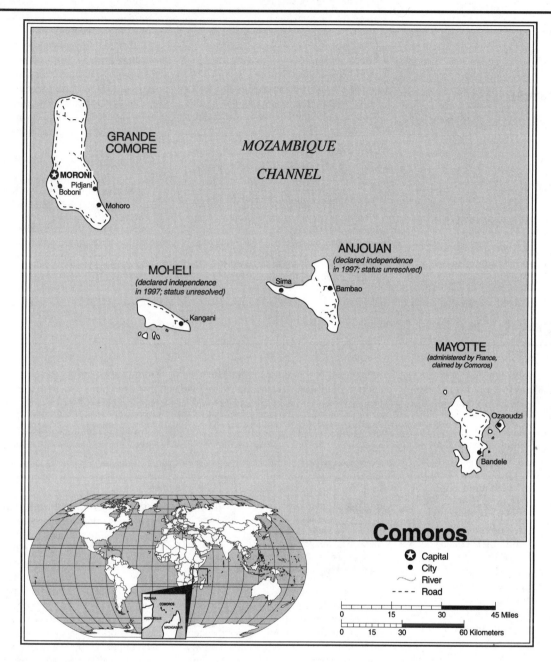

Comoros Statistics

GEOGRAPHY

Area in Square Miles (Kilometers): 838 (2,171) (about 12 times the size of Washington, D.C.)

Capital (Population): Moroni (49,000)

Environmental Concerns: soil degradation and erosion; deforestation

Geographical Features: volcanic islands; interiors vary from steep mountains to low hills

Climate: tropical marine

PEOPLE

Population

Total: 615,000

Annual Growth Rate: 2.99%

Rural/Urban Population Ratio: 69/31

Major Languages: Arabic; French; Comoran

Ethnic Makeup: Antalote; Cafre; Makoa; Oimatsaha; Sakalava

Religions: 98% Sunni Muslim; 2% Roman Catholic

Health

Life Expectancy at Birth: 58 years (male); 62 years (female)

Infant Mortality: 81.7/1,000 live births

Physicians Available: 1/6,600 people

HIV/AIDS Rate in Adults: 0.12%

Education
Adult Literacy Rate: 57.3%
Compulsory (Ages): 7–16

COMMUNICATION
Telephones: 7,000 main lines
Internet Users: 1,500 (2001)

TRANSPORTATION
Highways in Miles (Kilometers): 522 (870)
Railroads in Miles (Kilometers): none
Usable Airfields: 4

GOVERNMENT
Type: republic
Independence Date: July 6, 1975 (from France)
Head of State/Government: President Azali Assoumani; Prime Minister Hamada Madi Bolero

Political Parties: Rassemblement National pour le Development; Front National pour la Justice
Suffrage: universal at 18

MILITARY
Military Expenditures (% of GDP): 3%
Current Disputes: Comoros claims the French-administered island of Mayotte; Moheli and Anjouan seek independence

ECONOMY
Currency ($ U.S. Equivalent): 461 francs = $1
Per Capita Income/GDP: $710/$424 million
GDP Growth Rate: 1%
Inflation Rate: 3.5%
Unemployment Rate: 20%; extreme underemployment

Labor Force by Occupation: 80% agriculture
Population Below Poverty Line: 60%
Natural Resources: negligible
Agriculture: perfume essences; copra; coconuts; cloves; vanilla; bananas; cassava
Industry: tourism; perfume distillation
Exports: $35.3 million (primary partners France, United States, Singapore)
Imports: $44.9 million (primary partners France, South Africa, Kenya)

SUGGESTED WEB SITES
http://www.arabji.com/Comoros/
  index.htm
http://www.sas.upenn.edu/
  African_Studies/
  Country_Specific/Comoros.html
http://www.africaindex.
  africainfo.no/africaindex/
  africaindex1/countries/
  comoros.html

Comoros Country Report

A small archipelago consisting of three main islands—Grande Comore, Moheli, and Anjouan (a fourth island, Mayotte, has voluntarily remained under French rule)—in recent years Comoros has struggled to maintain its fragile unity. In 1997, separatists seized control of Anjouan and Moheli, subsequently declaring independence. But after years of failed mediation efforts by other African states, in December 2001 voters throughout Comoros were able to overwhelmingly agree on a new Constitution designed to reunite their country as a loose federation. This followed the seizure of power by a "military committee" on Anjouan that was committed to reunification. In April 2002, Azali Assoumani, who had previously seized power on Grande Comore, was sworn in as the president of the new "Union of Comoros." But his authority was soon challenged by Mze Abdou Soule Elbak, who a month later was elected as the president of Grande Comore. A military standoff thereafter developed on Grande Comore between followers of Azali and Elbak, further threatening the islands' prospects of ever achieving political stability.

The years since independence from France, in 1975, have not been kind to Comoros, which has been consistently listed by the United Nations as one of the world's least-developed countries. Lack of economic development has been compounded at times by natural disasters, eccentric and authoritarian leadership, political violence,

and external interventions. The 1990 restoration of multiparty democracy, along with subsequent elections in 1992–1993, has so far failed to provide a basis for national consensus.

DEVELOPMENT

One of the major projects undertaken since independence has been the ongoing expansion of the port at Mutsamundu, to allow large ships to visit the islands. Vessels of up to 25,000 tons can now dock at the harbor. In recent years, there has been a significant expansion of tourism to Comoros.

Meanwhile, the entire archipelago remains impoverished. While many Comorans remain underemployed as subsistence farmers, more than half of the country's food is imported. As a result, many Comorans have questioned the wisdom of independence, but appeals by Anjouan and Moheli islanders for a return of French control have been rejected by Paris.

The Comoros archipelago was populated by a number of Indian Ocean peoples, who—by the time of the arrival of Europeans during the early 1500s—had combined to form the predominantly Muslim, Swahili-speaking society found on the islands today. In 1886, the French proclaimed a protectorate over the three main islands that currently constitute the Union of Comoros (France had ruled Mayotte since 1843). Throughout the colonial period, Co-

moros was especially valued by the French for strategic reasons. A local elite of large landholders prospered from the production of cash crops. Life for most Comorans, however, remained one of extreme poverty.

A month after independence, the first Comoran government, led by Ahmed Abdullah Abderemane, was overthrown by mercenaries, who installed Ali Soilih in power. He promised a socialist transformation of the nation and began to implement land reform, but he rapidly lost support both at home and abroad—under his leadership, gangs of undisciplined youths terrorized society, while the basic institutions and services of government all but disappeared. In 1977, the situation was made even worse by a major volcanic eruption, which left 20,000 people homeless, and by the arrival of 16,000 Comoran refugees following massacres in neighboring Madagascar.

FREEDOM

Freedom was abridged after independence under both Ahmed Abdullah and Ali Soilih. The government elected in 1990 ended human-rights abuses.

In 1978, another band of mercenaries—this time led by the notorious Bob Denard, whose previous exploits in Zaire (present-day Democratic Republic of the Congo or

D.R.C.), Togo, and elsewhere had made his name infamous throughout Africa—overthrew Soilih and restored Abdullah to power. Denard, however, remained the true power behind the throne.

The Denard–Abdullah government enjoyed close ties with influential right-wing elements in France and South Africa. Connections with Pretoria were manifested through the use of Comoros as a major conduit for South African supplies to the Renamo rebels in Mozambique. Economic ties with South Africa, especially in tourism and sanctions-busting, also grew. The government also established good relations with Saudi Arabia, Kuwait, and other conservative Arab governments while attracting significant additional aid from the international donor agencies.

HEALTH/WELFARE

Health statistics improved during the 1980s, but a recent World Health Organization survey estimated that 10% of Comoran children ages 3 to 6 years are seriously malnourished and another 37% are moderately malnourished.

In 1982, the country legally became a one-party state. Attempted coups in 1985 and 1987 aggravated political tensions. Many Comorans particularly resented the overbearing influence of Denard and his men. By November 1989, this group included President Abdullah himself. With the personal backing of President François Mitterand of France and President F. W. de Klerk of South Africa, Abdullah moved to replace Denard's mercenaries with a French-approved security unit. But before this move could be implemented, Abdullah was murdered following a meeting with Denard.

ACHIEVEMENTS

Comoros has long been the world's leading exporter of ylang-ylang, an essence used to make perfume. It is also the second-leading producer of vanilla and a major grower of cloves. Together, these cash crops account for more than 95% of export earnings. Unfortunately, the international prices of these crops have been low for the past 2 decades.

The head of the Supreme Court, Said Mohamed Djohar, was appointed interim president in the wake of the assassination. After a period of some confusion, during which popular protests against Denard swelled, Djohar quietly sought French intervention to oust the mercenaries. With both Paris and Pretoria united against him, Denard agreed to relinquish power, in exchange for safe passage to South Africa. The removal of Denard and temporary stationing of a French peacekeeping force was accompanied by the lifting of political restrictions in preparation for presidential elections. In 1990, a runoff resulted in a 55 percent electoral mandate for Djohar.

In September 1995, Denard's men returned to overthrow Djohar. But the mercenaries were soon forced to surrender to French forces, who installed Caambi el Yachourtu, rather than Djohar, as acting president. At the end of 1996, Mohamed Taki Abdulkarim replaced Yachourtu as president. In November 1998, Taki died suddenly and was replaced by Tadjiddine Ben Said Massounde as the head of a ruling military committee. Massounde's government was overthrown in a bloodless coup on April 30, 1999. Azali Assoumani was subsequently installed as president.

Timeline: PAST

1500s
Various groups settle in the islands, which become part of a Swahili trading network

1886
A French protectorate over the remaining Comoros islands is proclaimed

1914–1946
The islands are ruled as part of the French colony of Madagascar

1975
Independence is followed by a mercenary coup, which installs Ali Soilih

1978
Ali Soilih is overthrown by mercenaries; Ahmed Abdullah is restored

1980s
Abdullah proclaims a one-party state; real power remains in the hands of mercenary leader Bob Denard

1990s
The assassination of Abdullah leads to the removal of Denard and to multiparty elections

PRESENT

2000s
The country is renamed "Union of Comoros"

Despite the name change, Comoros's political unity has not been achieved

Djibouti (Republic of Djibouti)

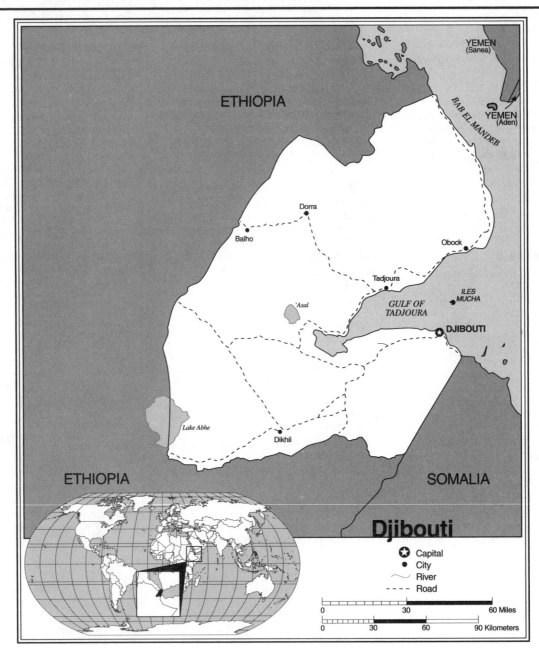

Djibouti Statistics

GEOGRAPHY

Area in Square Miles (Kilometers): 8,492 (22,000) (about the size of Massachusetts)

Capital (Population): Djibouti (542,000)

Environmental Concerns: insufficient potable water; desertification

Geographical Features: coastal plain and plateau, separated by central mountains

Climate: desert

PEOPLE

Population

Total: 473,000

Annual Growth Rate: 2.59%

Rural/Urban Population Ratio: 17/83

Major Languages: French; Arabic; Somali; Afar

Ethnic Makeup: 60% Issa/Somali; 35% Afar; 5% French, Arab, Ethiopian, Italian

Religions: 94% Muslim; 6% Christian

Health

Life Expectancy at Birth: 50 years (male); 53 years (female)
Infant Mortality: 99.7/1,000 live births
Physicians Available: 1/3,790 people
HIV/AIDS Rate in Adults: 11.75%

Education

Adult Literacy Rate: 46.2%

COMMUNICATION

Telephones: 10,000 main lines
Televisions: 43/1,000 people
Internet Users: 3,300 (2002)

TRANSPORTATION

Highways in Miles (Kilometers): 1,801 (2,906)
Railroads in Miles (Kilometers): 60 (97)
Usable Airfields: 12
Motor Vehicles in Use: 16,000

GOVERNMENT

Type: republic
Independence Date: June 27, 1977 (from France)
Head of State/Government: President Ismail Omar Guellah; Prime Minister Dileita Mohamed Dileita
Political Parties: People's Progress Assembly; Democratic Renewal Party; Democratic National Party; others
Suffrage: universal for adults

MILITARY

Military Expenditures (% of GDP): 4.4%
Current Disputes: ethnic conflict; border clashes with Eritrea

ECONOMY

Currency ($ U.S. Equivalent): 180 francs = $1

Per Capita Income/GDP: $1,400/$586 million
GDP Growth Rate: 0%
Inflation Rate: 2%
Unemployment Rate: 50%
Population Below Poverty Line: 50%
Natural Resources: geothermal areas
Agriculture: livestock; fruits; vegetables
Industry: port and maritime support; construction
Exports: $260 million (primary partners Somalia, Yemen, Ethiopia)
Imports: $440 million (primary partners France, Ethiopia, Italy)

SUGGESTED WEB SITES

```
http://www.sas.upenn.edu/
  African_Studies/
  Country_Specific/Djibouti.html
http://www.192.203.180.62/mlas/
  djibouti.html
http://www.republique-
  djibouti.com
```

Djibouti Country Report

After a decade of civil unrest, Djibouti has settled down under the leadership of its second president, Ismail Omar Guellah. In April 1999, Guellah succeeded the aging Hassan Gouled Aptidon, who stepped down due to ill health. While Guellah's main opponent, Musa Ahmed Idriss, was arrested after claiming massive electoral fraud, the new president has since consolidated his authority by building on the process of national reconciliation that had begun under his predecessor. In February 2000, this resulted in a peace agreement between the government and armed rebel holdouts of the Front for the Restoration of Democratic Unity (FRUD), resulting in the return of the rebel leader Ahmad Dini. A subsequent coup attempt, allegedly orchestrated by the chief of police, was crushed.

DEVELOPMENT

Recent discoveries of natural-gas reserves in Djibouti could result in a surplus for export. A number of small-scale irrigation schemes have been established. There is also a growing, though still quite small, fishing industry.

Political conflict in Djibouti has mirrored the country's ethnic tensions between the Somali-speaking Issas and the Afar-speakers. An earlier, 1997, power-sharing agreement was reached between the long-ruling, Issa-dominated Popular Rally for Progress Party (RPP) and a more moderate faction of the Afar-dominated FRUD. Although the FRUD moderates went on to win all 65 seats in December 1997 legislative elections, more radical elements of FRUD continued their armed resistance to President Aptidon. While the conflict has now ended, suspicions between Afars and Issas continue to threaten Djibouti's fragile political unity.

FREEDOM

The government continues to harass and detain its critics. Prison conditions are harsh, with the sexual assault of female prisoners being commonplace.

Since achieving its independence from France, Djibouti has also had to strike a cautious balance between the competing interests of its larger neighbors, Ethiopia and Somalia. In the past, Somalia has claimed ownership of the territory, based on the numerical preponderance of Djibouti's Somali population, variously estimated at 50 to 70 percent. However, local Somalis as well as Afars also have strong ties to communities in Ethiopia. Furthermore, Djibouti's location at the crossroads of Africa and Eurasia has made it a focus of continuing strategic concern to nonregional powers, particularly France, which maintains a large military presence in the country.

Modern Djibouti's colonial genesis is a product of mid-nineteenth-century European rivalry over control of the Red Sea. In 1862, France occupied the town of Obock, across the harbor from the city of Djibouti. This move was taken in anticipation of the 1869 opening of the Suez Canal, which transformed the Red Sea into the major shipping route between Asia, East Africa, and Europe. In 1888, Paris, having acquired Djibouti city and its hinterland, proclaimed its authority over French Somaliland, the modern territory of Djibouti.

HEALTH/WELFARE

Progress has been made in reducing infant mortality, but health services are strained in this very poor country. However, on the positive side, school enrollment has expanded by nearly one third since 1987.

The independence of France's other mainland African colonies by 1960, along with the formation in that year of the Somali Democratic Republic, led to local agitation for an end to French rule. To counter the effects of Somali nationalism, the French began to favor the Afar minority in local politics and employment. French president Charles de Gaulle's 1966 visit was accompanied by large, mainly

Somali, pro-independence demonstrations. As a result, a referendum was held on the question of independence. Colonial control of voter registration assured a predominantly Afar electorate, who, fearful of Somali domination, opted for continued French rule. French Somaliland was then transformed into the self-governing "Territory of Afars and Issas." The name reflected a continuing colonial policy of divide-and-rule; members of the Issas clan constituted just over half of the area's Somali-speakers.

ACHIEVEMENTS

 Besides feeding its own refugees, the government of Djibouti has played a major role in assisting international efforts to relieve the effects of recurrent famines in Ethiopia, Somalia, and Sudan.

By the 1970s, neither Ethiopia nor France was opposed to Djibouti's independence but, for their own strategic reasons, both countries backed the Afar community in its desire for assurances that the territory would not be incorporated into Somalia. An ethnic power-sharing arrangement was established that in effect acknowledged local Somali preponderance. The empowerment of local Somalis, in particular Issas, was accompanied by diminished pan-Somali sentiment. On June 27, 1977, the Republic of Djibouti became independent. French troops remained in the country, however, supposedly as a guarantee of its sovereignty. Internally, political power was divided by means of ethnically balanced cabinets.

War broke out between Ethiopia and Somalia a few months after Djibouti's independence. Djibouti remained neutral, but ethnic tensions mounted with the arrival of Somali refugees. In 1981, the Afar-dominated Djiboutian Popular Movement was outlawed. The Issa-dominated Popular Rally for Progress (RPP) then became the country's sole legal party.

Refugees have poured into Djibouti for years now, fleeing conflict and famine in Ethiopia, Somalia, and Sudan. The influx has swelled the country's population by about one third and has deepened Djibouti's dependence on external food aid. Massive unemployment among Djibouti's largely urban population remains a critical problem.

Timeline: PAST

1862
France buys the port of Obock

1888
France acquires the port of Djibouti

1917
The Addis Ababa-Djibouti Railroad is completed

1958
Djibouti votes to remain part of Overseas France

1977
Independence; the Ogaden War

1980s
The underground Union of Movements for Democracy is formed as an interethnic, antigovernment coalition

1990s
Civil war rends the country; Ismail Omar Guellah is elected to replace President Hassan Gouled Aptidon

PRESENT

2000s
Ethnic conflict continues

Eritrea (State of Eritrea)

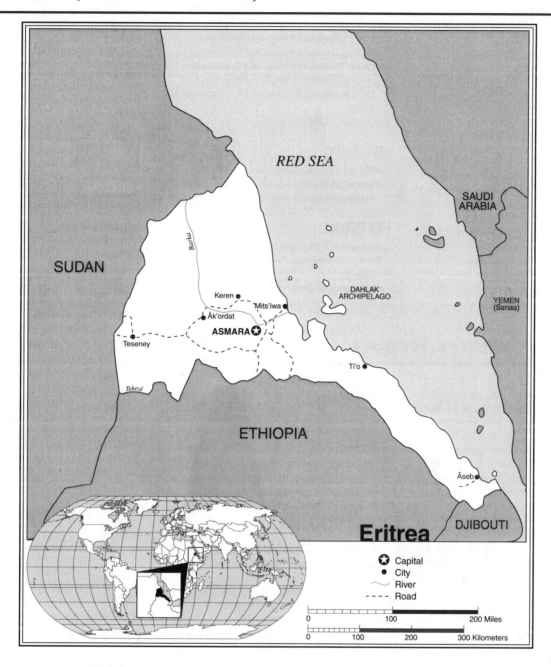

Eritrea Statistics

GEOGRAPHY

Area in Square Miles (Kilometers): 46,829 (121,320) (about the size of Pennsylvania)

Capital (Population): Asmara (503,000)

Environmental Concerns: deforestation; desertification; soil erosion; overgrazing

Geographical Features: north-south–trending highlands, descending on the east to a coastal desert plain, on the northwest to hilly terrain, and on the southwest to flat-to-rolling plains

Climate: hot, dry desert on the seacoast; cooler and wetter in the central highlands; semiarid in western hills and lowlands

PEOPLE

Population

Total: 4,466,000

Annual Growth Rate: 3.86%

Rural/Urban Population Ratio: 82/18

Major Languages: various, including Tigrinya, Tigre, Kunama, Arabic, Amharic, and Afar

Ethnic Makeup: 50% ethnic Tigrinya; 40% Tigre and Kunama; 4% Afar; 3% Saho; 3% others

Religions: Muslim; Coptic Christian; Roman Catholic; Protestant

Health

Life Expectancy at Birth: 54 years (male);
  59 years (female)
Infant Mortality: 73.6/1,000 live births
Physicians Available: 1/36,000 people
HIV/AIDS Rate in Adults: 2.87%

Education

Adult Literacy Rate: 25%
Compulsory (Ages): 7–13; free

COMMUNICATION

Telephones: 30,000 main lines
Televisions: 6/1,000 people
Internet Users: 12,000 (2001)

TRANSPORTATION

Highways in Miles (Kilometers): 2,436
  (3,930)
Railroads in Miles (Kilometers): 191
  (307)
Usable Airfields: 21

GOVERNMENT

Type: transitional government
Independence Date: May 24, 1993 (from
  Ethiopia)
Head of State/Government: President
  Isaias Aferweki (or Afworki) is both
  head of state and head of government
Political Parties: People's Front for
  Democracy and Justice (only recognized
  party)
Suffrage: universal at 18

MILITARY

Military Expenditures (% of GDP): 19.8%
Current Disputes: border disputes with
  Yemen; uneasy cease-fire with Ethiopia

ECONOMY

Currency ($ U.S. Equivalent): n/a
Per Capita Income/GDP: $740/$3.2
  billion
GDP Growth Rate: 7%

Inflation Rate: 15%
Labor Force by Occupation: 80%
  agriculture
Natural Resources: gold; potash; zinc;
  copper; salt; possibly petroleum and
  natural gas; fish
Agriculture: sorghum; lentils; vegetables;
  maize; cotton; tobacco; coffee; sisal;
  livestock; fish
Industry: food processing; beverages;
  clothing and textiles
Exports: $34.8 million (primary partners
  Sudan, Ethiopia, Japan)
Imports: $470 million (primary partners
  Italy, United Arab Emirates, Germany)

SUGGESTED WEB SITES

http://www.sas.upenn.edu/
  African_Studies/
  Country_Specific/Eritrea.html
http://www.eritrea.org

Eritrea Country Report

In 1998, a border dispute between Eritrea and Ethiopia, around the town of Badme, erupted into open war. This formally ended with a cease-fire agreement in June 2000, but not before leaving thousands of soldiers dead on both sides. In December of that year, Eritrea and Ethiopia signed a further agreement establishing commissions to mark the border, exchange prisoners, return displaced people, and hear compensation claims. On February 6, 2001, Eritrea accepted the United Nations' plan for a temporary demilitarized zone along its border with Ethiopia. By the end of the month, Ethiopia had completed its troop withdrawal. A key provision of the peace agreement was met in April when Eritrea announced that its forces had pulled out of the border zone with Ethiopia. In May, Eritrea and Ethiopia agreed on a UN-proposed mediator to try to demarcate their disputed border. The completion of this task in 2002 has resulted in what will hopefully be a lasting peace between Eritrea and Ethiopia, after decades of conflict.

Eritrea became Africa's newest nation in May 1993, ending 41 years of union with Ethiopia. The origins of Eritrea's separation date back to September 1961, when a small group of armed men calling themselves the Eritrean Liberation Front (ELF) began a bitter independence struggle that would last for three decades.

Between 60,000 and 70,000 people perished as a result of that war, while another

DEVELOPMENT

Since liberation, the government has concentrated its efforts on restoring agricultural and communications infrastructure. The railway and ports of Assab and Massawa are being rehabilitated. In 1991, 80% of the country was dependent on food aid, but subsequent good rains helped boost crop production.

700,000—then about one fifth of the total population—went into exile. What had been one of the continent's most sophisticated light-industrial infrastructures was largely reduced to ruins. Yet the war has also left a positive legacy, in the spirit of unity, self-reliance, and sacrifice that it engendered among Eritreans.

There is no clear-cut reason why a nationalist sentiment should have emerged in Eritrea. Like most African countries, the boundaries of Eritrea are an artificial product of the late-nineteenth-century European scramble for colonies. Between 1869 and 1889, the territory fell under the rule of Italy. Italian influence survives today, especially in the overcrowded but elegant capital city of Asmara, which was developed as a showcase of neo-Roman imperialism. Italian rule came to an abrupt end in 1941, when British troops occupied the territory in World War II. The British withdrew only in 1952. In accordance with the wishes of the UN Security Council, the ter-

ritory was then federated as an autonomous state within the "Empire of Ethiopia."

The federation did not come about through the wishes of the Eritreans. It was, rather, based on the dubious Ethiopian claim that Eritrea was an integral part of the Empire that had been alienated by the Italians. Among the Christians, there were historic cultural ties with their Ethiopian coreligionists, though the Tigrinya-speaking Copts of Eritrea were ethnically distinct from the Empire's then–politically dominant Amharic-speakers. The Muslim lowland areas had never been under any form of Ethiopian control. But, perhaps more important, developments under Italian rule had laid the basis for a sense that Eritrea had its own identity.

FREEDOM

The Eritrean government has pledged to uphold a bill of rights. While the government is dominated by the former EPLF, other parties and organizations participate in the 105-seat Provisional Council. Multiparty elections in 1997 confirmed former EPLF leader Isaias Aferweki as president.

In the face of growing dissatisfaction inside the territory, Ethiopia's emperor, Haile Selassie, ended Eritrea's autonomous status in 1962. Fighting intensified in the early 1970s, after a faction ultimately known as the Eritrean Popular Liberation

Front (EPLF) split from the ELF. The 1974 overthrow of Selassie briefly brought hopes of a peaceful settlement. But Ethiopia's new military rulers, known as the Dergue, committed themselves to securing the area by force. The ELF faded as the EPLF became increasingly effective in pinning down larger numbers of Ethiopian troops. In a major break with tradition, a large proportion of the EPLF's "Liberation Army," including many in command positions, was made up of women. In areas liberated by the EPLF, women were given the right to own land and choose their husbands, while the practice of female circumcision was discouraged.

HEALTH/WELFARE

A major challenge for the government has been the repatriation of hundreds of thousands of war refugees, mostly from neighboring Sudan. Rebuilding efforts were spearheaded by ex-combatants of the Liberation Army, who continued to work for virtually no pay. The EPLF established its own medical and educational services during the war.

Had it not been for the massive military support that the Dergue received from the Soviet Union and its allies, the conflict would have ended sooner. In the late 1980s, the EPLF began to work more closely with other groups inside Ethiopia proper that had taken up arms against their government. This resulted in an alliance between the EPLF and the Ethiopian People's Revolutionary Democratic Front

(EPRDF), which was facilitated by the fact that leading members of both groups spoke Tigrinya. In May 1991, the Dergue collapsed, with the EPLF taking Asmara in the same month that EPRDF troops entered the Ethiopian capital of Addis Ababa. In July, the new EPRDF government agreed in principle to Eritrea's right to self-determination.

ACHIEVEMENTS

Eritrea's independence struggle and on-going national development efforts have been carried out against overwhelming odds, and with very little external support. During the war, self-reliance was manifested in the fact that most weapons and ammunition used by the EPLF were captured from Ethiopian forces.

In 1997, the EPLF, transformed as the People's Front for Democracy and Justice, claimed an overwhelming mandate in elections in which there was little effective opposition. A number of smaller parties, including remnants of the ELF, had joined the Front. Former EPLF leader Isaias Aferweki was confirmed as president of Eritrea.

The renewal of war in Ethiopia has had a devastating effect on Eritrea's economy, which had been making significant progress in the years following independence. The rehabilitation of the port of Massawa and other infrastructure had boosted trade. Light industries, mostly based in Asmara, had recovered. International investors have shown increased interest in the country's mineral wealth,

especially offshore oil. In July 1997, the country introduced its own currency, the nakfa, which replaced the Ethiopian birr. Resulting exchange disputes between the two nations led to a souring of relations prior to the outbreak of the border war.

Timeline: PAST

1869
Italians occupy the Eritrean port of Assab

1889
Italians occupy all of Eritrea

1935–1936
Italians use Eritrea as a springboard for conquest of Ethiopia

1941–1952
Great Britain occupies Eritrea

1952
Eritrea is federated with Ethiopia

1961
The ELF begins the liberation struggle

1962
Federation ends; Eritrea is a province of Ethiopia

1990s
99.8% vote yes for Eritrea's independence; Isaias Aferweki becomes the newly independent nation's first president

PRESENT

2000s
Eritrea and Ethiopia try to forge a lasting peace

Ethiopia (Federal Democratic Republic of Ethiopia)

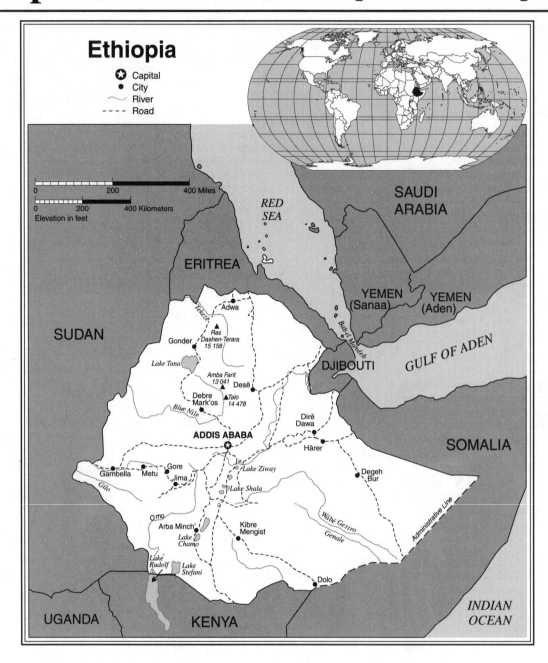

Ethiopia Statistics

GEOGRAPHY

Area in Square Miles (Kilometers): 435,071 (1,127,127) (about twice the size of Texas)

Capital (Population): Addis Ababa (2,753,000)

Environmental Concerns: deforestation; overgrazing; soil erosion; desertification

Geographical Features: a high plateau with a central mountain range divided by the Great Rift Valley; landlocked

Climate: tropical monsoon with wide topographic-induced variation

PEOPLE

Population

Total: 67,700,000

Annual Growth Rate: 2.64%

Rural/Urban Population Ratio: 83/17

Major Languages: Amharic; Tigrinya; Oromo; Somali; Arabic; Italian; English

Ethnic Makeup: 40% Oromo; 32% Amhara and Tigre; 9% Sidamo; 19% others

Religions: 45%–50% Muslims; 35%–40% Ethiopian Orthodox Christian; 12% animist; remainder others

Health

Life Expectancy at Birth: 43 years (male);
44 years (female)
Infant Mortality: 98.6/1,000 live births
Physicians Available: 1/36,600 people
HIV/AIDS Rate in Adults: 10.63%

Education

Adult Literacy Rate: 35.5%
Compulsory (Ages): 7–13; free

COMMUNICATION

Telephones: 232,000 main lines
Televisions: 4/1,000 people
Internet Users 20,000 (2002)

TRANSPORTATION

Highways in Miles (Kilometers): 17,016
(28,360)
Railroads in Miles (Kilometers): 425
(681)
Usable Airfields: 86
Motor Vehicles in Use: 66,000

GOVERNMENT

Type: federal republic

Independence Date: oldest independent
country in Africa (at least 2,000 years)
Head of State/Government: President
Girma Woldegiorgis; Prime Minister
Meles Zenawi
Political Parties: Ethiopian People's
Revolutionary Democratic Front; many
others
Suffrage: universal at 18

MILITARY

Military Expenditures (% of GDP): 12.6%
Current Disputes: border conflicts with
Somalia; uneasy cease-fire with Eritrea

ECONOMY

Currency ($ U.S. Equivalent): 8.72 birrs =
$1
Per Capita Income/GDP: $700/$46
billion
GDP Growth Rate: 7.3%
Inflation Rate: 6.8%
Labor Force by Occupation: 80%
agriculture; 12% government and
services; 8% industry
Population Below Poverty Line: 64%

Natural Resources: gold; platinum,
copper; potash; natural gas; hydropower
Agriculture: cereals; pulses; coffee;
oilseed; sugarcane; vegetables; livestock
Industry: food processing; beverages;
textiles; cement; building materials;
hydropower
Exports: $442 million (primary partners
Germany, Japan, Djibouti)
Imports: $1.54 billion (primary partners
Saudi Arabia, United States, Japan)

SUGGESTED WEB SITES

```
http://www.ethiopians.com
http://www.ethiopianembassy.org
http://www.ethiopianonline.net
http://www.ethiopianspokes.net
http://www.ethiopiadaily.com
http://www.state.gov/www/
    background_notes/
    ethiopia_0398_bgn.html
http://www.sas.upenn.edu/
    African_Studies/
    Country_Specific/Ethiopia.html
```

Ethiopia Country Report

The end of the year 2000 witnessed two major events in Ethiopia. In November, Haile Selassie, the former emperor of Ethiopia, was officially buried in Addis Ababa's Trinity Cathedral. And in December, Ethiopia and Eritrea signed a peace agreement in Algeria, formally ending two years of bloody armed conflict. The agreement, which followed a successful Ethiopian military offensive, established commissions to delineate the disputed border between the countries and provided for the exchange of prisoners and the return of displaced people. Subsequently the two countries officially accepted a new common border, drawn up by an independent commission in The Hague. But both sides made claims to the town of Badme.

DEVELOPMENT

There has been some progress in the country's industrial sector in recent years, after a sharp decline during the 1970s. Soviet-bloc investment resulted in the establishment of new enterprises in such areas as cement, textiles, and farm machinery.

In April 2001, thousands of demonstrators clashed with police in Addis Ababa, in

protest against police brutality and in support of calls for political and academic freedom. A year earlier, President Meles Zenawi's Ethiopian People's Revolutionary Democratic Front (EPRDF) had won an easy victory in legislative elections against some 25 opposition parties. In almost half of the constituencies, EPRDF candidates ran unopposed.

FREEDOM

Despite its public commitment to freedom of speech and association, the EPRDF government has resorted to authoritarian measures against its critics. In 1995, Ethiopia had the highest number of jailed journalists in Africa. Basic freedoms are also compromised.

Following equally overwhelming and controversial election victories in 1994–1995, the EPRDF remains the country's dominant political group. The movement initially came to power in 1991 as an armed movement, following the successful overthrow of Ethiopia's Marxist-oriented military dictatorship, after years of struggle.

Since coming to power, the EPRDF has faced a wide spectrum of opponents. Some

critics see its transformation of Ethiopia into a multiethnic federation of 14 self-governing regions as a threat to national unity. Others contend that its devolutionary structures are a sham designed to obscure its own determination to rule from the center as a virtual one-party state. International as well as domestic supporters, however, see Ethiopia's new Constitution as a bold experiment in institutionalizing a new model of multiethnic statehood.

The EPRDF had emerged in the 1980s as an umbrella movement fighting to liberate Ethiopia from the repressive misrule by the Provisional Military Administrative Council, popularly known as the *Dergue* (Amharic for "Committee"). The Dergue had come to power through a popular uprising against the country's former imperial order. It is still uncertain whether Ethiopia's second revolution in two decades will succeed where its first one failed.

Political instability has reduced Ethiopia from a developing breadbasket to a famine-ridden basket case. Interethnic conflict among an increasingly desperate population, many of whom have long had better access to arms and ammunition than to food and medicine, could lead to the state's disintegration. In the late 1990s, the problems facing the EPRDF government

(United Nations photo by Muldoon)

From 1916 to 1974, Ethiopia was ruled by Haile Selassie, also known as Ras Tafari (from which today's term *Rastafarian* is derived). He is pictured above, on the left, shaking hands with the now infamous Idi Amin of Uganda.

were intensified by the outbreak of the border war with Eritrea, which compromised the landlocked country's access to the sea.

AN IMPERIAL PAST

Ethiopia rivals Egypt as Africa's oldest country. For centuries, its kings claimed direct descent from the biblical King Solomon and the Queen of Sheba. Whether Ethiopia was the site of Sheba is uncertain, as is the local claim that, prior to the birth of Christ, the country became the final resting place of the Ark of the Covenant holding the original Ten Commandments given to Moses (the Ark is said to survive in a local monastery).

Local history is better established from the time of the Axum Empire, which prospered from the first century. During the fourth century, the Axumite court adopted the Coptic Christian faith, which has remained central to the culture of Ethiopia's highland region. The Church still uses the Geez, the ancient Axumite tongue from which the modern Ethiopian languages of Amharic and Tigrinya are derived, in its services.

From the eighth century A.D., much of the area surrounding the highlands fell under Muslim control, all but cutting off the

Copts from their European coreligionists. (Today, most Muslim Ethiopians live in the lowlands.) For many centuries, Ethiopia's history was characterized by struggles among the groups inhabiting these two regions and religions. Occasionally a powerful ruler would succeed in making himself truly "King of Kings" by uniting the Christian highlands and expanding into the lowlands. At other times, the mountains would be divided into weak polities that were vulnerable to the raids of both Muslim and non-Muslim lowlanders.

HEALTH/WELFARE

 Ethiopia's progress in increasing literacy during the 1970s was undermined by the severe dislocations of the 1980s. By 1991, Ethiopia had some 500 government soldiers for every teacher.

MODERN HISTORY

Modern Ethiopian history began in the nineteenth century, when the highlands became politically reunited by a series of kings, culminating in Menilik II, who built up power by importing European armaments. Once the Coptic core of his kingdom was intact, Menilik began to spread

his authority across the lowlands, thus uniting most of contemporary Ethiopia. In 1889 and 1896, Menilik also defeated invading Italian armies, thus preserving his empire's independence during the European partition of Africa.

From 1916 to 1974, Ethiopia was ruled by Ras Tafari (from which is derived the term *Rasta,* or *Rastafarian*), who, in 1930, was crowned Emperor Haile Selassie. The late Selassie remains a controversial figure. For many decades, he was seen both at home and abroad as a reformer who was modernizing his state. In 1936, after his kingdom had been occupied by Benito Mussolini, the leader of Italy, he made a memorable speech before the League of Nations, warning the world of the price it would inevitably pay for appeasing Fascist aggression. At the time, many African-Americans and Africans outside of Ethiopia saw Selassie as a great hero in the struggle of black peoples everywhere for dignity and empowerment. Selassie returned to his throne in 1941 and thereafter served as an elder statesman to the African nationalists of the 1950s and 1960s. However, by the latter decade, his own domestic authority was increasingly being questioned.

In his later years, Selassie could not, or would not, move against the forces that

(United Nations photo by John Isaac)

Ethiopians experienced a brutal civil war from 1974 to 1991. The continuous fighting displaced millions of people. The problems of this forced migration were compounded by drought and starvation. The drought victims pictured above are gathered at one of the many relief camps.

were undermining his empire. Despite its trappings of progress, the Ethiopian state remained quasi-feudal in character. Many of the best lands were controlled by the nobility and the Church, whose leading members lived privileged lives at the expense of the peasantry. Many educated people grew disenchanted with what they perceived as a reactionary monarchy and social order. Urban workers resented being paid low wages by often foreign owners. Within the junior ranks of the army and civil service, there was also great dissatisfaction with the way in which their superiors were able to siphon off state revenues for personal enrichment. But the empire's greatest weakness was its inability to accommodate the aspirations of the various ethnic, regional, and sectarian groupings living within its borders.

Ethiopia is a multiethnic state. Since the time of Menilik, the dominant group has been the Coptic Amhara-speakers, whose preeminence has been resented by their Tigrinya coreligionists as well as by predominantly non-Coptic groups such as the Afars, Gurages, Oromo, and Somalis. In recent years, movements fighting for ethnoregional autonomy have emerged among the Tigrinya of Tigray, the Oromo, and, to a lesser extent, the Afars, while many Somalis in Ethiopia's Ogaden region have long struggled for union with neighboring Somalia. Somali irredentism led to

open warfare between the two principal Horn of Africa states in 1963–1964 and again in 1977–1978.

ACHIEVEMENTS

 With a history spanning 2 millennia, the cultural achievements of Ethiopia are vast. Today, Addis Ababa is the site of the headquarters of the Organization of African Unity. Ethiopia's Kefe Province is the home of the coffee plant, from whence it takes its name.

The former northern coastal province of Eritrea was a special case. From the late nineteenth century until World War II, it was an Italian colony. After the war, it was integrated into Selassie's empire. Thereafter, a local independence movement, largely united as the Eritrean People's Liberation Front (EPLF), waged a successful armed struggle, which led to Eritrea's full independence in 1993.

REVOLUTION AND REPRESSION

In 1974, Haile Selassie was overthrown by the military, after months of mounting unrest throughout the country. A major factor triggering the coup was the government's inaction in 1972–1974, when famine swept across the northern provinces, claiming

200,000 lives. Some accused the Amhara government of using the famine as a way of weakening the predominantly Tigrinya areas of the empire. Others saw the tragedy simply as proof of the venal incompetence of Selassie's administration.

The overthrow of the old order was welcomed by most Ethiopians. Unfortunately, what began as a promising revolutionary transformation quickly degenerated into a repressive dictatorship, which pushed the nation into chronic instability and distress. By the end of 1974, after the first in a series of bloody purges within its ranks, the Dergue had embraced Marxism as its guiding philosophy. Revolutionary measures followed. Companies and lands were nationalized. Students were sent into the countryside to assist in land reforms and to teach literacy. Peasants and workers were organized into cooperative associations, called *kebeles*. Initial steps were also taken to end Amhara hegemony within the state.

Progressive aspects of the Ethiopian revolution were offset by the murderous nature of the regime. Power struggles within the Dergue, as well as its determination to eliminate all alternatives to its authority, contributed to periods of "red terror," during which thousands of supporters of the revolution as well as those associated with the old regime were killed. By 1977, the Dergue itself had been transformed from a collective decision-making body to a small clique loyal to Colonel Mengistu Haile Mariam, who became a presidential dictator.

Mengistu sought for years to legitimize his rule through a commitment to Marxist-Leninism. He formally presided over a Commission for Organizing the Party of the Working People of Ethiopia, which, in 1984, announced the formation of a single-party state, led by the new Workers' Party. But real power remained in the hands of Mengistu's Dergue.

CIVIL WAR

From 1974 to 1991, Ethiopians suffered through civil war. In the face of oppressive central authority, ethnic-based resistance movements became increasingly effective in their struggles throughout much of the country. In the late 1970s, the Mengistu regime began to receive massive military aid from the Soviet bloc in its campaigns against the Eritreans and Somalis. Some 17,000 Cuban troops and thousands of other military personnel from the Warsaw Pact countries allowed the government temporarily to gain the upper hand in the fighting. The Ethiopian Army grew to more than 300,000 men under arms at any given time, the largest military force on the continent. Throughout the 1980s, military expenditures claimed more than half of the national budget.

Despite the massive domestic and international commitment on the side of the Mengistu regime, the rebels gradually gained the upper hand. Before 1991, almost all of northern Eritrea, except its besieged capital city of Asmara, had fallen to the EPLF, which had built up its own powerful arsenal, largely from captured government equipment. Local rebels had also liberated the province of Tigray and, as part of the EPRDF coalition, pushed south toward Ethiopia's capital city of Addis Ababa. In the south, independent Oromo and Somali rebels challenged government authority. There was also resistance to Mengistu from within the ranks of the national army. A major rebellion against his authority in 1989 was crushed, devastating military morale in the process. The regime was further undermined by the withdrawal of remaining Cuban and Soviet-bloc support.

Ethiopians have paid a terrible price for their nation's conflicts. Tens of thousands have been killed in combat, while many more have died from the side effects of war. In 1984–1985, the conscience of the world was moved by the images of mass starvation in the northern war zone. (At the time, however, the global media and concerned groups like Band Aid paid rela-tively little attention to the nonenvironmen-tal factors that contributed to the crisis.) Up to 1 million lives were lost before adequate relief supplies reached the famine areas. Although drought and other environmental factors, such as soil erosion, contributed to the catastrophe, the fact that people contin-ued to starve despite the availability of inter-national relief can be attributed only to the use of food as a weapon of war.

There were other political constraints on local crop production. Having seized the lands of the old ruling class, the Mengistu regime, in accordance with its Marxist-Leninist precepts, invested most of its agri-cultural inputs in large state farms, whose productivity was abysmal. Peasant produc-tion also fell in nondrought areas, due to in-secure tenure, poor producer prices, lack of credit, and an absence of consumer goods. Ethiopia's rural areas were further dis-rupted by the government's heavy-handed villagization and relocation schemes. In 1984–1985, thousands died when the gov-ernment moved some 600,000 northerners to what were supposedly more fertile re-gions in the southwest. Many considered the scheme to be part of the central govern-ment's war effort against local communi-ties resistant to its authority. By the same token, villagization has long been associ-ated with counterinsurgency efforts; con-centrated settlements allow occupying armies to exert greater control over poten-tially hostile populations.

UNCERTAIN PROSPECTS

The Dergue's demise has not as yet been accompanied by national reconciliation. Opposition to the EPRDF's attempt to transform Ethiopia into a multiethnic fed-eration has been especially strong among Amharas, many of whom support the All-Amhara People's Organization. Others ac-cuse the EPRDF—or, more especially, former Stalinists within the Tigrean Peo-ple's Liberation Front (TPLF), which has been its predominant element—of trying to create its own monopoly of power.

In 1992, fighting broke out between the EPRDF and the forces of its former rebel partner, the Oromo Liberation Front (OLF), which claims to represent Ethio-pia's largest ethnic group (Oromos consti-tute 40 percent of the population). The OLF was prominent among those who boy-cotted June 1992 local-government elec-tions, which were further marred by allegations of vote-rigging and intimida-tion on behalf of the EPRDF. In December 1993, the OLF joined a number of other movements in a Council of Alternative Forces for Peace and Democracy (CAFPD). At its inaugural meeting, seven CAFPD delegates were detained for alleg-edly advocating the armed overthrow of the government. In April, another antigov-ernment coalition, the Ethiopian National Democratic party, was formed. Both movements called for an election boycott.

Another source of resistance to the EPRDF was the Ogadeni National Libera-tion Front (ONLF), which won strong sup-port from Ogadeni Somalis in the June 1992 elections. In April, the Transitional Government removed ONLF's Hassan Jireh from power as the Ogaden region's elected administrator; in May, he was ar-rested. As a result, clashes occurred be-tween the ONLF and EPRDF in the area, with the former boycotting the subsequent polls. Smaller uprisings and acts of terror, such as a January 1996 bombing of an Ad-dis Ababa hotel, have posed further chal-lenges for the government, which has also had to cope with drought. But while the ex-tent of its electoral mandate is disputed, the EPRDF has demonstrated that it retains strong popular support, while its opposi-tion is divided.

Timeline: PAST

1855
Emperor Tewodros begins the conquest and development of modern Ethiopia

1896
Ethiopia defeats Italian invaders at the Battle of Adowa

1936
Fascist Italy invades Ethiopia and rules until 1941

1961
The Eritrean liberation struggle begins

1972–1973
Famines in Tigray and Welo Provinces result in up to 200,000 deaths

1974
Emperor Haile Selassie is overthrown; the PMAC is established

1977
Diplomatic realignment and a new arms agreement with the Soviet Union

1980s
Massive famine, resulting from both drought and warfare

1990s
The Mengistu regime is overthrown by EPRDF rebels; Eritrea achieves independence of Ethiopia

PRESENT

2000s
Interethnic political tensions continue

Ethiopia and Eritrea work to forge a lasting peace

Kenya (Republic of Kenya)

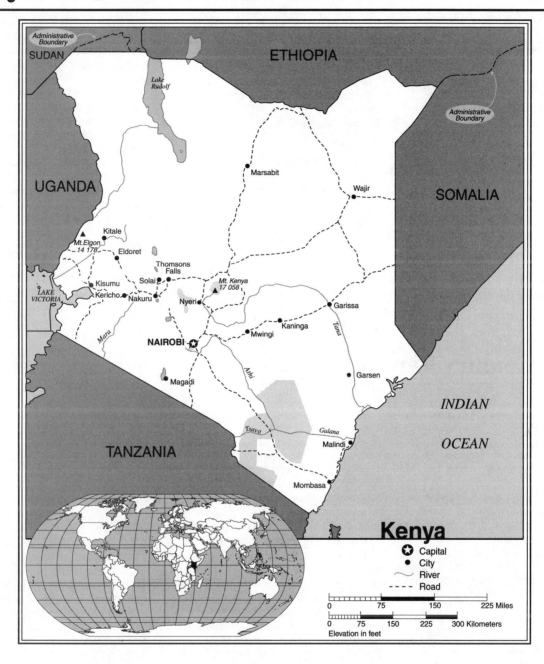

Kenya Statistics

GEOGRAPHY

Area in Square Miles (Kilometers): 224,900 (582,488) (twice the size of Nevada)

Capital (Population): Nairobi (2,343,000)

Environmental Concerns: water pollution; deforestation; soil erosion; poaching; water-hyacinth infestation in Lake Victoria

Geographical Features: low plains rise to central highlands bisected by the Great Rift Valley; fertile plateau in the west

Climate: tropical to arid

PEOPLE

Population

Total: 31,140,000

Annual Growth Rate: 1.15%

Rural/Urban Population Ratio: 68/32

Major Languages: English; Kiswahili; Maasai; others

Ethnic Makeup: 22% Kikuyu; 14% Luhya; 13% Luo; 12% Kalenjin; 11% Kamba; 28% others

Religions: 38% Protestant; 28% Catholic; 26% indigenous beliefs; 8% others

77

Health

Life Expectancy at Birth: 46 years (male);
48 years (female)
Infant Mortality: 67.2/1,000 live births
Physicians Available: 1/5,999 people
HIV/AIDS Rate in Adults: 13.5%

Education

Adult Literacy Rate: 78%
Compulsory (Ages): 6–14; free

COMMUNICATION

Telephones: 310,000 main lines
Televisions: 18/1,000 people
Internet Users: 250,000 (2001)

TRANSPORTATION

Highways in Miles (Kilometers): 38,198
(63,663)
Railroads in Miles (Kilometers): 1,654
(2,650)
Usable Airfields: 231
Motor Vehicles in Use: 357,000

GOVERNMENT

Type: republic
Independence Date: December 12, 1963
(from the United Kingdom)
Head of State/Government: President
Mwai Kibaki is both head of state and
head of government
Political Parties: Kenya African National
Union; Forum for the Restoration of
Democracy; Democratic Party of Kenya;
others
Suffrage: universal at 18

MILITARY

Military Expenditures (% of GDP): 1.8%
Current Disputes: border conflict with
Sudan; civil unrest and interethnic
violence; tensions with Somalia

ECONOMY

Currency ($ U.S. Equivalent): 77.2
shillings = $1
Per Capita Income/GDP: $1,000/$31
billion
GDP Growth Rate: 1%

Inflation Rate: 3.3%
Unemployment Rate: 40%
Labor Force by Occupation: 75%–80%
agriculture
Population Below Poverty Line: 50%
Natural Resources: gold; limestone; soda
ash; salt barites; rubies; fluorspar;
garnets; wildlife; hydropower
Agriculture: coffee; tea; corn; wheat;
sugarcane; fruit; vegetables; livestock
and dairy products
Industry: small-scale consumer goods;
agricultural processing; oil refining;
cement; tourism
Exports: $1.8 billion (primary partners
Uganda, United Kingdom, Tanzania)
Imports: $3.1 billion (primary partners
United Kingdom, United Arab Emirates,
Japan)

SUGGESTED WEB SITES

http://www.kenyaweb.com
http://www.kenyaembassy.com
http://www.kentimes.com

Kenya Country Report

In December 2002, the 24-year rule of Kenya's second president, Daniel arap Moi, ended when Kenyans elected a new leader. Opposition-party candidate Mwai Kibaki won by a landslide.

Late in 2002, Moi had informed Kenyans that Uhuru Kenyatta, the son of Kenya's first president, Jomo Kenyatta, would serve as the ruling Kenyan African National Union (KANU) party presidential candidate. Uhuru Kenyatta had entered politics a year earlier, when Moi appointed him to both the Parliament and the cabinet.

The designation of Kenyatta as Moi's successor caused a number of resignations by members of cabinet and other senior KANU leaders, who then formed a "National Rainbow Coalition." In October, the grouping combined with most of the other established opposition parties in rallying around former vice-president Kibaki as its presidential candidate.

DEVELOPMENT

Under its director Richard Leakey, the Kenyan Wildlife Service during the early 1990s cracked down on poachers while placing itself in the forefront of the global campaign to ban all ivory trading. As a result, local wildlife populations began to recover.

The elections took place against a backdrop of economic adversity. With a major portion of its economy devoted to tourism, since the terrorist attacks in the United States on September 11, 2001, Kenya has been faced with the worst recession since independence. In addition, years of drought in northern and central Kenya have left more than 1 million people dependent on food supplies from the government.

The designation of Uhuru Kenyatta as his successor was but the latest in a series of surprise political moves by Moi. In June 2001, he reshuffled the cabinet, appointing opposition-party leader Raila Odinga as energy minister in the first coalition government in Kenya's history. Earlier, in 1999, President Moi had surprised many by appointing one of his best-known critics, the internationally renowned archaeologist and conservationist Richard Leakey, as head of Kenya's civil service, with an apparent mandate to restore the country's flagging economic fortunes by rooting out endemic corruption and lethargy in the public sector. While Leakey and his "Recovery Team" could claim some progress in shaking up the system, entrenched interests had largely frustrated its efforts.

In August 1998, the world's attention had focused on the bombing of the U.S. Embassy building in Nairobi, Kenya's capital city. Scores of people in the vicinity of

FREEDOM

In 1997–1998, Kenya's already poor human-rights record deteriorated, with extra-judicial killings, beatings, and torture by the police. The National Youth Service, intended to provide young Kenyans with vocational training, was co-opted to block opposition political meetings. The worst atrocities, however, were linked to interethnic violence.

the blast were killed or injured. For Kenyans, it was an unprecedented act of international terrorism on their soil. It was also a further blow to the country's already troubled tourist industry, which had been hard hit by a continuing upward spiral of both criminal and political violence. The strength of the tourist industry—a critical foreign-exchange earner—coupled with years of growth in manufacturing and services had made Kenya, and especially Nairobi, the commercial center of East Africa. But today, after four decades of independence, most Kenyans remain the impoverished citizens of a state struggling to develop as a nation. And the recent restoration of multiparty politics has so far served only to intensify interethnic conflict.

In the precolonial past, Kenyan communities belonged to relatively small-scale, but economically interlinked, societies.

Predominantly pastoral groups, such as the Maasai and Turkana, exchanged cattle for the crops of the Kalinjin, Kamba, Kikuyu, Luo, and others. Swahili city-states developed on the coast. In the 1800s, caravans of Arab as well as Swahili traders stimulated economic and political changes. However, the outsiders who had the greatest modern impact on the Kenyan interior were European settlers, who began to arrive in the first decade of the twentieth century. By the 1930s, much of the temperate hill country around Nairobi had become the "White Highlands." More than 6 million acres of land—Maasai pasture and Kikuyu and Kamba farms—were stolen by the settlers. African communities were often displaced to increasingly overcrowded reserves. Laborers, mostly Kikuyu migrants from the reserves, worked for the new European owners, sometimes on lands that they had once farmed for themselves.

HEALTH/WELFARE

Kenya's social infrastructure has been burdened by the influx of some 300,000 refugees from the neighboring states of Ethiopia, Somalia, and Sudan. Circumcision of girls under age 17 was banned in 2001.

By the 1950s, African grievances had been heightened by increased European settlement and the growing removal of African "squatters" from their estates. There were also growing class and ideological differences among Africans, leading to tensions between educated Christians with middle-class aspirations and displaced members of the rural underclass. Many members of the latter group, in particular, began to mobilize themselves in largely Kikuyu oathing societies, which coalesced into the Mau Mau movement.

Armed resistance by Mau Mau guerrillas began in 1951, with isolated attacks on white settlers. In response, the British proclaimed a state of emergency, which lasted for 10 years. Without any outside support, the Mau Mau held out for years by making effective use of the highland forests as sanctuaries. Nonetheless, by 1955, the uprising had largely been crushed. Although the name Mau Mau became for many outsiders synonymous with antiwhite terrorism, only 32 European civilians actually lost their lives during the rebellion. In contrast, at least 13,000 Kikuyu were killed. Another 80,000 Africans were detained by the colonial authorities, and more than 1 million were resettled in controlled villages. While the Mau Mau were overwhelmed by often ruthless counterinsurgency measures, they achieved an important victory: The British

realized that the preservation of Kenya as a white-settler–dominated colony was militarily and politically untenable.

In the aftermath of the emergency, the person who emerged as the charismatic leader of Kenya's nationalist movement was Jomo Kenyatta, who had been detained and accused by the British—without any evidence—of leading the resistance movement. At independence, in 1963, he became the president. He held the office until his death in 1978.

ACHIEVEMENTS

Each year, Kenya devotes about half of its government expenditures to education. Most Kenyan students can now expect 12 years of schooling. Tertiary education is also expanding.

To many, the situation in Kenya under Kenyatta looked promising. His government encouraged racial harmony, and the slogan Harambee (Swahili for "Pull together") became a call for people of all ethnic groups and races to work together for development. Land reforms provided plots to 1.5 million farmers. A policy of Africanization created business opportunities for local entrepreneurs, and industry grew. Although the Kenya African National Union was supposedly guided by a policy of "African Socialism," the nation was seen by many as a showcase of capitalist development.

POLITICAL DEVELOPMENT

Kenyatta's Kenya quickly became a de facto one-party state. In 1966, the country's first vice-president, Oginga Odinga, resigned to form an opposition party, the Kenyan People's Union (KPU). Three years later, however, the party was banned and its leaders, including Odinga, were imprisoned. Thereafter KANU became the focus of political competition, and voters were allowed to remove sitting members of Parliament, including cabinet ministers. But politics was marred by intimidation and violence, including the assassinations of prominent critics within government, most notably Economic Development Minister Tom Mboya, in 1969, and Foreign Affairs Minister J. M. Kariuki, in 1975. Constraints on freedom of association were justified in the interest of preventing ethnic conflict—much of the KPU support came from the Luo group. However, ethnicity has always been important in shaping struggles within KANU itself.

Under Daniel arap Moi, Kenyatta's successor, the political climate grew steadily more repressive. In 1982, his government

was badly shaken by a failed coup attempt, in which about 250 people died and approximately 1,500 others were detained. The air force was disbanded and the university, whose students came out in support of the coup-makers, was temporarily closed.

Timeline: PAST

1895
The British East African Protectorate is proclaimed

1900–1910
British colonists begin to settle in the Highlands area

1951
Mau Mau, a predominantly Kikuyu movement, resists colonial rule

1963
Kenya gains independence under the leadership of Jomo Kenyatta

1978
Daniel arap Moi becomes president upon the death of Kenyatta

1980s
A coup attempt by members of the Kenyan Air Force is crushed; political repression grows

1990s
Prodemocracy agitation leads to a return of multiparty politics; interethnic violence threatens democratic transition; President Daniel arap Moi is reelected; the U.S. Embassy building in Nairobi is bombed

PRESENT

2000s
Kenya seeks to root out public corruption

Kenya works to strengthen its economy and international reputation

Opposition candidate Mwai Kibaki wins the presidency, ending KANU rule

In the aftermath of the coup, all parties other than KANU were formally outlawed. Moi followed this step by declaring, in 1986, that KANU was above the government, the Parliament, and the judiciary. Press restrictions, detentions, and blatant acts of intimidation became common. Those members of Parliament brave enough to be critical of Moi's imperial presidency were removed from KANU and thus Parliament. Political tensions were blamed on the local agents of an ever-growing list of outside forces, including Christian missionaries and Muslim fundamentalists, foreign academics and the news media, and Libyan and U.S. meddlers.

A number of underground opposition groups emerged during the mid-1980s, most notably the socialist-oriented Mwakenya movement, whose ranks included such prominent exiles as the writer Ngugi wa Thiong'o. In 1987, many of these groups came together to form the United Movement for Democracy, or UMOJA (Swahili for "unity"). But in the immediate aftermath of the 1989 KANU elections, which in many areas were blatantly rigged, Moi's grip on power appeared strong.

The early months of 1990, however, witnessed an upsurge in antigovernment unrest. In February, the murder of Foreign Minister Robert Ouko touched off rioting in Nairobi and his home city of Kisumu. Another riot occurred when squatters were forcibly evicted from the Nairobi shantytown of Muoroto. Growing calls for the restoration of multiparty democracy fueled a cycle of unrest and repression. The detention in July of two former cabinet ministers, Kenneth Matiba and Charles Rubia, for their part in the democracy agitation sparked nationwide rioting, which left at least 28 people dead and 1,000 arrested. Opposition movements, most notably the Forum for the Restoration of Democracy (FORD), began to emerge in defiance of the government's ban on their activities.

Under mounting external pressure from Western donor countries as well as from his internal opponents, Moi finally agreed to the legalization of opposition parties, in December 1991. Unfortunately, this move failed to diffuse Kenya's increasingly violent political, social, and ethnic tensions.

Continued police harassment of the opposition triggered renewed rioting throughout the country. There was also a rise in interethnic clashes in both rural and urban areas, which many, even within KANU, attributed to government incitement. In the Rift Valley, armed members of Moi's own Kalinjin grouping attacked other groups for supposedly settling on their land. Hundreds were killed and thousands injured and displaced in the worst violence since the Mau Mau era. In the face of the government's cynical resort to divide-and-rule tactics, the fledgling opposition movement betrayed the hopes of many of its supporters by becoming hopelessly splintered. New groups, such as the Islamic Party of Kenya, openly appealed for support along ethnoreligious lines. More significantly, a leadership struggle in the main FORD grouping between Matiba and the veteran Odinga split the party into two, while a proposed alliance with Mwai Kibaki's Democratic Party failed to materialize. Although motivated as much by personal ambitions and lingering mistrust between KANU defectors and long-term KANU opponents, the FORD split soon took on an ethnic dimension, with many Kikuyu backing Matiba, while Luo remained solidly loyal to Odinga.

Taking skillful advantage of his opponents' disarray, Moi called elections in December 1992, which resulted in his plurality victory, with 36 percent of the vote. KANU was able to capture 95 of the 188 parliamentary seats that were up for grabs. The two FORD factions won 31 seats each, while the Democratic Party captured 23 seats. Notwithstanding voting irregularities, most independent observers blamed the divided opposition for sowing the seeds of its own defeat. The death of the widely respected Odinga in January 1994 coincided with renewed attempts to form a united opposition to KANU. In 1996, a group of opposition Members of Parliament was established.

Kenya's politics reflects class as well as ethnic divisions. The richest 10 percent of the population own an estimated 40 percent of the wealth, while the poorest 30 percent own only 10 percent. Past economic growth has failed to alleviate poverty. Kenya's relatively large middle class has grown resentful of increased repression and the evident corruption at the very top, but it is also fearful about perceived anarchy from below.

Although its rate of growth has declined during much of the past two decades, the Kenyan economy has expanded since independence. Nairobi is now the leading center of industrial and commercial activity in East Africa. While foreign capital has played an important role in industrial development, the largest share of investment has come from government and the local private sector.

A significant percentage of foreign-exchange earnings has come from agriculture. A wide variety of cash crops is exported, a diversity that has buffered the nation's economy to some degree from the uncertainties associated with single-commodity dependence. While large plantations—now often owned. by wealthy Kenyans—have survived from the colonial era, much of the commercial production is carried out by small landholders.

A major challenge to Kenya's well-being has been its rapidly expanding population. Although there are some hopeful signs that women are beginning to plan for fewer children than in the past, the nation has been plagued with one of the highest population growth rates in the world, until very recently hovering around 3 percent per year (it is now estimated at 1.15 percent). More than half of all Kenyans are under age 15. Pressure on arable land is enormous. It will be difficult to create nonagricultural employment for the burgeoning rural-turned-urban workforce, even in the context of democratic stability and renewed economic growth.

Madagascar (Republic of Madagascar)

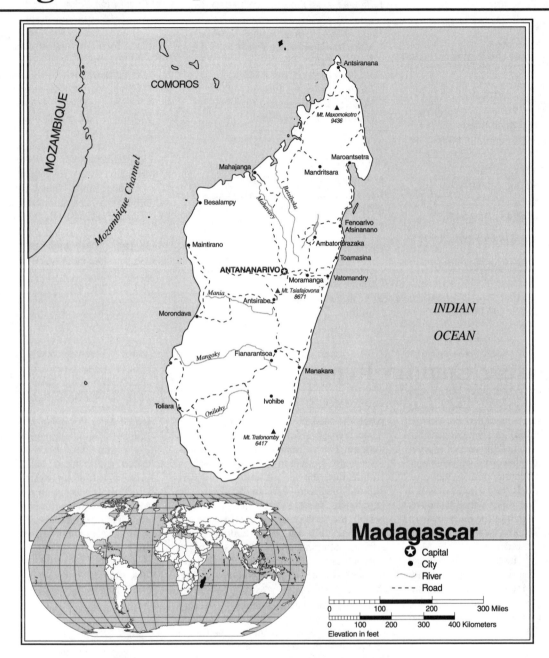

Madagascar

- ✪ Capital
- ● City
- ～ River
- - - - Road

0   100   200   300 Miles

0   100   200   300   400 Kilometers

Elevation in feet

Madagascar Statistics

GEOGRAPHY

Area in Square Miles (Kilometers): 226,658 (587,041) (about twice the size of Arizona)

Capital (Population): Antananarivo (1,689,000)

Environmental Concerns: soil erosion resulting from deforestation and overgrazing; desertification; water contamination; endangered species

Geographical Features: narrow coastal plain; high plateau and mountains in the center; the world's fourth-largest island

Climate: tropical along the coast; temperate inland; arid in the south

PEOPLE

Population

Total: 16,474,000

Annual Growth Rate: 3.02%

Rural/Urban Population Ratio: 71/29

Major Languages: Malagasy; French

Ethnic Makeup: Malayo Indonesian; Cotiers; French; Indian; Creole; Comoran

Religions: 52% indigenous beliefs; 41% Christian; 7% Muslim

Health

Life Expectancy at Birth: 53 years (male); 58 years (female)

Infant Mortality: 81.9/1,000 live births
Physicians Available: 1/8,628 people
HIV/AIDS Rate in Adults: 0.15%

Education

Adult Literacy Rate: 46%
Compulsory (Ages): for 5 years between 6–13

COMMUNICATION

Telephones: 55,000 main lines
Daily Newspaper Circulation: 4/1,000 people
Televisions: 20/1,000 people
Internet Users: 30,000 (2000)

TRANSPORTATION

Highways in Miles (Kilometers): 29,900 (49,837)
Railroads in Miles (Kilometers): 530 (883)
Usable Airfields: 130
Motor Vehicles in Use: 76,000

GOVERNMENT

Type: republic
Independence Date: June 26, 1960 (from France)
Head of State/Government: President Marc Ravalomanana; Prime Minister Jacques Sylla
Political Parties: I Love Madagascar; Movement for the Progress of Madagascar; Renewal of the Social Democratic Party; others
Suffrage: universal at 18

MILITARY

Military Expenditures (% of GDP): 1.2%
Current Disputes: territorial disputes over islands administered by France; civil strife

ECONOMY

Currency ($ U.S. Equivalent): 6,161 francs = $1
Per Capita Income/GDP: $870/$14 billion

GDP Growth Rate: 5%
Inflation Rate: 7%
Population Below Poverty Line: 70%
Natural Resources: graphite; chromite; coal; bauxite; salt; quartz; tar sands; semiprecious stones; mica; fish; hydropower
Agriculture: coffee; vanilla; sugarcane; cloves; cocoa; rice; cassava (tapioca); beans; bananas; livestock products
Industry: meat processing; soap; breweries; tanneries; sugar; textiles; glassware; cement; automobile assembly; paper; petroleum; tourism
Exports: $680 million (primary partners France, United States, Germany)
Imports: $919 million (primary partners France, Hong Kong, China)

SUGGESTED WEB SITES

http://www.embassy.org
http://www.madagascarnews.com
http://www.sas.upenn.edu/
  African_Studies/
  Country_Specific/Madagascar.html

Madagascar Country Report

Madagascar has been called the "smallest continent"; indeed, many geologists believe that it once formed the core of a larger landmass, whose other principal remnants are the Indian subcontinent and Australia. The world's fourth-largest island remains a world unto itself in other ways. Botanists and zoologists know it as the home of flora and fauna not found elsewhere. The island's culture is also distinctive. The Malagasy language, which with dialectical variations is spoken throughout the island, is related to the Malay tongues of distant Indonesia. But despite their geographic separation from the African mainland and their Asiatic roots, the Malagasy are very aware of their African identity.

DEVELOPMENT

In 1989, "Export Processing Zones" were established to attract foreign investment through tax and currency incentives. The government especially hopes to attract business from neighboring Mauritius, whose success with such zones has led to labor shortages and a shift toward more value-added production.

While the early history of Madagascar is the subject of much scholarly debate, it is clear that by the year A.D. 500, the island was being settled by Malay-speaking peoples, who may have migrated via the East African coast rather than directly from Indonesia. The cultural imprint of Southeast Asia is also evident in such aspects as local architecture, music, cosmology, and agricultural practices. African influences are equally apparent. During the precolonial period, the peoples of Madagascar were in communication with communities on the African mainland, and the waves of migration across the Mozambique channel contributed to the island's modern ethnic diversity.

During the early nineteenth century, most of Madagascar was united by the rulers of Merina. In seeking to build up their realm and preserve the island's independence, the Merina kings and queens welcomed European (mostly English) missionaries, who helped introduce new ideas and technologies. As a result, many Malagasy, including the royalty, adopted Christianity. The kingdom had established diplomatic relations with the United States and various European powers and was thus a recognized member of the international community. Foreign businesspeople were attracted to invest in the island's growing economy, while the rapid spread of schools and medical services, increasingly staffed by Malagasy, brought profound changes to the society.

The Merina court hoped that its "Christian civilization" and modernizing army would deter European aggression. But the French were determined to rule the island. The 1884–1885 Franco–Malagasy War ended in a stalemate, but a French invasion in 1895 led to the Merina kingdom's destruction. It was not an easy conquest. The Malagasy Army, with its artillery and modern fortifications, held out for many months; eventually, however, it was outgunned by the invaders. French sovereignty was proclaimed in 1896, but "pacification" campaigns continued for another decade.

FREEDOM

Respect for human rights has improved since 1993, and there has been little political violence since the 1996 election. There are isolated reports of police brutality against criminal suspects and detainees, as well as instances of arbitrary arrest and detention. Prison conditions are often life threatening, with women experiencing abuse, including rape. New judges are being appointed in an effort to relieve the overburdened judiciary.

French rule reduced what had been a prospering state into a colonial backwater. The pace of development slowed as the local economy was restructured to serve the interests of French settlers, whose numbers had swelled to 60,000 by the time of World

(United Nations photo by L. Rajaonina)

Madagascar has a unique ethnic diversity, created by migrations from the African mainland and Southeast Asia. The varied ethnic makeup of the population can be seen in the faces of these schoolchildren.

War II. Probably the most important French contribution to Madagascar was the encouragement their misrule gave to the growth of local nationalism. By the 1940s, a strong sense of Malagasy identity had been forged through common hatred of the colonialists.

HEALTH/WELFARE

Primary-school enrollment is now universal. Thirty-six percent of the appropriate age group attend secondary school, while 5% of those ages 20 to 24 are in tertiary institutions. Malaria remains a major health challenge. Madagascar's health and education facilities are underfunded.

The local overthrow of Vichy power by the British in 1943 created an opening for Malagasy nationalists to organize themselves into mass parties, the most prominent of which was the Malagasy Movement for Democratic Renewal (MRDM). In 1946, the MRDM elected two overseas deputies to the French National Assembly, on the basis of its call for immediate independence. France responded by instructing its administrators to "fight the MRDM by every means." Arrests led to resistance. In March 1947, a general insurrection began. Peasant rebels, using whatever weapons they could find, liberated large areas from French control. French troops countered by destroying crops and blockading rebel areas, in an effort to starve the insurrectionists into surrendering. Thousands of Malagasy were massacred. By the end of the year, the rebellion had been largely crushed, although a state of siege was maintained until 1956. No one knows precisely how many Malagasy lost their lives in the uprising, but contemporary estimates indicate about 90,000.

INDEPENDENCE AND REVOLUTION

Madagascar gained its independence in 1960. However, many viewed the new government, led by Philibert Tsiranana of the Social Democratic Party (PSD), as a vehicle for continuing French influence; memories of 1947 were still strong. Lack of economic and social reform led to a peasant uprising in 1971. This Maoist-inspired rebellion was suppressed, but the government was left weakened. In 1972, new unrest, this time spearheaded by students and workers in the towns, led to Tsiranana's overthrow by the military. After a period of confusion, radical forces consolidated power around Lieutenant Commander Didier Ratsiraka, who assumed the presidency in 1975.

Under Ratsiraka, a new Constitution was adopted that allowed for a controlled process of multiparty competition, in which all parties were grouped within the National Front. Within this framework, the largest party was Ratsiraka's Vanguard of the Malagasy Revolution (AREMA). Initially, all parties were expected to support the president's Charter of the Malagasy Revolution, which called for a Marxist-oriented socialist transformation. In accordance with the Charter, foreign-owned banks and financial institutions were nationalized. A series of state enterprises were also established to promote industrial development, but few proved viable.

Although 80 percent of the Malagasy were employed in agriculture, investment in rural areas and concerns was modest. The government attempted to work through *fokonolas* (indigenous village-management bodies). State farms and collectives were also established on land expropriated from French settlers. While these efforts led to some improvements, such as increased mechanization, state marketing monopolies and planning constraints contributed to shortfalls. Efforts to keep consumer prices low were blamed for a drop in rice production, the Malagasy staple, while cash-crop production, primarily coffee, vanilla, and cloves, suffered from falling world prices.

ACHIEVEMENTS

 A recently established wildlife preserve will allow the unique animals of Madagascar to survive and develop. Sixty-six species of land animals are found nowhere else on earth, including the aye-aye, a nocturnal lemur that has bat ears, beaver teeth, and an elongated clawed finger, all of which serve the aye-aye in finding food.

Since 1980, Madagascar has experienced grave economic difficulties, which have given rise to political instability. Food shortages in towns have led to rioting, while frustrated peasants have abandoned their fields. Ratsiraka's government turned increasingly from socialism to a greater reliance on market economics. But the economy has remained impoverished.

In 1985, having abandoned attempts to make the National Front into a vehicle for a single-party state, Ratsiraka presided over a loosening of his once-authoritarian control. In February 1990, most remaining restrictions on multiparty politics were lifted. But the regime's opponents, including a revived PSD, became militant in their demands for a new constitution. After six months of crippling strikes and protests, Ratsiraka formally ceded many of his powers to a transitional government, headed by Albert Zafy, in November 1991. In February 1993, Zafy won the presidency by a large margin. But subsequent divisions with Parliament over his rejection of an International Monetary Fund austerity plan, accompanied by alle-

gations of financial irregularities, led to his impeachment in August 1996. In elections held at the end of the year, Ratsiraka made a comeback, narrowly defeating Zafy. More than half of the population, however, stayed away from the polls.

Marc Ravalomanana shocked most outside observers when, under the banner of his newly organized I Love Madagascar (TIM) party, he claimed outright victory in presidential elections in December 2001. A bitter seven-month struggle for power ensued with the president of 27 years, Didier Ratsiraka.

Ravalomanana argued that, having won a majority of the votes in the first round of the elections, there need not be another round of voting. Ratsiraka, on the other hand, sought solace in the Constitution and demanded another round of voting. since many saw the elections as rigged in favor of Ratsiraka, it was widely perceived in Madagascar that Ravalomanana was the winner and that Ratsiraka was merely stalling for time and acting as an impediment to hold onto power. In April 2002, after a recount, the High Constitutional Court named Ravalomanana the winner of the December 2001 polls, but Ratsiraka ignored the verdict.

Once the demand for new elections was turned down, Ratsiraka mobilized the army and took over various areas of Madagascar. Without Madagascar Army support, Ravalomanana organized a motley crew of reserves and attacked the army-held positions. The result was chaotic. For seven months the country had two presidents and two capitals. Widespread demonstrations and worker strikes paralyzed trade, commerce, and industry, with political discourse becoming intransigent. The dispute caused immediate economic depression, as Ratsiraka's allies put up economic barricades throughout much of the country. Gaoline and medical supplies were in short supply, as foreign reserves held in U.S. banks were frozen.

In July 2002, shortly after the United States and France recognized Ravalomanana as the legitimate leader, Ratsiraka and various family members flew into exile in the Seychelles, and his forces on the is-

land either surrendered or switched sides. Ravalomanana has promised to use his entrepreneurial flair to fight the poverty and unemployment that afflict many Madagascans. But he has inherited an economy that is suffering after months of economic disruption and political violence.

The African Union has so far refused to accept Ravalomanana's presidency and did not admit him to the initiation of the organization in September 2002. The AU had demanded that new elections be held to resolve the issue. The lack of AU recognition, however, has been offset by recognition from France, Britain, and the United States.

Timeline: PAST

1828
Merina rulers gain sovereignty over other peoples of the island

1884–1885
Franco– Malagasy War

1904
The French complete the conquest of the island

1947–1948
A revolt is suppressed by the French, with great loss of life

1960
Independence from France; Philibert Tsiranana becomes the first president

1972
A coup leads to the fall of the First Malagasy Republic

1975
Didier Ratsiraka becomes president by military appointment

1980s
Economic problems intensify

1990
Elections in 1989–1990 strengthen multiparty democracy

PRESENT

2000s
Ratsiraka finally cedes power to Marc Ravalomanana

Mauritius (Republic of Mauritius)

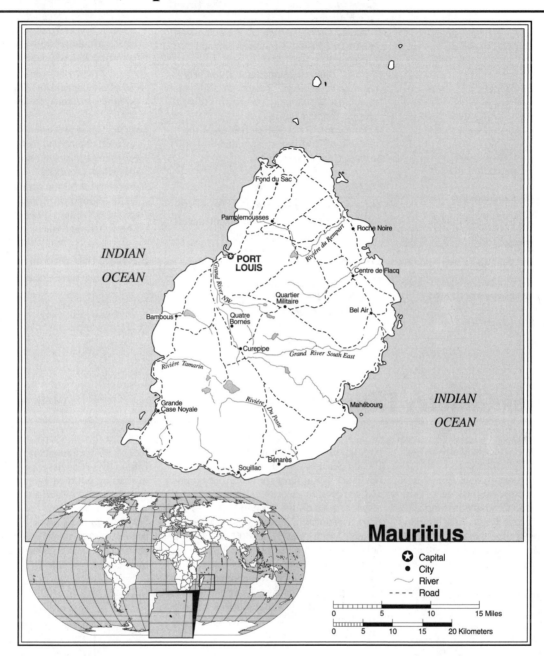

Mauritius

- ⭐ Capital
- ● City
- River
- --- Road

0   5   10   15 Miles
0   5   10   15   20 Kilometers

Mauritius Statistics

GEOGRAPHY

Area in Square Miles (Kilometers): 720
(1,865) (about 11 times the size of
Washington, D.C.)

Capital (Population): Port Louis
(176,000)

Environmental Concerns: water pollution;
population pressures on land and water
resources

Geographical Features: a small coastal
plain rising to discontinuous mountains
encircling a central plateau

Climate: tropical

PEOPLE

Population

Total: 1,200,000

Annual Growth Rate: 0.86%

Rural/Urban Population Ratio: 59/41

Major Languages: English; Creole;
French; Hindi; Urdu; Hakka; Bojpoori

Ethnic Makeup: 68% Indo-Mauritian;
27% Creole; 3% Sino-Mauritian; 2%
Franco-Mauritian

Religions: 52% Hindu; 28% Christian;
17% Muslim; 3% others

Health

Life Expectancy at Birth: 67 years (male); 75 years (female)
Infant Mortality: 16.6/1,000 live births
Physicians Available: 1/1,182 people
HIV/AIDS Rate in Adults: 0.08%

Education

Adult Literacy Rate: 83%
Compulsory (Ages): 5–12

COMMUNICATION

Telephones: 245,000 main lines
Daily Newspaper Circulation: 49/1,000 people
Televisions: 150/1,000 people
Internet Users: 87,000 (2001)

TRANSPORTATION

Highways in Miles (Kilometers): 1,126 (1,877)
Railroads in Miles (Kilometers): none
Usable Airfields: 5
Motor Vehicles in Use: 82,500

GOVERNMENT

Type: parliamentary democracy
Independence Date: March 12, 1968 (from the United Kingdom)
Head of State/Government: President Karl Offmann; Prime Minister Aneerood Jugnauth
Political Parties: Mauritian Labor Party; Mauritian Militant Movement; Militant Socialist Movement; Mauritian Militant Renaissance; Mauritian Social Democratic Party; Organization of the People of Rodrigues; Hizbullah; others
Suffrage: universal at 18

MILITARY

Military Expenditures (% of GDP): 0.2%
Current Disputes: territorial disputes with France and the United Kingdom

ECONOMY

Currency ($ U.S. equivalent): 27.4 rupees = $1
Per Capita Income/GDP: $10,800/$12.9 billion
GDP Growth Rate: 5.2%
Inflation Rate: 4.2%
Unemployment Rate: 8.6%
Labor Force by Occupation: 36% construction and industry; 24% services; 14% agriculture; 26% others
Population Below Poverty Line: 10%
Natural Resources: arable land; fish
Agriculture: sugarcane; tea; corn; potatoes; bananas; pulses; cattle; goats; fish
Industry: food processing; textiles; apparel; chemicals; metal products; transport equipment; nonelectrical machinery; tourism
Exports: $1.6 billion (primary partners United Kingdom, France, United States)
Imports: $2 billion (primary partners South Africa, France, India)

SUGGESTED WEB SITES

http://www.mauritius-info.com
http://www.ncb.intnet.ma/govt/
http://www.sas.upenn.edu/
  African_Studies/
  Country_Specific/Mauritius.html

Mauritius Country Report

Although it was not permanently settled until 1722, today Mauritius is home to 1.2 million people of South Asian, Euro-African, Chinese, and other origins. Out of this extraordinary human diversity has emerged a society that in recent decades has become a model of democratic stability and economic growth as well as ethnic, racial, and sectarian tolerance.

DEVELOPMENT

The success of the Mauritian EPZ along with the export-led growth of various Asian economies has encouraged a growing number of other African countries, such as Botswana, Cape Verde, and Madagascar, to launch their own export zones.

Mauritius was first settled by the French, some of whom achieved great wealth by setting up sugar plantations. From the beginning, the plantations prospered through their exploitation of slave labor imported from the African mainland. Over time, the European and African communities merged into a common Creole culture; that membership currently accounts for one quarter of the Mauritian population. A small number claim pure French descent. For decades, members of this latter group have formed an economic and social elite. More than half the sugar acreage remains the property of 21 large Franco–Mauritian plantations; the rest is divided among nearly 28,000 small landholdings. French cultural influence remains strong. Most of the newspapers on the island are published in French, which shares official-language status with English. Most Mauritians also speak a local, French-influenced, Creole language. Most Mauritian Creoles are Roman Catholics.

In 1810, Mauritius was occupied by the British; they ruled the island until 1968. (After years of debate, in 1992, the country cut its ties with Great Britain to become a republic.) When the British abolished slavery, in 1835, the plantation owners turned to large-scale use of indentured labor from what was then British India. Today nearly two thirds of the population are of South Asian descent and have maintained their home languages. Most are Hindu, but a substantial minority are Muslim. Other faiths, such as Buddhism, are also represented.

Although the majority of Mauritians gained the right to vote after World War II, the island has maintained an uninterrupted record of parliamentary rule since 1886. Ethnic divisions have long been important in shaping political allegiances. But ethnic constituency-building has not led, in recent years, to communal polarization. Other factors—such as class, ideology, and opportunism—have also been influential. All postindependence governments have been multiethnic coalitions.

FREEDOM

Political pluralism and human rights are respected on Mauritius, but problem areas remain. There are reports of police abuse of suspects and delayed access to defense counsel. Child labor exists. Legislation outlawing domestic violence has been passed recently. The nation has more than 30 political parties, of which about a half dozen are important at any given time. The Mauritian labor movement is one of the strongest in all of Africa.

While government in Mauritius has been characterized by shifting coalitions, with no single party winning a majority of seats, there has been relative stability in leadership at the top. Over the past half-century, Mauritius has had only three prime ministers. In September 2000, a coalition led by former prime minister Sir Aneerood Jugnauth won a landslide victory, ousting the rival coalition of Navin Ramgoolam, who had ousted Jugnauth five years earlier. Jugnauth had first come to power in 1982 by defeating Seewoosagar Ramgoolam, Navin's father.

HEALTH/WELFARE

Medical and most educational expenses are free. Food prices are heavily subsidized. Rising government deficits, however, threaten future social spending. Mauritius has a high life expectancy rate and a low infant mortality rate. Human-rights education has been introduced in secondary schools.

Although most major political parties have in the past espoused various shades of socialism, Mauritius's economic success in recent decades has created a strong consensus in favor of export-oriented market economics. Until the 1970s, the Mauritian economy was almost entirely dependent on sugar. While 45 percent of the island's total landmass continues to be planted with the crop, sugar now ranks below textiles and tourism in its contribution to export earnings and gross domestic product. The transformation of Mauritius from monocrop dependency into a fledging industrial state with a strong service sector has made it one of the major economic success stories of the developing world. Mauritian growth has been built on a foundation of export-oriented manufacturing. At the core of the Mauritian take-off is its island-wide Export Processing Zone (EPZ), which has attracted industrial investment through a combination of low wages, tax breaks, and other financial incentives. Although most of the EPZ output has been in the field of

cheap textiles, the economy has begun to diversify into more capital- and skill-intensive production. In 1989, Mauritius also entered the international financial services market by launching Africa's first offshore banking center.

ACHIEVEMENTS

Perhaps Mauritius's most important modern achievement has been its successful efforts to reduce its birth rate. This has been brought about by government-backed family planning as well as by increased economic opportunities for women.

The success of the Mauritian economy is measured in relative terms. Mauritius is still considered a middle-income country. In reality, however, there are, as with most developing societies, great disparities in the distribution of wealth. Nonetheless, quality-of-life indicators confirm a rising standard of living for the population as a whole. While great progress has been made toward eliminating poverty and disease, concern has also grown about the environmental capacity of the small, crowded country to sustain its current rate of development. There is also a general recognition that Mauritian prosperity is—and will for the foreseeable future remain—extremely vulnerable to global-market forces. This was demonstrated after the September 11, 2001, terrorist attacks in the United States, when tourism took a precipitous dive, and previously with the declining level of sugar production caused by the cyclone in 1999. The much-acclaimed ethnic diversity of Mauritius also came under examination as there were racially inspired riots in 2000 between Creoles and Hindu communities after the death while in police custody of a Creole Rastafarian pop star, Rasta Karya.

Timeline: PAST

1600s
The Dutch claim, but abandon, Mauritius

1722
French settlers arrive, and slaves imported from the African mainland

1814
The Treaty of Paris formally cedes Mauritius to the British

1835
Slavery is abolished; South Asians arrive

1937
Rioting on sugar estates shakes the political control of the Franco-Mauritian elite

1948
An expanded franchise allows greater democracy

1968
Independence

1979
A cyclone destroys homes as well as much of the sugar crop

1982
Aneerood Jugnauth replaces Seewoosagar Ramgoolam as prime minister

1990s
Mauritius becomes a republic; Mauritius becomes the 12th member of the SADC

PRESENT

2000s
Jugnauth regains the prime ministership in a landslide electoral victory

Mauritius continues its focus on market economics

Rwanda (Rwandese Republic)

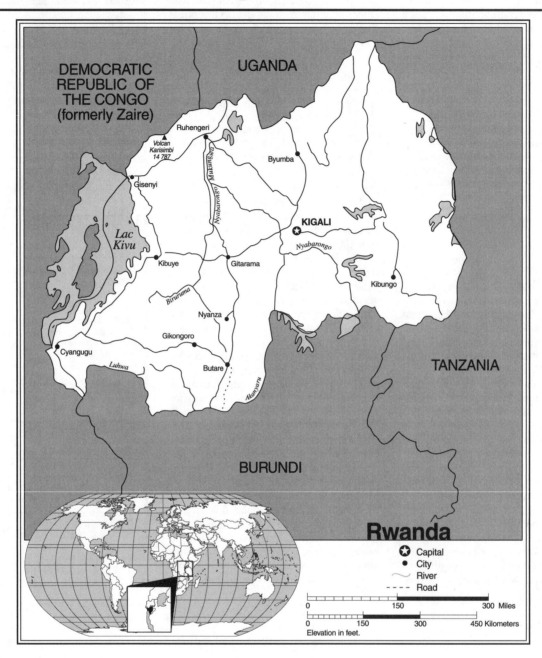

Rwanda Statistics

GEOGRAPHY

Area in Square Miles (Kilometers): 10,169 (26,338) (about the size of Maryland)

Capital (Population): Kigali (412,000)

Environmental Concerns: deforestation; overgrazing; soil exhaustion and erosion; poaching

Geographical Features: mostly grassy uplands and hills; landlocked

Climate: temperate

PEOPLE

Population

Total: 7,398,000

Annual Growth Rate: 1.16%

Rural/Urban Population Ratio: 94/6

Major Languages: Kinyarwanda; French; Kiswahili; English

Ethnic Makeup: 84% Hutu; 15% Tutsi; 1% Twa

Religions: 95% Christian; 4% Muslim; 2% none or other

Health

Life Expectancy at Birth: 38 years (male); 39 years (female)

Infant Mortality: 118/1,000 live births

Physicians Available: 1/50,000 people

HIV/AIDS Rate in Adults: 11.21%

Education

Adult Literacy Rate: 60.5%
Compulsory (Ages): 7–14

COMMUNICATION

Telephones: 11,000 main lines
Internet Users: 5,000 (2001)

TRANSPORTATION

Highways in Miles (Kilometers): 7,200 (12,000)
Railroads in Miles (Kilometers): none
Usable Airfields: 8
Motor Vehicles in Use: 28,000

GOVERNMENT

Type: republic
Independence Date: July 1, 1962 (from Belgian-administered UN trusteeship)

Head of State/Government: President Paul Kagame; Prime Minister Bernard Makuza
Political Parties: Rwanda Patriotic Front; Democratic Republican Movement; Liberal Party; Democratic Socialist Party; others
Suffrage: universal at 18

MILITARY

Military Expenditures (% of GDP): 3.1%
Current Disputes: continued internal ethnic violence; conflict in the Democratic Republic of the Congo

ECONOMY

Currency ($ U.S. equivalent): n/a
Per Capita Income/GDP: $1,000/$7.2 billion
GDP Growth Rate: 5%
Inflation Rate: 5%

Labor Force by Occupation: 90% agriculture
Population Below Poverty Line: 70%
Natural Resources: tungsten ore; tin; cassiterite; methane; hydropower; arable land
Agriculture: coffee; tea; pyrethrum; bananas; sorghum; beans; potatoes; livestock
Industry: agricultural-products processing; mining; light consumer goods; cement
Exports: $61 million (primary partners Europe, Pakistan, United States)
Imports: $248 million (primary partners Kenya, Europe, United States)

SUGGESTED WEB SITES

http://www.rwanda1.com/government/
http://www.rwandanews.org
http://www.sas.upenn.edu/African_Studies/Country_Specific/Rwanda.html

Rwanda Country Report

In July 2002, the presidents of Rwanda and the Democratic Republic of the Congo (D.R.C.) signed a peace agreement to end their four-year war. With South Africa and the United Nations acting as guarantors, the deal committed Rwanda to the withdrawal of troops from eastern D.R.C., and the D.R.C. to helping disarm Rwandan Hutu refugees in its territory who were responsible for the 1994 genocide that targeted Rwanda's Tutsi minority. By October 2002, Rwanda had honored its commitment, bringing cautious hope for a lasting peace.

DEVELOPMENT

Hydroelectric stations meet much of the country's energy needs. Before the 1994 genocide, plans were being made to exploit methane-gas reserves under Lake Kivu.

Rwandans have been living a nightmare. The 1994 genocide ranks as one of the world's greatest tragedies. A small country, only about the size of Maryland, Rwanda had 8 million inhabitants early in 1994, making it continental Africa's most densely populated state. This population was divided into three groups: the Hutu majority (89 percent); the Tutsis (10 percent); and the Twa, commonly stereotyped as "Pygmies" (1 percent). By September 1994, civil war involving genocidal conflict between Hutus and Tutsis had nearly halved the country's resident population while radically altering its group demography. Up to 1 million people, including women and children, were killed and an additional 2 million displaced. The international community, especially France, the United States, and the United Nations, did little to stop the genocide.

Rebel forces of the ousted government that initiated the genocide have continued to threaten Rwanda, from the D.R.C. In response, since 1997 Rwanda has intervened against its opponents in the D.R.C. as well as in Rwanda itself. In the process, much of eastern D.R.C. has come under Rwandan occupation. Rwanda's ability to send and sustain some 50,000 well-armed troops into the D.R.C. would probably not have been possible without the tacit support of its donors. A UN report released in October 2002 confirmed the involvement of the Rwandan, as well as the Ugandan and Zimbabwean, militaries in the exploitation of the D.R.C.'s natural resources.

The dominant figure in Rwanda over the past decade has been Paul Kagame, who in April 2000 was inaugurated as president, following the resignation of Pasteur Bizimungu. The change at the top had little immediate effect. As vice-president and minister of defense, Kagame, through his largely Tutsi-controlled Rwanda Patriotic Front (RPF), had dominated the "Government of National Unity" since its installation in 1994. While the resignation of Bizimungu and other members of the mainly Hutu Republican Democratic Movement (RDM) raised eyebrows, the RDM itself opted to remain within the government.

FREEDOM

The Rwandan government and its opponents continue to be responsible for serious human-rights abuses, including massacres. More than 120,000 prisoners are in overcrowded jails, most accused of participating in the 1994 genocide. Genocide trials, which began at the end of 1996, have made little progress and are expected to take years to complete. Hutu death squads, composed of members of the defeated former Rwandan Armed Forces and Interahamwe genocide gangs, continue to target Tutsis and foreigners.

At the end of 2001, voting to elect members of traditional *gacaca* (courts) began. The courts, in which ordinary Rwandans judge their peers, aim to clear the backlog of 1994 genocide cases. Just as important, justice can be seen to be done by the people themselves, based on their traditional standards.

THE 1994 GENOCIDE

The genocide began within a half-hour of the April 6, 1994, death of the country's democratizing dictator, President Juvenal Habyarimana. Along with president Cyprien Ntaryamira of neighboring Burundi, Habyarimana was killed when his plane was shot down. While the identity of the culprits remains a matter of speculation, Belgian troops reported that rockets were fired from Kanombe military base, which was then controlled by the country's Presidential Guard, known locally as the Akuza. In broadcasts over the independent Radio Libre Mille Collins, Hutu extremists then openly called for the destruction of the Tutsis—"The graves are only half full, who will help us fill them up?" Because the tirades were in idiomatic Kinyarwanda, the national language, they initially escaped the attention of most international journalists on the spot. Meanwhile, Akuza and regular army units set up roadblocks and began systematically to massacre Tutsi citizens in the capital, Kigali. Even greater numbers perished in the countryside, because their names appeared on death lists that had been prepared with the help of local Hutu chiefs.

By July, more than 500,000 people had been murdered. While most were Tutsis, Hutus who had supported moves toward ethnic reconciliation and democratization had also been targeted. In addition to elements within the Hutu-dominated military, the killings were carried out by youth-wing militias of the ruling party, the National Revolutionary Movement for Development (MRND) and the Coalition for Defense of Freedom (CDR). Known respectively as the Interahamwe and Impuzamugambi, the ranks of these two all-Hutu militias had mushroomed in the aftermath of an August 1993 agreement designed to return the country to multiparty rule. The extent of the killings was apparent in neighboring Tanzania, where thousands of corpses were televised being carried downstream by the Kagara River. At a rate of 80 an hour, they entered Lake Victoria, more than 100 miles from the Rwandan border.

SYSTEMATIC PLANS FOR MASS MURDER

Preparations for the genocide had been going on for months. According to Amnesty International, Hutu "Zero Network" death squads had already murdered some 2,300 people in the months leading up to the crisis. Although this information was the subject of press reports, no action was taken by the 2,500 peacekeeping troops who had been stationed in the country since June 1993 as the United Nations Assistance Mission to Rwanda (UNAMIR). Once the crisis began, most of UNAMIR's personnel were hastily withdrawn. French and Belgian paratroopers arrived for a brief time to evacuate their nationals.

Rwanda's genocide did not end with the destruction of a third or more of the Tutsi minority. Enraged by the massacres of their brethren, the 14,000-man Tutsi-dominated Rwanda Patriotic Front, which had been waging an armed struggle against the Habyarimana regime since October, launched a massive offensive. The 35,000-man regular army, along with the militias, crumbled. In July 1994, the RPF took full control of Kigali and drove the remnants of the government and its army eastward into Zaire (since 1997, the D.R.C.). Two million panic-stricken Hutu civilians also fled across the border. By then, about 1 million Rwandans were already in exile. Another 2.5 million people were crowded into a "safe zone" created by the French military. As the French prepared to pull out, the fate of these refugees was uncertain.

In depopulated Kigali, the RPF set up a "Provisional Government" with a Hutu president, Pasteur Bizimungu, and prime minister Faustin Twagiramungu. Its most powerful figure, however, was the RPF commander, Major General Paul Kagame, who became both the vice-president and minister of defense.

HUTUS AND TUTSIS

The roots of Hutu–Tutsi animosity in Rwanda (as well as Burundi) run deep. Yet it is not easy for an outsider to differentiate between the two groups. Their members both speak Kinyarwanda and look the same physically, notwithstanding the stereotype of the Tutsis being exceptionally tall; intermarriage between the two groups has taken place for centuries. By some accounts, the Tutsi arrived as northern Nilotic conquerors, perhaps in the fifteenth century. But others believe that the two groups have always been defined by class or caste rather than by ethnicity.

In the beginning, according to one epic Kinyarwanda poem, the godlike ruler Kigwa fashioned a test to choose his successor. He gave each of his sons a bowl of milk to guard during the night. His son

Gatwa drank the milk. Yahutu slept and spilled the milk. Only Gatutsi guarded it well. The myth justifies the old Rwandan social order, in which the Twas were the outcasts, the Hutus servants, and the Tutsis aristocrats. Historically, Hutu serfs herded cattle and performed various other services for their Tutsi "protectors." At the top of the hierarchy was the Mwami, or king.

THE COLONIAL ERA: HUTU AND TUTSI ANIMOSITIES CONTINUE

Rwanda's feudal system survived into the colonial era. German and, later, Belgian administrators opted to rule through the existing order. But the social order was subtly destabilized by the new ideas emanating from the Catholic mission schools and by the colonialists' encouragement of the predominantly Hutu peasantry to grow cash crops, especially coffee. Discontent also grew due to the ever-increasing pressure of people and herds on already crowded lands.

In the late 1950s, under UN pressure, Belgium began to devolve political power to Rwandans. The death of the Mwami in 1959 sparked a bloody Hutu uprising against the Tutsi aristocracy. Tens of thousands, if not hundreds of thousands, were killed. Against this violent backdrop, pre-independence elections were held in 1961. These resulted in a victory for the first president, Gregoire Kayibanda's, Hutu Emancipation Movement (better known as Parmehutu). Thus, at independence, in 1962, Rwanda's traditionally Tutsi-dominated society was suddenly under a Hutu-dominated government.

In 1963 and 1964, the continued inter-ethnic competition for power exploded into more violence, which resulted in the flight of hundreds of thousands of ethnic Tutsis to neighboring Burundi, Tanzania, and Uganda. Along with their descendants, this refugee population today numbers about 1 million. Successive Hutu-dominated governments have barred their return, questioning their citizenship and citing extreme land pressure as barriers to their re-absorption. But the implied hope that the refugees would integrate into their host societies has failed to materialize. The RPF was originally formed in Uganda by Tutsi exiles, many of whom were hardened veterans of that country's past conflicts. The repatriation of all Rwandan Tutsis has been a key RPF demand.

HABYARIMANA TAKES POWER

Major General Juvenal Habyarimana, a Hutu from the north, seized power in a military coup in 1973. Two years later, he institutionalized his still army-dominated regime as a one-party state under the MRND, in the name of overcoming ethnic divisions. Yet hostility between the Hutus and Tutsis remained. Inside the country, a system of ethnic quotas was introduced, which formally limited the remaining Tutsi minority to a maximum of 14 percent of the positions in schools and at the workplace. In reality, the Tutsis were often allocated less, while the MRND's critics maintained that the best opportunities were reserved for Hutus from Habyarimana's northern home area of Kisenyi.

POPULAR DISCONTENT

In the 1980s, many Hutus, as well as Tutsis, grew impatient with their government's corrupt authoritarianism. The post-1987 international collapse of coffee prices, Rwanda's major export-earner, led to an economic decline, further fueling popular discontent. Even before the armed challenge of the RPF, the MRND had agreed to give up its monopoly of power, though this pledge was compromised by continued repression. Prominent among the new parties that then emerged were the Democratic Republic Movement (MDR), the Social Democrats (PSD), and the Liberals (PL). The PL and PSD were able to attract both Hutu and Tutsi support. As a result, many of

their Hutu as well as Tutsi members were killed in 1994. The MDR was associated with southern-regional Hutu resentment at the MRND's supposed northern bias.

A political breakthrough occurred in March 1992 with the formation of a "Transitional Coalition Government," headed by the MDR's Dismas Nsengiyaremye, which also included MRND, PSD, and PL ministers. Habyarimana remained as president. With French military assistance, including the participation of several hundred French "advisers," Habyarimana's interim government of national unity was able to halt the RPF's advance in 1992. A series of ceasefires was negotiated with the RPF, leading up to the promise of (but never-realized) UN–supervised elections. But from the beginning, progress toward national reconciliation was compromised by hard-line Hutus within the ruling military/MRND establishment and the extremist CDR. Ironically, these elements, who conspired to carry out the anti-Tutsi genocide in order to maintain control, were pushed out of the country by the RPF. These Hutu officials, soldiers, and militarymen, thought to be responsible for massacring hundreds of thousands of Tutsis and moderate Hutus during the 1994 Civil War, were exiled to camps in Zaire (D.R.C.) and Tanzania. Soon these militants returned and began a two-month wave of killings in western Rwanda in an apparent attempt to stop the Tutsis from testifying at genocide trials being conducted by the Rwandan government and the United Nations.

As Rwanda civil unrest intensified, refugees continued to flow into strife-torn Zaire. Estimates are that between 100,000 and 350,000 Hutu citizens are still in camps there. In March 1997, some 70,000 Rwandan Hutu refugees were gathered in Ubunda, a town 80 miles south of Kisan-

gani on the Zaire River. But civil war in the D.R.C. has caused these people to be pushed back into Rwanda, where they are faced with the persistent and serious unrest in their home country.

Timeline: PAST

A.D. 1860–1895
Mwami Kigeri Rwabugiri expands and consolidates the kingdom

1916
Belgium rules Rwanda as a mandate of the League of Nations

1959
The Hutu rebellion

1962
Rwanda becomes independent; Gregoire Kayibana is president

1973
Juvenal Habyarimana seizes power

1975
The National Revolutionary Movement for Development is formed

1978
A new Constitution is approved in a nationwide referendum; Habyarimana is reelected president

1990s
Genocidal conflict results in a dramatic drop in the country's resident population; millions are killed or displaced; French relief workers withdraw from Rwanda; Tutsi massacre of Hutu at Kibeho refugee camp

PRESENT

2000s
Paul Kagame becomes president

Rwandans occupy much of eastern Democratic Republic of the Congo

Seychelles (Republic of Seychelles)

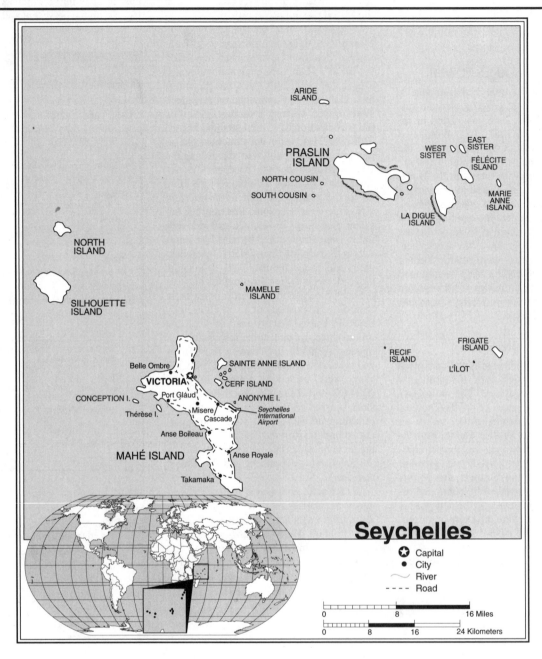

MAP LABELS:

ARIDE ISLAND

PRASLIN ISLAND

WEST SISTER · EAST SISTER · FÉLÉCITE ISLAND

NORTH COUSIN

SOUTH COUSIN

MARIE ANNE ISLAND

LA DIGUE ISLAND

NORTH ISLAND

SILHOUETTE ISLAND

MAMELLE ISLAND

FRIGATE ISLAND

RECIF ISLAND

L'ÎLOT

Belle Ombre

VICTORIA

Port Glaud

SAINTE ANNE ISLAND

CERF ISLAND

CONCEPTION I.

Misere

ANONYME I.

Thérèse I.

Cascade

Seychelles International Airport

Anse Boileau

MAHÉ ISLAND

Anse Royale

Takamaka

Seychelles

⭐ Capital
• City
〜 River
--- Road

0   8   16 Miles

0   8   16   24 Kilometers

Seychelles Statistics

GEOGRAPHY

Area in Square Miles (Kilometers): 185 (455) (2.5 times the size of Washington, D.C.)

Capital (Population): Victoria (30,000)

Environmental Concerns: uncertain freshwater supply

Geographical Features: Mahé Group is granitic, narrow coastal strip; rocky, hilly; other islands are coral, flat, elevated reefs

Climate: tropical marine

PEOPLE

Population

Total: 80,100

Annual Growth Rate: 0.5%

Rural/Urban Population Ratio: 37/63

Major Languages: English; French; Creole

Ethnic Makeup: Seychellois (mixture of Asians, Africans, and Europeans)

Religions: 96% Christian; 4% others

Health

Life Expectancy at Birth: 65 years (male); 76 years (female)

Infant Mortality: 16.8/1,000 live births

Physicians Available: 1/906 people

HIV/AIDS Rate in Adults: na

Education

Adult Literacy Rate: 58%

Compulsory (Ages): 6–15; free

COMMUNICATION

Telephones: 19,700 main lines

Daily Newspaper Circulation: 41/1,000 people

Televisions: 173/1,000 people

Internet Users: 6,000 (2001)

TRANSPORTATION

Highways in Miles (Kilometers): 162 (270)

Railroads in Miles (Kilometers): none

Usable Airfields: 14

Motor Vehicles in Use: 8,600

GOVERNMENT

Type: republic

Independence Date: June 29, 1976 (from the United Kingdom)

Head of State/Government: President France Albert René is both head of state and head of government

Political Parties: Seychelles People's Progressive Front; Seychelles National Party; Democratic Party

Suffrage: universal at 17

MILITARY

Military Expenditures (% of GDP): 1.8%

Current Disputes: claims Chagos Archipelago

ECONOMY

Currency ($ U.S. Equivalent): 5.66 rupees = $1

Per Capita Income/GDP: $7,600/$605 million

GDP Growth Rate: 1.5%

Inflation Rate: 6.1%

Labor Force by Occupation: 71% services; 19% industry; 10% agriculture

Natural Resources: fish; copra; cinnamon trees

Agriculture: vanilla; coconuts; sweet potatoes; cinnamon; cassava; bananas; chickens; fish

Industry: tourism; fishing; copra and vanilla processing; coconut oil; boat building; printing; furniture; beverages

Exports: $182.6 million (primary partners United Kingdom, Italy, France)

Imports: $360 million (primary partners Italy, South Africa, France)

SUGGESTED WEB SITES

http://www.seychelles-online. com/sc

http://www.sas.upenn.edu/ African_Studies/ Country_Specific/Seychelles.html

http://www.tbc.gov.bc.ca/cwgames/ country/Seychelles/ seychelles.html

Seychelles Country Report

Africa's smallest country in terms of both size and population, the Republic of Seychelles consists of a number of widely scattered archipelagos off the coast of East Africa. Over the last quarter-century, Seychellois have enjoyed enormous economic and social progress. According to the United Nations, today they enjoy the highest standard of living in Africa.

DEVELOPMENT

Seychelles has declared an Exclusive Economic Zone of 200 miles around all of its islands in order to promote the local fishing industry. Most of the zone's catch is harvested by foreign boats, which are supposed to pay licensing fees to Seychelles.

But for many years, the country's politics was bitterly polarized between supporters of President James Mancham and his successor, Albert René. The holding of multiparty elections in 1992 and 1993, after 15 years of single-party rule under René, has been accompanied by a significant degree of reconciliation between the partisans of these two long-time rivals.

FREEDOM

Since the restoration of multiparty democracy, there has been greater political freedom in Seychelles. The opposition nonetheless continues to complain of police harassment and to protest about the government's control over the broadcast media.

The roots of Seychelles' modern political economy go back to 1963, when Mancham's Democratic Party and René's People's United Party were established. The former originally favored private enterprise and the retention of the British imperial connection, while the latter advocated an independent socialist state. Electoral victories in 1970 and 1974 allowed Mancham to pursue his dream of turning Seychelles into a tourist paradise and a financial and trading center by aggressively seeking outside investment. Tourism began to flourish following the opening of an international airport on the main island of Mahe in 1971, fueling an economic boom. Between 1970 and 1980,

per capita income rose from nearly $150 to $1,700 (today it is about $7,600).

In 1974, Mancham, in an about-face, joined René in advocating the islands' independence. The Democratic Party, despite its modest electoral and overwhelming parliamentary majority, set up a coalition government with the People's United Party. On June 29, 1976, Seychelles became independent, with Mancham as president and René as prime minister.

HEALTH/WELFARE

A national health program has been established; private practice has been abolished. Free-lunch programs have raised nutritional levels among the young. Education is also free, up to age 15.

On June 5, 1977, with Mancham out of the country, René's supporters, with Tanzanian assistance, staged a successful coup in which several people were killed. Thereafter René assumed the presidency and suspended the Constitution. A period of rule by decree gave way in 1979, without the benefit of referendum, to a new constitutional framework in which the People's Progressive Front Seychelles (SPPF), successor to the People's United Party, was recognized as the nation's sole political voice. The first years of one-party government were characterized by continued economic growth, which allowed for an impressive expansion of social-welfare programs.

ACHIEVEMENTS

 Seychelles has become a world leader in wildlife preservation. An important aspect of the nation's conservation efforts has been the designation of one island as an international wildlife refuge.

Political power since the coup has largely remained concentrated in the hands of René. The early years of his regime, however, were marked by unrest. In 1978, the first in a series of unsuccessful countercoups was followed, several months later, by violent protests against the government's attempts to impose a compulsory National Youth Service, which would have removed the nation's 16- and 17-year-olds from their families in order to foster their sociopolitical indoctrination in accordance with the René government's socialist ideals. Another major incident occurred in 1981, when a group of international mercenaries, who had the backing of authorities in Kenya and South Africa as well as exiled Seychellois, were forced to flee in a hijacked jet after an airport shootout with local security forces. Following this attempt, Tanzanian troops were sent to the islands. A year later, the Tanzanians were instrumental in crushing a mutiny of Seychellois soldiers.

Despite its success in creating a model welfare state, which undoubtedly strengthened its popular acceptance, for years René continued to govern in a repressive manner. Internal opposition was not tolerated by his government, and exiled activists were largely neutralized. About one fifth of the islands' population now live overseas (not all of these people left the country, however, for political reasons).

In 1991, René gave in to rising internal and external pressures for a return to multiparty democracy. In July 1992, his party won 58 percent of the vote for a commission to rewrite the Constitution. Mancham's Democrats received just over a third of the vote. But in November 1992, voters heeded Mancham's call, rejecting the revised constitution proposed by the pro-René commission. Faced with a possible deadlock, the two parties reached consensus on new proposals, which were ratified in a June 1993 referendum. Presidential and parliamentary elections held the following month confirmed majority support for René's party.

In elections in 1998 and 2001, the SPPF gained renewed mandates. These latter contests are perhaps more significant for the emergence of the United Opposition coalition, led by Wavel Ramkalawan, as the main opposition party in place of the aging Mancham's Democrats. Increasing support of the United Opposition coalition was apparent in September 2001, when President René won another term in office, with 54 percent of the votes, against 45 percent for Ramkalawan.

Timeline: PAST

1771
French settlement begins

1814
British rule is established

1830
The British end slavery

1903
Seychelles is detached from Mauritius by the British and made a Crown colony

1948
Legislative Council with qualified suffrage is introduced

1967
Universal suffrage

1976
Independence

1977
An Albert René coup against James Mancham

1980s
An Amnesty International report alleges government fabrication of drug-possession cases for political reasons

1990s
René agrees to a multiparty system; René and his party are approved in presidential and parliamentary elections

PRESENT

2000s
The tourism sector accounts for 30% of Seychellois employment and 70% of hard-currency earnings

The government seeks to diversify the economy

Somalia

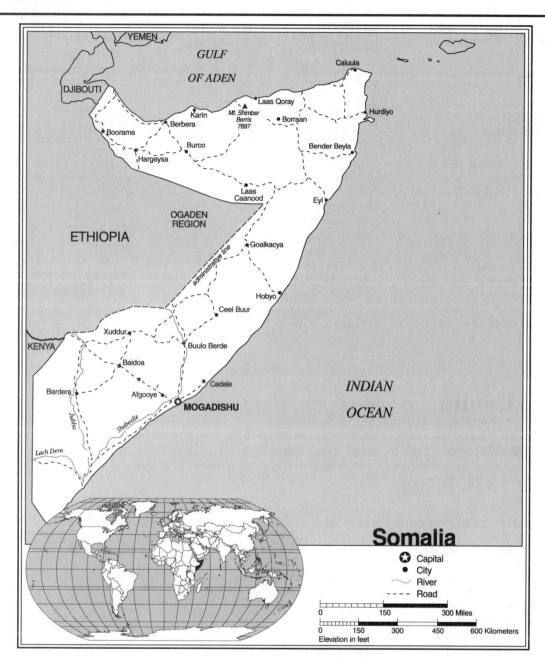

Somalia Statistics

GEOGRAPHY

Area in Square Miles (Kilometers): 246,331 (638,000) (about the size of Texas)

Capital (Population): Mogadishu (1,212,000)

Environmental Concerns: famine; contaminated water; deforestation; overgrazing; soil erosion; desertification

Geographical Features: principally desert; mostly flat to undulating plain, rising to hills in the north

Climate: arid to semiarid

PEOPLE

Population

Total: 7,754,000*

Annual Growth Rate: 3.46%

Rural/Urban Population Ratio: 73/27

Major Languages: Somali; Arabic; Italian; English

Ethnic Makeup: 85% Somali; Bantu; Arab

Religion: Sunni Muslim

Health

Life Expectancy at Birth: 45 years (male); 47 years (female)

Infant Mortality: 122/1,000 live births

Physicians Available: 1/19,071 people

Education

Adult Literacy Rate: 38%
Compulsory (Ages): 6–14; free

COMMUNICATION

Telephones: 15,000 main lines
Televisions: 18/1,000 people
Internet Users: 200 (2000)

TRANSPORTATION

Highways in Miles (Kilometers): 13,702 (22,100)
Railroads in Miles (Kilometers): none
Usable Airfields: 54
Motor Vehicles in Use: 20,000

GOVERNMENT

Type: "Transitional National Government"
Independence Date: July 1, 1960 (from a merger of British Somaliland and Italian Somaliland)
Head of State/Government: Interim President Abdiqassim Salah Hassan; Prime Minister Hassan Abshir Farah

Political Parties: none
Suffrage: universal at 18

MILITARY

Military Expenditures (% of GDP): 0.9%
Current Disputes: civil war; border and territorial disputes with Ethiopia

ECONOMY

Currency ($ U.S. Equivalent): 2,607 shillings = $1
Per Capita Income/GDP: $550/$4.1 billion
Inflation Rate: 100%
Labor Force by Occupation: 71% agriculture; 29% industry and services
Natural Resources: uranium; iron ore; tin; gypsum; bauxite; copper; salt
Agriculture: livestock; bananas; sugarcane; cotton; cereals; corn; sorghum; mangoes; fish
Industry: sugar refining; textiles; limited petroleum refining

Exports: $186 million (primary partners Saudi Arabia, United Arab Emirates, Yemen)
Imports: $314 million (primary partners Djibouti, Kenya, India)

SUGGESTED WEB SITES

http://www.unsomalia.org
http://www.somalianews.com
http://somalinet.com
http://somaliawatch.org
http://www.state.gov/www/
  background_notes/
  somalia_0798_bgn.html
http://www.cs.indiana.edu/hyplan/
  dmulholl/somalia/somalia.html
http://www.sas.upenn.edu/
  African_Studies/
  Country_SpecificSomalia.html

---

*Note: Population statistics in Somalia are complicated by the large number of nomads and by refugee movements in response to famine and clan warfare.

Somalia Country Report

Somalia has in effect been without a central government since 1991, when after the overthrow of President Siad Barre, the country entered a period of chaos from which it has never recovered. The country has been dominated by "warlords" reponsible for small fiefdoms supported by heavily armed militias under their control. The resulting intermilitia fighting, added to the inability to deal with famine and disease, has led to the deaths of up to 1 million people. There appears to be little hope for an early resolution to the continuing conflicts in Somalia. The international donor community has long vanished, fundamentally giving up efforts to work out a peaceful solution to the problems of the nation. To compound matters, the northern portion of the country has broken away from the south and is now called Somaliland.

A "Transitional National Government" led by Abdiqassim Salah Hassan has been in place since August 2000, but so far it has had little influence outside of the capital, Mogadishu. Hassan was elected by a "Transitional National Assembly," which had been formed at an internationally backed peace conference in neighboring Djibouti. After months of negotiations, the new government was accepted by a broad cross-section of Somali society. But a number of

key groupings, in addition to still-powerful military leaders, remained outside of the accord. The northern "Puntland" and "Somaliland" governments were among those that boycotted the talks. The latter administration has declared itself an independent state, despite international opposition.

DEVELOPMENT

 Most development projects have ended. Somalia's material infrastructure has largely been destroyed by war and neglect, though some local rebuilding efforts are under way, especially in the more peaceful central and northern parts of the country. In 1996, the European Union agreed to finance the reconstruction of the port of Berbera.

For much of the outside world, Somalia has become a symbol of failure of both international peacekeeping operations and the postcolonial African state. For the Somalis themselves, Somalia is an ideal that has ceased to exist—but may yet be re-created. Literally hundreds of thousands of Somalis starved to death in 1991–1992 before a massive U.S.–led United Nations intervention—officially known as UNITAF but labeled "Operation Restore Hope" by

the Americans—assured the delivery of relief supplies. The 1994 withdrawal of most UN forces (the last token units left in March 1995), following UNITAF's failure to disarm local militias while supporting the creation of a "Transitional National Council," led to the termination of relief efforts in many areas, but widespread famine was averted in 1994–1995. Repeated attempts to reach a settlement between the various armed factions have continued to fail, despite the death in August 1996 of Somalia's most powerful warlord, General Mohammed Farah Aideed.

FREEDOM

 Plagued by persistent hunger and internal violence, and with the continuing threat of governance by the anarchic greed of the warlords, the living have no true freedom in Somalia.

SOMALI SOCIETY

The roots of Somalia's suffering run deep. Somalis have lived with the threat of famine for centuries, as the climate is arid even in good years. Traditionally, most Somalis were nomadic pastoralists, but in recent

years, this way of life has declined dramatically. Prior to the 1990s crisis, about half the population were still almost entirely reliant on livestock. Somali herds have sometimes been quite big: In the early 1980s, more than 1 million animals, mostly goats and sheep, were exported annually. Large numbers of cattle and camels have also been kept. But hundreds of thousands of animals were lost due to lack of rain during the mid-1980s; and since 1983, reports of rinderpest led to a sharp drop in exports, due to the closing of the once-lucrative Saudi Arabian market to East African animals.

A quarter of the Somali population have long combined livestock-keeping with agriculture. Cultivation is possible in the area between the Juba and Shebelle Rivers and in portions of the north. Although up to 15 percent of the country is potentially arable, only about 1 percent of the land has been put to plow at any given time. Bananas, cotton, and frankincense have been major cash crops, while maize and sorghum are subsistence crops. Like Somali pastoralists, farmers walk a thin line between abundance and scarcity, for locusts as well as drought are common visitors.

HEALTH/WELFARE

Somalia's small health service has almost completely disappeared, leaving the country reliant on a handful of international health teams. By 1986, education's share of the national budget had fallen to 2%. Somalia had 525 troops per teacher, the highest such ratio in Africa.

The delicate nature of Somali agriculture helps to explain recent urbanization. One out of every four Somalis lives in the large towns and cities. The principal urban center is Mogadishu, which, despite being divided by war, still houses well over a million people. Unfortunately, as Somalis have migrated in from the countryside, they have found little employment. Even before the recent collapse, the country's manufacturing and service sectors were small. By 1990, more than 100,000 Somalis had become migrant workers in the Arab/Persian Gulf states. (In 1990–1991, many were repatriated as a result of the regional conflict over Kuwait.)

Until recently, many outsiders assumed that Somalia possessed a greater degree of national coherence than most other African states. Somalis do share a common language and a sense of cultural identity. Islam is also a binding feature. However, competing clan and subclan allegiances have long played a divisive political role in the society. Membership in all the current armed factions is congruent with blood loyalties. Traditionally, the clans were governed by experienced, wise men. But the authority of these elders has now largely given way to the power of younger men with a surplus of guns and a surfeit of education and a lack of moral decency.

ACHIEVEMENTS

Somalia has been described as a "nation of poets." Many scholars attribute the strength of the Somali poetic tradition not only to the nomadic way of life, which encourages oral arts, but to the role of poetry as a local social and political medium.

Past appeals to greater Somali nationalism have also been a source of conflict by encouraging irredentist sentiments against Somalia's neighbors. During the colonial era, contemporary Somalia was divided. For about 75 years, the northern region was governed by the British, while the southern portion was subject to Italian rule. These colonial legacies have complicated efforts at nation-building. Many northerners feel that their region has been neglected and would benefit from greater political autonomy or independence.

Somalia became independent on July 1, 1960, when the new national flag, a white, five-pointed star on a blue field, was raised in the former British and Italian territories. The star symbolized the five supposed branches of the Somali nation—that is, the now-united peoples of British and Italian Somalilands and the Somalis still living in French Somaliland (modern Djibouti), Ethiopia, and Kenya.

THE RISE AND FALL OF SIAD BARRE

Siad Barre came to power in 1969, through a coup promising radical change. As chairman of the military's Supreme Revolutionary Council, Barre combined Somali nationalism and Islam with a commitment to "scientific socialism." Some genuine efforts were made to restructure society through the development of new local councils and worker management committees. New civil and labor codes were written. The Somali Revolutionary Socialist Party was developed as the sole legal political party.

Initially, the new order seemed to be making progress. The Somali language was alphabetized in a modified form of Roman script, which allowed the government to launch mass-literacy campaigns. Various rural-development projects were also implemented. In particular, roads were built, which helped to break down isolation among regions.

The promise of Barre's early years in office gradually faded. Little was done to follow through the developments of the early 1970s, as Barre increasingly bypassed the participatory institutions that he had helped to create. His government became one of personal rule; he took on emergency powers, relieved members of the governing council of their duties, surrounded himself with members of his own Marehan branch of the Darod clan, and isolated himself from the public. Barre also isolated Somalia from the rest of Africa by pursuing irredentist policies in order to unite the other points of the Somali star under his rule. To accomplish this task, he began to encourage local guerrilla movements among the ethnic Somalis living in Kenya and Ethiopia.

Timeline: PAST

1886–1887
The British take control of northern regions of present-day Somalia

1889
Italy establishes a protectorate in the eastern areas of present-day Somalia

1960
Somalia is formed through a merger of former British and Italian colonies under UN Trusteeship

1969
Siad Barre comes to power through an army coup; the Supreme Revolutionary Council is established

1977–1978
The Ogaden war in Ethiopia results in Somalia's defeat

1980s
SNM rebels escalate their campaign in the north; government forces respond with genocidal attacks on the local Issaq population

1990s
The fall of Barre leaves Somalia without an effective central government; U.S.–led UN intervention feeds millions while attempting to restore order
In the face of mounting losses, foreign troops pull out of Somalia

PRESENT

2000s
Chaos still reigns in Somalia, causing untold suffering

Abdiqassim Salah Hassan is named interim president

In 1977, Barre sent his forces into the Ogaden region to assist the local rebels of the Western Somali Liberation Front. The invaders achieved initial military success against the Ethiopians, whose forces had been weakened by revolutionary strife and battles with Eritrean rebels. However, the intervention of some 17,000 Cuban troops and other Soviet-bloc personnel on the side of the Ethiopians quickly turned the tide of battle. At the same time, the Somali incursion was condemned by all members of the Organization of African Unity.

The intervention of the Soviet bloc on the side of the Ethiopians was a bitter disappointment to Barre, who had enjoyed Soviet support for his military buildup. In exchange, he had allowed the Soviets to establish a secret base at the strategic northern port of Berbera. However, in 1977, the Soviets decided to shift their allegiances decisively to the then–newly established revolutionary government in Ethiopia. Barre in turn tried to attract U.S. support with offers of basing rights at Berbera, but the Carter administration was unwilling to jeopardize its interests in either Ethiopia or Kenya by backing Barre's irredentist adventure. American–Somali relations became closer during the Reagan administration, which signed a 10-year pact giving U.S. forces access to air and naval facilities at Berbera, for which the United States increased its aid to Somalia, including limited arms supplies.

In 1988, Barre met with Ethiopian leader Mengistu Mariam. Together, they pledged to respect their mutual border. This understanding came about in the context of growing internal resistance to both regimes. By 1990, numerous clan-based armed resistance movements were enjoying success against Barre.

Growing resistance was accompanied by massive atrocities on the part of government forces. Human-rights concerns were cited by the U.S. and other governments in ending their assistance to Somalia. In March 1990, Barre called for national dialogue and spoke of a possible end to one-party rule. But continuing atrocities, including the killing of more than 100 protesters at the national stadium, fueled further armed resistance.

In January 1991, Barre fled Mogadishu, which was seized by resistance forces of the United Somali Congress (USC). The USC set up an interim administration, but its authority was not recognized by other groups. By the end of the year, the USC itself had split into two warring factions. A 12-faction "Manifesto Group" recognized Ali Mahdi as the country's president. But Mahdi's authority was repudiated by the four-faction Somali National Alliance (SNA), led by Farah Aideed. Much of Mogadishu was destroyed in inconclusive fighting between the two groupings. Other militias, including forces still loyal to Barre (that is, the Somali National Front, or SNF), also continued to fight one another. In the north, the Somali National Movement (SNM) declared its zone's sovereign independence as "Somaliland."

Continued fighting coincided with drought. As failed crops and dying livestock resulted in countrywide famine, international relief efforts were unable to supply sufficient quantities of outside food to those most in need, due to the prevailing state of lawlessness. In mid-1992, the International Red Cross estimated that, of southern Somalia's 4.5 million people, 1.5 million were in danger of starvation. Another 500,000 or so had fled the country. More than 300,000 children under age five were reported to have perished.

As Somalia's suffering grew and became publicized in the Western media, many observers suggested the need for the United Nations to intervene. A small UN presence, known as UNISOM, was established in August 1992, but its attempts to police the delivery of relief supplies proved to be ineffectual. Conceived as a massive U.S.–led military operation, initially consisting of 30,000 troops (22,000 Americans), UNITAF averted catastrophe by assuring the delivery of food and medical supplies to Somalia's starving millions. Still, the foreign troops' mission was unclear.

In 1993, a bloody clash between Aideed's SNA militia and Pakistani troops in Mogadishu led to full-scale armed conflict. Efforts by UNITAF forces to capture Aideed and neutralize his men were unsuccessful. After a U.S. helicopter was shot down in October 1993, President Bill Clinton decided to end American involvement in UNITAF by March 1994. By then, much higher losses had been suffered by several other nations participating in the UNITAF–UNISOM coalition, causing them also to reassess their commitments. Outgunned and demoralized, the remaining UN forces (officially labeled UNISOM II) remained largely confined to their compounds until their withdrawal.

Sudan (Republic of the Sudan)

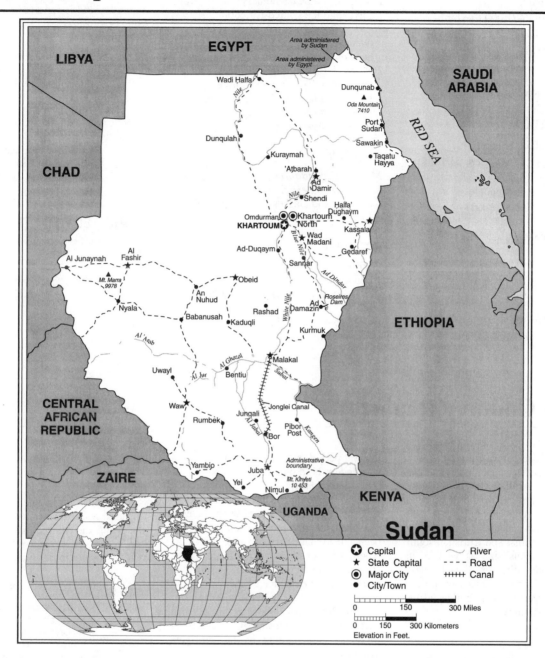

Sudan Statistics

GEOGRAPHY

Area in Square Miles (Kilometers): 967,247
(2,505,810) (about 1/4 the size of the
United States)

Capital (Population): Khartoum
(2,853,000)

Environmental Concerns: insufficient
potable water; excessive hunting of
wildlife; soil erosion; desertification

Geographical Features: generally flat,
featureless plain; mountains in the east
and west

Climate: arid desert to tropical

PEOPLE

Population

Total: 37,091,000

Annual Growth Rate: 2.73%

Rural/Urban Population Ratio: 65/35

Major Languages: Arabic; Sudanic
languages; Nubian; English; others

Ethnic Makeup: 52% black; 39% Arab;
6% Beja; 3% others

Religions: 70% Sunni Muslim, especially
in north; 25% indigenous beliefs; 5%
Christian

Health

Life Expectancy at Birth: 56 years (male); 58 years (female)
Infant Mortality: 67/1,000 live births
Physicians Available: 1/11,300 people
HIV/AIDS Rate in Adults: 0.99%

Education

Adult Literacy Rate: 46%

COMMUNICATION

Telephones: 400,000 main lines
Daily Newspaper Circulation: 21/1,000 people
Televisions: 8.2/1,000 people
Internet Users: 50,000 (2002)

TRANSPORTATION

Highways in Miles (Kilometers): 7,198 (11,610)
Railroads in Miles (Kilometers): 3,425 (5,516)
Usable Airfields: 65
Motor Vehicles in Use: 75,000

GOVERNMENT

Type: transitional
Independence Date: January 1, 1956 (from Egypt and the United Kingdom)
Head of State/Government: President Omar Hasan Ahmad al-Bashir is both head of state and head of government
Political Parties: National Congress Party; Popular National Congress; Umma; Sudan People's Liberation Movement (Army); others
Suffrage: universal at 17

MILITARY

Military Expenditures (% of GDP): 2.5%
Current Disputes: civil war; border disputes and clashes with Egypt and Kenya

ECONOMY

Currency ($ U.S. Equivalent): 2,582 pounds = $1
Per Capita Income/GDP: $1,360/$49.3 billion
GDP Growth Rate: 5.5%
Inflation Rate: 10%

Unemployment Rate: 19%
Labor Force by Occupation: 80% agriculture; 13% government; 7% industry and commerce
Natural Resources: petroleum; iron ore; chromium ore; copper; zinc; tungsten; mica; silver; gold; hydropower
Agriculture: cotton; sesame; gum arabic; sorghum; millet; wheat; sheep; groundnuts
Industry: textiles; cement; cotton ginning; edible oils; soap; sugar; shoes; petroleum refining
Exports: $2.1 billion (primary partners Japan, China, Saudi Arabia)
Imports: $1.6 billion (primary partners China, Saudi Arabia, United Kingdom)

SUGGESTED WEB SITES

```
http://www.sudan.net
http://sudanhome.com
http://www.sudmer.com
http://www.sunanews.net
http://www.sas.upenn.edu/
    African_Studies/
    Country_Specific/Sudan.html
http://www.sudan.net/
    contents.shtml
```

Sudan Country Report

The name Sudan comes from the Arabic *bilad al-sudan*, or "land of the blacks." Today, Sudan is Africa's largest country. Apart from an 11-year period of peace, it has been torn since its indpendence in 1956 by civil war between the mainly Muslim north and the animist and Christian south. Sudan's tremendous size as well as its great ethnic and religious diversity have frustrated the efforts of successive postindependence governments to build a lasting sense of national unity.

DEVELOPMENT

Many ambitious development plans have been launched since independence, but progress has been limited by political instability. The periodic introduction and redefinition of "Islamic" financial procedures have complicated long-term planning.

The current president, Omar Bashir, was reelected in 2001 for another five years. The elections were boycotted by the main opposition parties. The Machakos Protocol of July 2002, which was signed by both the government and the two largest southern rebel groups, calls for a six-year interval period, after which there will be a referendum held on self-determination for the south. However, the Muslim-led Sudanese government has continued to attack the southern rebels; through the age-old tactic of divide-and-conquer, it has been able to make periodic inroads into the rebels' African strongholds. Ethnic groups are pitted against one another. Meanwhile, there has been evidence of the widespread enslavement of blacks in the south. In December 2001, for example, more than 14,550 slaves, mainly blacks, were freed following campaigning by human-rights activists.

Listed by the U.S. government as a major supporter of terrorism, until 1997 Sudan's Islamic fundamentalist government provided refuge for Osama bin Laden. Since then, the government has been keen to overcome its image as a pariah state. The challenges facing Sudan have also been complicated by the discovery of major oil fields in the south. The government has sought to establish safe enclaves for the exploitation of the oil fields, at the cost of relocating people who were living in the area. The oil fields may make Sudan rich, but they remain a primary source of alienation, as funds generated by the government from oil revenues have been used to purchase weapons against the southern rebels.

The future looks like continued civil war until Sudan ceases to enslave its people, grants the southerners self-determination, and ceases trying to impose an Islamic state on its religiously mixed population.

FREEDOM

The current regime rules through massive repression. In 1992, Africa Watch accused it of practicing genocide against the Nuba people. Elsewhere, tales of massacres, forced relocations, enslavement, torture, and starvation are commonplace. The insurgent groups have also been responsible for numerous atrocities.

HISTORY

Sudan, like its northern neighbor Egypt, is a gift of the Nile. The river and its various branches snake across the country, providing water to most of the 80 percent of Sudanese who survive by farming. From ancient times, the Upper Nile region of northern Sudan has been the site of a series of civilizations whose histories are closely

(United Nations photo by Milton Grant)

Millions of Sudanese have been displaced by war and drought. The effects on the population have been devastating, and even the best efforts of the international community have met with only limited success.

intertwined with those of Egypt. There has been constant human interaction between the two zones. Some groups, such as the Nubians, expanded northward into the Egyptian lower Nile.

The last ruler to unite the Nile Valley politically was the nineteenth-century Turko–Egyptian ruler Muhammad Ali. After absorbing northern Sudan, by then predominantly Arabized Muslim, into his Egyptian state, Ali gradually expanded his authority to the south and west over non-Arabic and, in many cases, non-Muslim groups. This process, which was largely motivated by a desire for slave labor, united for the first time the diverse regions that today make up Sudan. In the 1880s, much of Sudan fell under the theocratic rule of the Mahdists, a local anti-Egyptian Islamic movement. The Mahdists were defeated by an Anglo–Egyptian force in 1898. Thereafter, the British dominated Sudan until its independence, in 1956.

Sudanese society has remained divided ever since. There has been strong pan-Arab sentiment in the north, but 60 percent of Sudanese, concentrated in the south and west, are non-Arab. About a third of Sudanese, especially in the south, are also non-Muslim. Despite this fact, many, but by no means all, Sudanese Muslims have favored the creation of an Islamic state. Ideological divisions among various socialist- and nonsocialist-oriented factions have also been important. Sudan has long had a strong Communist Party (whether legal or not), drawing on the support of organized labor, and an influential middle class.

The division between northern and southern Sudan has been especially deep. A mutiny by southern soldiers prior to independence escalated into a 17-year rebellion by southerners against what they perceived to be the hegemony of Muslim Arabs. Some 500,000 southerners perished before the Anya Nya rebels and the government reached a compromise settlement, recognizing southern autonomy in 1972.

HEALTH/WELFARE

 Civil strife and declining government expenditures have resulted in rising rates of infant mortality. Warfare has also prevented famine relief from reaching needy populations, resulting in instances of mass starvation.

In northern Sudan, the first 14 years of independence saw the rule of seven different civilian coalitions and six years of military rule. Despite this chronic instability, a tradition of liberal tolerance among political factions was generally maintained. Government became increasingly authoritarian during the administration of Jaafar Nimeiri, who came to power in a 1969 military coup.

Nimeiri quickly moved to consolidate his power by eliminating challenges to his government from the Islamic right and the Communist left. His greatest success was ending the Anya Nya revolt, but his subsequent tampering with the provisions of the peace agreement led to renewed resistance. In 1983, Nimeiri decided to impose Islamic law throughout Sudanese society. This led to the growth of the Sudanese People's Liberation Army (SPLA), under the leadership of John Garang, which quickly seized control of much of the southern Sudanese countryside. Opposition to Nimeiri had also been growing in the north, as more people became alienated by the regime's increasingly heavy-handed ways and inability to manage the declining economy. Finally, in 1985, he was toppled in a coup.

The holding of multiparty elections in 1986 seemed to presage a restoration of Sudan's tradition of pluralism. With the SPLA preventing voting in much of the south, the two largest parties were the northern-based Umma and Democratic Union (DUP). The National Islamic Front was the third-largest vote-getter, with eight other parties plus a number of independents gaining parliamentary seats. The major challenge facing the new coalition government, led by Umma, was reconciliation with the SPLA. Because the SPLA, unlike the earlier Anya Nya, was committed to national unity, the task did not appear insurmountable. However, arguments within the government over meeting key SPLA demands, such as the repeal of Islamic law, caused the war to drag on. A hard-line faction within Umma and the NIF sought to resist a return to secularism. In March 1989, a new government, made up of Umma and the DUP, committed itself to accommodating the SPLA. However, a month later, on the day the cabinet was to ratify an agreement with the rebels, there was a coup by pro-NIF officers.

ACHIEVEMENTS

Although his music is banned in his own country, Mohammed Wardi is probably Sudan's most popular musician. Now living in exile, he has been imprisoned and tortured for his songs against injustice, which also appeal to a large international audience, especially in North Africa and the Middle East.

Besides leading to a breakdown in all efforts to end the SPLA rebellion, the NIF/military regime has been responsible for establishing the most intolerant, repressive government in Sudan's modern history. Extra-judicial executions have become commonplace. Instances of pillaging and enslavement of non-Muslim communities by government-linked militias have increased. NIF-affiliated security groups have become a law unto themselves, striking out at their perceived enemies and intimidating Muslims and non-Muslims alike to conform to their fundamentalist norms. Islamic norms are also being invoked to justify a radical campaign to undermine the status of women.

In 1990, most of the now-banned political parties (including Umma, the DUP, and the Communists) aligned themselves with the SPLA as the National Democratic Alliance. But opposition by the northern-based parties proved ineffectual, leading to the formation of a new, Eritrean-based armed movement—the Sudan Alliance of Forces, headed by Abdul Azizi Khalid.

Beginning in 1991, the SPLA was weakened by a series of splits. Two factions—Kerubino Kuanyin Bol's SPLA–Bahr al-Ghazal group and Riek Macher's Southern Sudan Independence Army (SSIA)—accepted a government peace plan in April 1996. But the plan was rejected by John Garang's SPLA (Torit faction), which remains the most powerful southern group. After a number of years of being on the defensive, Garang's forces began making significant advances in 1996, partially as a result of increased support from neighboring countries that have come to look upon the Khartoum regime as a regional threat. (In June 1995, the regime was implicated in an attempt to assassinate Egyptian president Hosni Mubarak in Ethiopia, which resulted in the imposition of UN antiterrorism sanctions. Border clashes have since occurred with Eritrea, Kenya, and Uganda as well as Egypt.)

ECONOMIC PROSPECTS

Although it has great potential, political conflict has left Sudan one of the poorest nations in the world. Persistent warfare and lack of financing are blocking needed infrastructural improvements. Sudan's unwillingness to pay its foreign debt has led to calls for its expulsion from the International Monetary Fund.

Nearly 7 million Sudanese (out of a total population then of 23 million) had been displaced by 1988—more than 4 million by warfare, with drought and desertification contributing to the remainder. Sudan has been a major recipient of international emergency food aid for years, but warfare, corruption, and genocidal indifference have often blocked help from reaching the needy. In 1994, the United Nations estimated that 700,000 southern Sudanese faced the prospect of starvation.

Timeline: PAST

1820
Egypt invades northern Sudan

1881
The Mahdist Revolt begins

1956
Independence

1969
Jaafar Nimeiri comes to power

1972
Hostilities end in southern Sudan

1980s
Islamic law replaces the former penal code; renewed civil war in the south; Nimeiri is overthrown in a popular coup; an elected government is installed

1990s
The hard-line Islamic fundamentalist regime becomes increasingly repressive

PRESENT

2000s
Famine threatens large segments of the population

Omar al-Bashir claims victory in boycotted elections

Tanzania (United Republic of Tanzania)

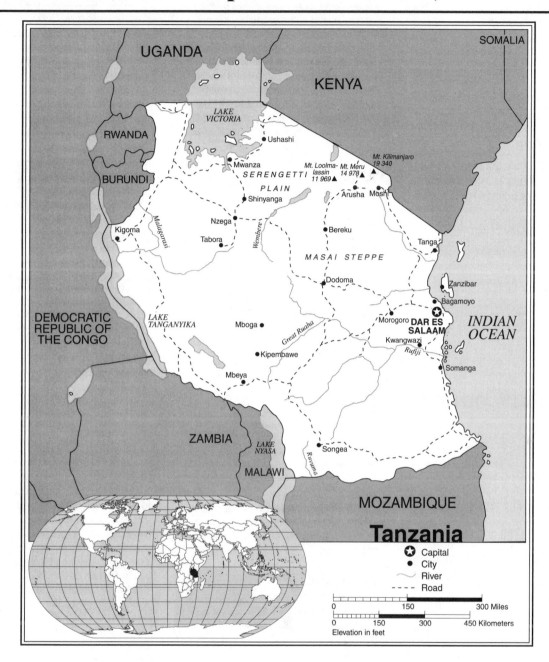

Tanzania Statistics

GEOGRAPHY

Area in Square Miles (Kilometers): 363,950 (939,652) (about twice the size of California)

Capital (Population): Dar es Salaam (2,347,000); Dodoma (to be new capital) (180,000)

Environmental Concerns: soil degradation; deforestation; desertification; destruction of coral reefs and marine environment

Geographical Features: plains along the coast; central plateau; highlands in the north and south

Climate: tropical to temperate

PEOPLE

Population

Total: 37,188,000

Annual Growth Rate: 2.6%

Rural/Urban Population Ratio: 68/32

Major Languages: Kiswahili; Chagga; Gogo; Ha; Haya; Luo; Maasai; English; others

Ethnic Makeup: 99% African; 1% others

Religions: indigenous beliefs; Muslim; Christian; Hindu

Health

Life Expectancy at Birth: 51 years (male);
53 years (female)
Infant Mortality: 77.8/1,000 live births
Physicians Available: 1/20,511 people
HIV/AIDS Rate in Adults: 8.09%

Education

Adult Literacy Rate: 68%
Compulsory (Ages): 7–14; free

COMMUNICATION

Telephones: 127,000 main lines
Televisions: 2.8/1,000 people
Internet Users: 115,000 (2001)

TRANSPORTATION

Highways in Miles (Kilometers): 52,800
(85,161)
Railroads in Miles (Kilometers): 2,141
(3,453)
Usable Airfields: 125
Motor Vehicles in Use: 134,000

GOVERNMENT

Type: republic
Independence Date: December 9, 1961
(from United Nations trusteeship)
Head of State/Government: President
Benjamin William Mkapa is both head of
state and head of government
Political Parties: Revolutionary Party;
National Convention for Construction
and Reform; Civic United Front; Union
for Multiparty Democracy; Democratic
Party; United Democratic Party; others
Suffrage: universal at 18

MILITARY

Military Expenditures (% of GDP): 0.2%
Current Disputes: boundary disputes with
Malawi; civil strife

ECONOMY

Currency ($ U.S. Equivalent): 1,068
shillings = $1
Per Capita Income/GDP: $610/$22.1
billion
GDP Growth Rate: 5%

Inflation Rate: 5%
Labor Force by Occupation: 80%
agriculture; 20% services and industry
Population Below Poverty Line: 51%
Natural Resources: hydropower; tin;
phosphates; iron ore; coal; diamonds;
gemstones; gold; natural gas; nickel
Agriculture: coffee; sisal; tea; cotton;
pyrethrum; cashews; tobacco; cloves;
wheat; fruits; vegetables; livestock
Industry: agricultural processing; mining;
oil refining; shoes; cement; textiles;
wood products; fertilizer; salt
Exports: $827 million (primary partners
United Kingdom, India, Germany)
Imports: $1.55 billion (primary partners
South Africa, Japan, United Kingdom)

SUGGESTED WEB SITES

```
http://www.tanzania.go.tz
http://www.tanzanianews.com
http://www.sas.upenn.edu/
   African_Studies/
   Country_Specific/Tanzania.html
http://www.tanzania_online.gov.uk
```

Tanzania Country Report

After winning a second term of office in 2000, taking nearly 72 percent of the vote, President Benjamin Mkapa and the ruling Chama Cha Mapinduzi (CCM) party were confronted with a series of challenges. The government banned opposition rallies, which were demanding new elections. Police staged a raid in January 2001 on the offices of the Civic United Front (CUF), the main opposition party in Zanzibar, and killed two in the process. Widespread protests in Zanzibar ensued, during which at least 31 people were killed. One hundred were arrested, and CUF chairman Ibrahim Lupimba was charged with unlawful assembly and disturbing the peace.

The government sent in troop reinforcements, but the solution to the unrest ultimately was to prove to be a political one. CCM and the CUF agreed in March 2001 to the formation of a joint committee to restore calm to Zanzibar, and also to encourage the return of around 2,000 refugees who had fled to Kenya.

Still, tensions remained high throughout Tanzania, and the opposition parties picked up a great deal of support. In April, opposition parties staged the first major political demonstrations against the government in more than 20 years. Some 50,000

supporters of the opposition marched in Dar es Salaam.

In November 2001 the presidents of Tanzania, Uganda, and Kenya launched a regional parliament and court of justice in Arusha. They will legislate on matters of common interest such as trade and immigration rules.

In October 1999, more than 3 million Tanzanians joined leaders from around the world in filing through a temporary mausoleum housing the body of the country's late first president, Julius Nyerere. It was an overwhelming tribute to the man who was known at home and abroad as *Mwalimu*—Swahili for "teacher." Nyerere voluntarily relinquished power in 1985, but the legacy of his nation-building efforts can be found throughout the country. After independence, until recently Tanzania enjoyed internal unity and an expansion of social services. The Kiswahili language helped bind the nation together. But economic

growth has remained modest; Tanzania continues to be one of the poorest nations in the world.

After a period of harsh German rule followed by paternalistic British trusteeship, the Tanzanian mainland gained its independence, as Tanganyika, in 1961. In 1964, it merged with the small island state of Zanzibar, which had been a British protectorate, to form the "United Republic of Tanzania." Political activity in Tanzania was restricted to the Chama Cha Mapinduzi party, which joined the former Tanganyika African National Union with its Zanzibar partner, the Afro-Shirazi Party.

In February 1992, the CCM agreed to compete with other "national parties"—provided they did not "divide the people

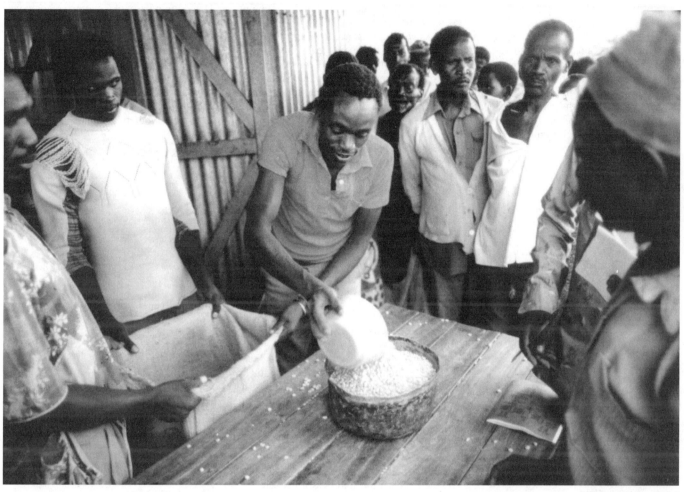

(UN photo by Ray Witlin)

The Tanzanian economy is primarily agriculture-based. However, rainfall for most of the country is sporadic. This, coupled with wide swings in world-market demand for its cash crops, has led to economic pressures. These men in the village of Lumeji are receiving seed grains needed to develop Tanzania's food production.

along tribal, religious or racial lines." Multiparty elections were held in October–December 1995, with the CCM claiming victory over a divided opposition in a poll characterized by massive irregularities. The disputed official results were CCM, 60 percent of the vote and 187 members of Parliament (MPs), including 128 seats where the results were still being legally contested months later; the National Committee for Constitutional Reform (NCCR), 25 percent of the vote and 15 MPs; and the Civic United Front, 24 MPs but only a small percentage of the vote, concentrated in Zanzibar and Pemba Islands. The new CCM leader, Ben Mkapa, replaced as president Ali Hassan Mwinyi, who was forced to step down after having served two terms. The dominant personality in the CCM, however, remained former president Nyerere.

By 1967, the CCM's predecessors had already eliminated legal opposition, when they proclaimed their commitment to the Arusha Declaration, a blueprint for "African Socialism." At the time, Tanzania was one of the least-developed countries in the world. It has remained so. Beyond this fact, there is much controversy over the degree to which the goals of the declaration have been achieved. To some critics, the Arusha experiment has been responsible for reducing a potentially well-off country to ruin. Supporters often counter that it has led to a stable society in which major strides have been made toward greater democracy, equality, and human development. Both sides exaggerate.

HEALTH/WELFARE

 The Tanzanian Development Plan calls for the government to give priority to health and education in its expenditures. This reflects a recognition that early progress in these areas has been undermined to some extent in recent years. Malnutrition remains a critical problem.

Like many African states, Tanzania has a primarily agrarian economy that is con-strained by a less than optimal environment. Although some 80 percent of the population are employed in agriculture, only 8 percent of the land is under cultivation. Rainfall for most of the country is low and erratic, and soil erosion and deforestation are critical problems in many areas. But geography and environmental problems are only one facet of Tanzania's low agricultural productivity. There has also been instability in world-market demand for the nation's principal cash crops: coffee, cotton, cloves, sisal, and tobacco. The cost of imported fuel, fertilizers, and other inputs has risen simultaneously.

ACHIEVEMENTS

 The government has had enormous success in its program of promoting the use of Kiswahili (Swahili) as the national language throughout society. Mass literacy in Kiswahili has facilitated the rise of a national culture, helping to make Tanzania one of the more cohesive African nations.

Government policies have also been responsible for underdevelopment. Perhaps the greatest policy disaster was the program of villagization. Tanzania hoped to relocate its rural and unemployed urban populations into *ujaama* (Swahili for "familyhood") villages, which were to become the basis for agrarian progress. In the early 1970s, coercive measures were adopted to force the pace of resettlement. Agricultural production is estimated to have fallen as much as 50 percent during the initial period of ujaama dislocation, transforming the nation from a grain exporter to a grain importer.

Another policy constraint was the exceedingly low official produce prices paid by the government to farmers. Many peasants withdrew from the official market, while others turned to black-market sales. Since 1985, the official market has been liberalized, and prices have risen. This has been accompanied by a modest rise in production, yet the lack of consumer goods in rural areas is widely seen as a disincentive to greater development.

All sectors of the Tanzanian economy have suffered from deteriorating infrastructure. Here again there are both external and internal causes. Balanced against rising imported-energy and equipment costs have been inefficiencies caused by poor planning, barriers to capital investment, and a relative neglect of communications and transport. Even when crops or goods are available, they often cannot reach their destination. Tanzania's few bituminized roads have long been in a chronic state of disrepair, and there have been frequent shutdowns of its railways. In particular, much of the southern third of the country is isolated from access to even inferior transport services.

Manufacturing declined from 10 to 4 percent of gross domestic product in the 1980s, with most sectors operating at less than half of their capacity. Inefficiencies also grew in the nation's mining sector. Diamonds, gold, iron, coal, and other minerals are exploited, but production has been generally falling and now accounts for less than 1 percent of GDP. Lack of capital investment has led to a deterioration of existing operations and an inability to open up new deposits.

As with agriculture, the Tanzanian government has in recent years increasingly abandoned socialism in favor of market economics, in its efforts to rehabilitate and expand the industrial and service sectors of the economy. A number of state enterprises are being privatized, and better opportunities are being offered to outside investors. Tourism is now being actively promoted, after decades of neglect.

Tanzania has made real progress in extending health, education, and other social services to its population since independence, though the statistical evidence is inadequate and official claims exaggerated. Some 1,700 health centers and dispensaries have been built since 1961, but they have long been plagued by shortages of medicines, equipment, and even basic supplies such as bandages and syringes. Although the country has a national health service, patients often end up paying for material costs.

. Much of the progress that has been made in human services is a function of outside donations. Despite the Arusha Declaration's emphasis on self-reliance, Tanzania has for decades been either at or near the top of the list of African countries in per capita receipt of international aid.

Even before the recent opening to multipartyism, Tanzania's politics was in a state of transition. Political life was dominated from the 1950s by Julius Nyerere, who was the driving personality behind the Arusha experiment. However, in 1985, he gave up the presidency in favor of Ali Hassan Mwinyi, and, in 1990, Nyerere resigned as chairman of the CCM, without having to give up his leading influence in the party.

The move to multiparty politics is complicated by the omnipresent CCM. The party has sought to control all organized social activity outside of religion. A network of community and workplace cells has assured that all Tanzanians have at least one party official responsible for monitoring their affairs.

In 1993, a dozen new opposition parties were registered, though others, notably the Democratic Party of Reverend Christopher Mtikila, remain banned. Opposition disunity contributed to subsequent CCM election victories. In addition, the CCM still enjoys a near media monopoly, occasionally invoking the National Security Act to harass independent journalists. Overt political repression has been most notable on Zanzibar and Pemba Islands.

Timeline: PAST

1820
The sultan of Oman transfers the capital to Zanzibar as Arab commercial activity increases

1885
Germany declares a protectorate over the area

1905–1906
The Maji Maji rebellion unites many ethnic groups against German rule

1919
Tanganyika becomes a League of Nations mandate under the United Kingdom

1961
Tanganyika becomes independent; Julius Nyerere is the leader

1964
Tanzania is formed of Tanganyika and Zanzibar

1967
The Arusha Declaration establishes a Tanzanian socialist program for development

1985
Nyerere retires; Ali Hassan Mwinyi succeeds as president

1990s
The CCM wins disputed multiparty elections

PRESENT

2000s
Tanzania continues to look for ways to diversify its economy

Tensions erupt between the government and the opposition

President Ben Mkapa is criticized for ordering a $21 million presidential plane

Uganda (Republic of Uganda)

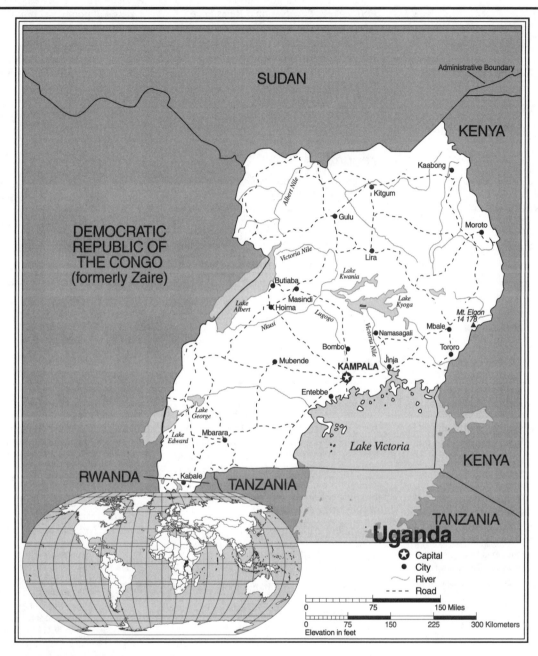

Uganda Statistics

GEOGRAPHY

Area in Square Miles (Kilometers):
91,076 (235,885) (about the size of Oregon)

Capital (Population): Kampala (1,274,000)

Environmental Concerns: draining of wetlands; deforestation; overgrazing; soil erosion; widespread poaching

Geographical Features: mostly plateau, with a rim of mountains

Climate: generally tropical, but semiarid in the northeast

PEOPLE

Population

Total: 24,700,000
Annual Growth Rate: 2.94%
Rural/Urban Population Ratio: 87/13
Major Languages: English; Swahili; Bantu languages; Nilotic languages

Ethnic Makeup: Bantu; Nilotic; Nilo-Hamitic; Sudanic

Religions: 66% Christian; 18% indigenous beliefs; 16% Muslim

Health

Life Expectancy at Birth: 39 years (male); 40 years (female)
Infant Mortality: 89.3/1,000 live births
Physicians Available: 1/20,700 people
HIV/AIDS Rate in Adults: 6.1%

Education

Adult Literacy Rate: 62%

COMMUNICATION

Telephones: 81,000 main lines
Televisions: 27/1,000 people
Internet Users: 25,000

TRANSPORTATION

Highways in Miles (Kilometers): 16,200
   (27,000)
Railroads in Miles (Kilometers): 745
   (1,241)
Usable Airfields: 27
Motor Vehicles in Use: 51,000

GOVERNMENT

Type: republic
Independence Date: October 9, 1962
   (from the United Kingdom)

Head of State/Government: President
   Yoweri Kaguta Museveni is both head of
   state and head of government
Political Parties: National Resistance
   Movement (only organization allowed to
   operate unfettered)
Suffrage: universal at 18

MILITARY

Military Expenditures (% of GDP): 2.1%
Current Disputes: continuing ethnic strife
   in the region

ECONOMY

Currency ($ U.S. Equivalent): 1,738
   Uganda shillings = $1
Per Capita Income/GDP: $1,200/$29
   billion
GDP Growth Rate: 5.1%
Inflation Rate: 3.5%
Labor Force by Occupation: 82%
   agriculture; 13% services; 5% industry
Population Below Poverty Line: 35%

Natural Resources: copper; cobalt; salt;
   limestone; hydropower; arable land
Agriculture: coffee; tea; cotton; tobacco;
   cassava; potatoes; corn; millet; pulses;
   livestock
Industry: sugar; brewing; tobacco; textiles;
   cement
Exports: $367 million (primary partner
   Europe)
Imports: $1.26 billion (primary partners
   Kenya, United States, India)

SUGGESTED WEB SITES

http://www.ugandaweb.com
http://www.government.go.ug
http://www.monitor.co.ug
http://www.mbendi.co.za/
   cyugcy.htm
http://www.sas.upenn.edu/
   African_Studies/
   Country_Specific/Uganda.html
http://www.state.gov/www/
   background_notes/
   uganda_0298_bgn.html
http://www.uganda.co.ug/

Uganda Country Report

In 2000, Ugandans voted to reject multi-party politics in favor of continuing President Yoweri Museveni's "no-party" system. This action was followed in 2001 by another victory for Museveni in the presidential elections, in which he defeated his rival, Kizza Besigye, by 69 percent to 28 percent. Museveni has remained in power since 1985 through his tight control over the military and governing institutions. But his position has been challenged by his involvement in regional conflicts as well as by domestic rebels, most notably the Lords Resistance Army (LRA), which continues to terrorize northern Uganda. This led to a series of signifiant diplomatic initiatives in 2002.

DEVELOPMENT

In the past few years, Uganda's economy has been growing at an average annual rate of about 5%, boosted by increased investment. Foreign economic assistance nonetheless accounts for approximately 29% of government spending.

In March 2002, Sudan and Uganda signed an agreement aimed at containing the LRA, which was active along their common border. The deal allowed Uganda to carry out limited operations in Sudan. Although the LRA thereafter came under

sustained attack in what the Ugandan military dubbed "Operation Iron Fist," the movement not only survived but also was able to regain control over areas in northern Uganda. In July, the Ugandan government broke off the offensive and agreed to begin negotiations with the LRA to end the war.

Finding common ground with the LRA may not be easy. A self-proclaimed "prophet" named Joseph Kony, who says he wants to run Uganda in conformity with the biblical Ten Commandments, leads the movement. But, in practice, the LRA has been responsible for the abduction of thousands of boys and girls. The boys are indoctrinated to kill and rape, while the girls are consigned to serve as laborers and targets of sexual gratification.

Over the past five years, Uganda's army has also been heavily involved in fighting in the Democratic Republic of the Congo (D.R.C., the former Zaire). In March 2001, Uganda classified Rwanda—its former ally in the D.R.C.'s civil war—as a hostile nation after several months of clashes between the two countries' armed forces over key areas of eastern D.R.C., which they had collectively occupied since 1998. Both countries, along with Zimbabwe, have been accused of pillaging the D.R.C.'s natural wealth. Tensions were eased between the two states in February 2002, when

Museveni met the Rwandan president, Paul Kagame, as part of an on-going, British-backed effort to defuse tensions.

In August 2002, Museveni signed a peace accord with D.R.C. president Joseph Kabila, which had been brokered by the Angolan president Jose Eduardo dos Santos. This resulted in the rapid withdrawal of Ugandan forces from Eastern D.R.C.

FREEDOM

The human-rights situation in Uganda remains poor, with government security forces linked to torture, extra-judicial executions, and other atrocities. Freedom of speech and association are curtailed. Insurgent groups are also associated with atrocities; the Lord's Resistance Army continues to kill, torture, maim, and abduct large numbers of civilians, enslaving numerous children.

Uganda's foreign and domestic conflicts pose a potential threat to the very real progress that the country has made since the coming to power of Museveni's National Resistance Movement (NRM). After years of repressive rule accompanied by massive interethnic violence, Uganda is still struggling for peace and reconciliation. A land rich in natural and human resources, Uganda suffered dreadfully during the despotic regimes of Milton Obote (1962–1971, 1980–1985) and Idi Amin (1971–1979). Under these two dictators, hundreds of thousands of Ugandans were murdered by the state.

HEALTH/WELFARE

Millions of Ugandans live below the poverty line. Uganda's traditionally strong school system was damaged but not completely destroyed under Amin and Obote. In 1986, some 70% of primary-school children attended classes. The killing and exiling of teachers have resulted in a serious drop of standards at all levels of the education system, but progress is under way. The adult literacy rate has risen to 62%.

The country had reached a state of general social and political collapse by 1986, when the NRM seized power. The new government soon made considerable progress in restoring a sense of normalcy in most of the country, except for the north. In May 1996, Museveni officially received 74 percent of the vote in a contested presidential poll. Despite charges of fraud by his closest rival, Paul Ssemogerere, most independent observers accepted the poll as an endorsement of Museveni's leadership, including his view that politics should remain organized on a nonparty basis. There has since, however, emerged growing international criticism of his intolerance of genuine political pluralism.

HISTORIC GEOGRAPHY

The breakdown of Uganda is an extreme example of the disruptive role of ethnic and sectarian competition, which was fostered by policies of both its colonial and postcolonial governments. Uganda consists of two major zones: the plains of the northeast and the southern highlands. It has been said that you can drop anything into the rich volcanic soils of the well-watered south and it will grow. Until the 1960s, the area was divided into four kingdoms—Buganda, Bunyoro, Ankole, and Toro—populated by peoples using related Bantu languages. The histories of these four states stretch back hundreds of years. European visitors of the nineteenth century were impressed by their sophisticated social orders, which the Europeans equated with the feudal monarchies of medieval Europe.

When the British took over, they integrated the ruling class of the southern highlands into a system of "indirect rule." By then, missionaries had already succeeded in converting many southerners to Christianity; indeed, civil war among Protestants, Catholics, and Muslims within Buganda had been the British pretext for establishing their overrule.

The Acholi, Langi, Karamojang, Teso, Madi, and Kakwa peoples, who are predominant in the northeast, lack the political heritage of hierarchical state-building found in the south. These groups are also linguistically separate, speaking either Nilotic or Nilo-Hamitic languages. The British united the two regions as the Uganda Protectorate during the 1890s (the word *Uganda,* which is a corruption of "Buganda," has since become the accepted name for the larger entity). But the zones developed separately under colonial rule.

Cash-crop farming, especially of cotton, by local peasants spurred an economic boom in the south. The Bugandan ruling class benefited in particular. Increasing levels of education and wealth led to the European stereotype of the "progressive" Bugandans as the "Japanese of Africa." A growing class of Asian entrepreneurs also played an important role in the local economy, although its prosperity, as well as that of the Bugandan elite, suffered from subordination to resident-British interests.

ACHIEVEMENTS

The Ugandan government was one of the first countries in Africa (and the world) to acknowledge the seriousness of the HIV/AIDS epidemic within its borders. It has instituted public-information campaigns and welcomed outside support.

The south's growing economy stood in sharp contrast to the relative neglect of the northeast. Forced to earn money to pay taxes, many northeasterners became migrant workers in the south. They were also recruited, almost exclusively, to serve in the colonial security forces.

As independence approached, many Bugandans feared that other groups would compromise their interests. Under the leadership of their king, Mutesa II, they sought to uphold their separate status. Other groups feared that Bugandan wealth and educational levels could lead to their dominance. A compromise federal structure was agreed to for the new state. At independence, the southern kingdoms retained their autonomous status within the "United Kingdom of Uganda." The first government was made up of Mutesa's royalist party and the United People's Congress (UPC), a largely non-Bugandan coalition, led by Milton Obote, a Langi. Mutesa was elected president and Obote prime minister.

THE REIGN OF TERROR

In 1966, the delicate balance of ethnic interests was upset when Obote used the army—still dominated by fellow northeasterners—to overthrow Mutesa and the Constitution. In the name of abolishing "tribalism," Obote established a one-party state and ruled in an increasingly dictatorial fashion. However, in 1971, he was overthrown by his army chief, Idi Amin. Amin began his regime with widespread public support but alienated himself by favoring fellow Muslims and Kakwa. He expelled the 40,000-member Asian community and distributed their property to his cronies. The Langi, suspected of being pro-Obote, were also early targets of his persecution, but his attacks soon spread to other members of Uganda's Christian community, at the time about 80 percent of the total population. Educated people in particular were purged. The number of Ugandans murdered by Amin's death squads is unknown; the most commonly cited figure is 300,000, but estimates range from 50,000 to 1 million. Many others went into exile. Throughout the world, Amin's name became synonymous with despotic rule.

A Ugandan military incursion into Tanzania led to war between the two countries in 1979. Many Ugandans joined with the Tanzanians in defeating Amin's army and its Libyan allies.

Unfortunately, the overthrow of Amin, who fled into exile, did not lead to better times. In 1980, Obote was returned to power, through a fraudulent vote count. His second administration was characterized by a continuation of the violence of the Amin years. Obote's security forces massacred an estimated 300,000 people, mostly southerners, while an equal number fled the country. Much of the killing occurred in the Bugandan area known as the Luwero triangle, which was completely depopulated; its fields are still full of skeletons today. As the killings escalated, so did the resistance of Museveni's NRM guerrillas, who had taken to the bush in the aftermath of the failed election. In 1985, a split between Ancholi and Langi officers led to Obote's overthrow and yet another pattern of interethnic recrimination. Finally, in 1986, the NRM gained the upper hand.

Thereafter a new political order began to emerge based on Museveni's vision of a "no-party government." His position was strengthened in March 1994, when elections to a Constituent Assembly resulted in his supporters' capturing more than two thirds of the seats. In another controversial initiative, Museveni allowed the restoration of traditional offices, including Bugandan kingship.

THE STRUGGLE CONTINUES

Museveni's National Resistance Movement administration has faced enormous challenges in trying to bring about national reconstruction. The task has been complicated by continued warfare in the northeast by armed factions representing elements of the former regimes, independent Karamojong communities, and followers of prophetic religious movements. In 1987, an uprising of the Holy Spirit rebels of Alice Lakwena was crushed, at the cost of some 15,000 lives.

The restoration of peace to most of the country has promoted economic growth. Western-backed economic reforms produced an annual growth rate of 13 percent between 1990 and 1998. The rate of inflation also improved, falling from 200 to 7 percent in the same period. On the regional level, Museveni has championed the formation of a new "East African Community" (EAC), which is intended to lay the groundwork for economic and ultimately monetary integration with Tanzania and Kenya.

While rebuilding their shattered country, Ugandans have had to cope with an especially severe outbreak of HIV/AIDS. Thousands have died of the disease in the last decade; it is believed that literally hundreds of thousands of Ugandans are HIV-positive. The government's bold acknowledgment and proactive efforts to address the crisis, however, have been credited with helping to contain the pandemic.

North Africa

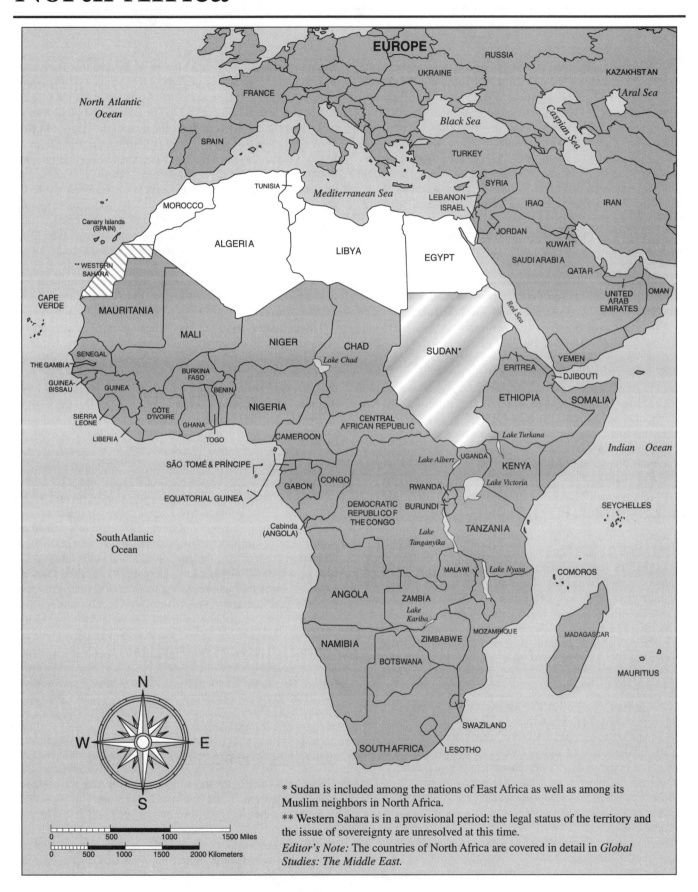

* Sudan is included among the nations of East Africa as well as among its Muslim neighbors in North Africa.

** Western Sahara is in a provisional period: the legal status of the territory and the issue of sovereignty are unresolved at this time.

Editor's Note: The countries of North Africa are covered in detail in *Global Studies: The Middle East.*

North Africa: The Crossroads of the Continent

Located at the geographical and cultural crossroads between Europe, Asia, and the rest of Africa, North Africa has served since ancient times as a link between the civilizations of sub-Saharan Africa and the rest of the world. Traders historically carried the continent's products northward, either across the Sahara Desert or up the Nile River and Red Sea, to the great port cities of the Mediterranean coast. Goods also flowed southward. In addition, the trade networks carried ideas: Islam, for example, spread from coastal North Africa across much of the rest of the continent to become the religion of at least one third of all Africans.

North Africa's role as the continent's principal window to the world gradually declined after the year A.D. 1500, as the trans-Atlantic trade increased. (The history of East Africa's participation in Indian Ocean trade goes back much further.) However, the countries of North Africa have continued to play an important role in the greater continent's development.

The countries of North Africa—Morocco, Algeria, Tunisia, Libya, and Egypt—and their millions of people differ from one another, but they share a predominant, overarching Arab-Islamic culture that both distinguishes them from the rest of Africa and unites them with the Arabic-speaking nations of the Middle East. To begin to understand the societies of North Africa and their role in the rest of the continent, it is helpful to examine the area's geography. The region's diverse environment has long encouraged its inhabitants to engage in a broad variety of economic activities: pastoralism, agriculture, trading, crafts, and, later, industry.

GEOGRAPHY AND POPULATION

Except for Tunisia, which is relatively small, the countries of North Africa are sprawling nations. Algeria, Libya, and Egypt are among the biggest countries on the African continent, and Morocco is not far behind. Their size can be misleading, for much of their territories is encompassed by the largely barren Sahara Desert. The approximate populations of the five states today range from Egypt's 71 million people to Libya's 6 million; Morocco has 31 million, Algeria 32 million, and Tunisia almost 10 million citizens. All these populations are increasing at a rapid rate; indeed, well over half of the region's citizens are under age 21.

Due to its scarcity, water is the region's most precious resource, so most people live either in valleys near the Mediterranean coast or along the Nile. The latter courses through the desert for thousands of miles, creating a narrow green ribbon that is the home of the 95 percent of Egypt's population who live within 12 miles of its banks. More than 90 percent of the people of Algeria, Libya, Morocco, and Tunisia live within 200 miles of either the Mediterranean or, in the case of Morocco, the Atlantic coast.

Besides determining where people live, the temperate, if often too dry, climate of North Africa has always influenced local economies and lifestyles. There is intensive agriculture along the coasts and rivers. Algeria, Morocco, and Tunisia are well known for their tree and vine crops, notably citrus fruits, olives, and wine grapes. The intensively irrigated Nile Valley has been a leading source of high-quality cotton as well as locally consumed foodstuffs since the time of the American Civil War, which temporarily removed U.S.–produced fiber from the world market. In the oases that dot the Sahara Desert, date palms are grown for their sweet fruits, which are almost a regional staple. Throughout the steppelands between the fertile coasts and the desert, pastoralists follow flocks of sheep and goats or herds of cattle and camels in constant search of pasture. Although now few in number, it was these nomads who in the past developed the trans-Saharan trade. As paved roads and airports have replaced their caravan routes, long-distance nomadism has declined. But the traditions it bred, including a love of independence, remain an important part of North Africa's cultural heritage.

Urban culture has flourished in North Africa since the ancient times of the Egyptian pharaohs and the mercantilist rulers of Carthage. Supported by trade and local industries, the region's medieval cities, such as Cairo, Fez, and Kairouan, were the administrative centers of great Islamic empires, whose civilizations shined during Europe's dark ages. In the modern era, the urban areas are bustling industrial centers, ports, and political capitals.

Geography—or, more precisely, geology—has helped to fuel economic growth in recent decades. Although agriculture continues to provide employment in Algeria and Libya for as much as a third of the labor force, discoveries of oil and natural gas in the 1950s dramatically altered these two nations' economic structures. Between 1960 and 1980, Libya's annual per capita income jumped from $50 to almost $10,000, transforming it from among the poorest to among the richest countries in the world. Algeria has also greatly benefited from the exploitation of hydrocarbons, although less dramatically than Libya. Egypt and Tunisia have developed much smaller oil industries, which nonetheless provide for their domestic energy needs and generate much-needed foreign exchange. The decline in oil prices during the 1980s, however, reduced revenues, increased unemployment, and contributed to social and political unrest, especially in Algeria. While it has no oil, Morocco profits from its possession of much of the world's phosphate production, which is concentrated in the Moroccan-occupied territory of Western Sahara.

CULTURAL AND POLITICAL HERITAGES

The vast majority of the inhabitants of North Africa are Arabic-speaking Muslims. Islam and Arabic both became established in the region between the seventh and eleventh centuries A.D. Thus, by the time of the Crusades in the eastern Mediterranean, the societies of North Africa were thoroughly incorporated into the Muslim world, even though the area had earlier been the home of many Christian scholars. Except for Egypt, where

(United Nations photo)

Geography has been less of a barrier to regional cohesiveness in North Africa than have politics and ideology.

about 5 percent of the population remain loyal to the Coptic Church, there is virtually no Christianity among modern North Africans. Until recently, important Jewish communities existed in all the region's countries, but their numbers have dwindled as a result of mass emigration to Israel.

With Islam came Arabic, the language of the Koran—the holy book of Islam. Today, Egypt and Libya are almost exclusively Arabic-speaking. In Algeria, Morocco, and Tunisia, Arabic coexists with various local minority languages, which are collectively known as Berber (from which the term *Barbary*, as in Barbary Coast, was derived). As many as a third of the Moroccans speak a form of Berber as their first language. Centuries of interaction between the Arabs and Berbers as well as their common adherence to Islam have promoted a sense of cultural unity between the two communities, although ethnic disputes have developed in Algeria and Morocco over demands that Berber be included in local school curriculums. As was the case almost everywhere else on the continent, the linguistic situation in North Africa was further complicated by the introduction of European languages during the colonial era. Today, French is particularly important as a language of technology and administration in Algeria, Morocco, and Tunisia.

By the seventeenth century, all the countries of North Africa, except Morocco, were autonomous provinces of the Ottoman Empire, which was based in present-day Turkey and also incorporated most of the Middle East. Morocco was an independent state; indeed, it was one of the earliest to recognize the independence of the United States from Britain. From 1830, the European powers gradually encroached upon the Ottoman Empire's North African realm. Thus, like most of their sub-Saharan counterparts, all the states of North Africa fell under European imperial control. Algeria's conquest by the French began in 1830 but took decades to accomplish, due to fierce local resistance. France also seized Tunisia in 1881 and, along with Spain, partitioned Morocco in 1912. Britain occupied Egypt in 1882, and Italy invaded Libya in 1911, although anti-Italian resistance continued until World War II, when the area was liberated by Allied troops.

The differing natures of their European occupations have influenced the political and social characters of each North African state. Algeria, which was directly incorporated into France as a province for 120 years, did not win its independence until 1962, after a protracted war of liberation. Morocco, by contrast, was accorded independence in 1956, after only 44 years of Franco–Spanish administration, during which the local monarchy continued to reign. Tunisia's 75 years of French rule also ended in 1956, as a strong nationalist party took the reins of power. Egypt, although formally independent of Great Britain, did not win genuine self-determination until 1952, when a group of nationalist army officers came to power by overthrowing the British-supported monarchy. Libya became a temporary ward of the United Nations after Italy was deprived of its colonial empire during World War II. The nation was granted independence by the United Nations in 1951, under a monarch whose religious followers had led much of the anti-Italian resistance.

113

NATIONAL POLITICS

Egypt

Egypt reemerged as an important actor on the world stage soon after Gamal Abdel Nasser came to power, in the aftermath of the overthrow of the monarchy. One of the major figures in the post-World War II Non-aligned Movement, Nasser gave voice to the aspirations of millions in the Arab world and Africa, through his championing of pan-Arab and pan-African anti-imperialist sentiments. Faced with the problems of his nation's burgeoning population and limited natural resources, Nasser nonetheless refused to let his government become dependent on a single foreign power. Domestically, he adopted a policy of developmental socialism.

Because of mounting debts, spurred by enormous military spending, and increasing economic problems, many Egyptians had already begun to reassess some aspects of Nasser's policies by the time of his death in 1970. His successor, Anwar al-Sadat, reopened Egypt to foreign investment in hopes of attracting much-needed capital and technology. In 1979, Sadat drew Egypt closer to the United States by signing the Camp David Accords, which ended more than three decades of war with Israel. Egypt has since been one of the largest recipients of U.S. economic and military aid.

Sadat's increasingly authoritarian rule, as well as his abandonment of socialism and foreign policy initiatives, made him a target of domestic discontent, and in 1981, he was assassinated. His successor, Hosni Mubarak, has modestly liberalized Egyptian politics and pursued what are essentially moderate internal and external policies. While maintaining peace with Israel, Mubarak has succeeded in reconciling Egypt with other Arab countries, which had strongly objected to the Camp David agreement. In 1990–1991, he took a leading role among the majority of Arab leaders opposed to Iraq's seizure of Kuwait. However, rapid urbanization, declining per capita revenues, debt, and unemployment, all linked to explosive population growth, have continued to strain the Egyptian economy and fuel popular discontent. Some of this discontent has in recent years been channeled into violence by extremist Islamic groups, which now threaten to destroy the traditional tolerance that has existed between Egypt's Muslim majority and Christian minority. Domestic terrorism has also had a negative impact on Egypt's tourism industry, the country's largest foreign-exchange earner. Particularly detrimental was a 1997 massacre by Islamic extremists of 58 foreign tourists at the ancient ruins of Luxor. The September 11, 2001, terrorist attacks in the United States also played a role in the reduction of tourism for Egypt. The industry has not fully recovered.

Libya

Libya was ruled for years by a pious, autocratic king whose domestic legitimacy was always in question. After 1963, the nation came under the heavy influence of foreign oil companies, which discovered and produced the country's only substantial resource. In 1969, members of the military, led by Colonel Muammar al-Qadhafi, overthrew the monarchy. Believed to be about age 27 at the time of the coup, Qadhafi was an ardent admirer of Nasser's vision of pan-Arab nationalism and anti-imperialism. Qadhafi invested billions of dollars, earned from oil,

(United Nations photo by Y. Nagata)

The Egyptian president, Hosni Mubarak, continues the legacy of his predecessor, Anwar al-Sadat.

in ambitious domestic development projects, successfully ensuring universal health care, housing, and education for his people by the end of the 1970s. He also spent billions more on military equipment and aid to what he deemed "nationalist movements" throughout the world. Considered a maverick, he came into conflict with many African and Arab rulers as well as with outside powers like the United States. Despite Qadhafi's persistent efforts to forge regional alliances, political differences, economic pressures, and the expulsion of expatriate workers (due to declining oil revenues) have increased tensions between Libya and some of its neighbors as well as between the country's own military and middle class.

Strained relations between Libya and the United States over Qadhafi's activist foreign policy and support for international terrorists culminated in a U.S. air raid on Tripoli in 1986. In that year, the United States required American businesses and citizens to leave Libya and since then has sought other ways to undermine Qadhafi's ambitions. With the support of the United States and other powers, the Hisséne Habré government of Chad was able in 1987 to expel the Libyan military from its northern provinces. In 1991, Libya came under greater international pressure when the UN Security Council backed American and British demands for the extradition of two alleged Libyan agents suspected of complicity in the 1987 blowing up of a Pan Am passenger jet over Lockerbie, Scotland. The Qadhafi government's initial failure to submit to this decision led to the imposition of international sanctions barring other countries from maintaining air links with or selling arms to Libya.

Since 1998, Qadhafi has broken out of his regime's international isolation. UN sanctions on Libya were lifted after Qadhafi turned over the Lockerbie suspects for trial under Scottish law at The Hague, Netherlands, seat of the International Court of Justice. A special Scottish court in the Netherlands found one of the two Libyans accused of the Lockerbie bombing, Abdelbaset Ali Mohamed al-Megrahi, guilty and sentenced him to life imprisonment.

Qadhafi has also put himself forward as a promoter of African integration. In 1999, he hosted a special Organization of African Unity conference, where his call for a new "African Union" was adopted in principle. The African Union held its first meeting in 2002, with Qadhafi playing an instrumental role in its organization. His influence was strengthened when Libya paid all the arrears owed by impoverished members of the OAU.

On the domestic front, there are increasing signs of disillusionment with the status quo, particularly among the nation's youth, many of whom face unemployment. A sign of the volubility of the Libyan youth took place in September 2000, when dozens of African immigrants were killed by Libyan mobs who were said to be angry at the large number of African laborers coming into the country.

Tunisia

Although having the fewest natural resources of the North African countries, Tunisia enjoyed a high degree of political stability and economic development during the first three decades that followed the restoration of its independence, in 1956. Habib Bourguiba, leader of the local nationalist party known as the Neo-Destour, led the country to independence while retaining cordial economic and political ties with France as well as other Western countries. Bourguiba's government was a model of pragmatic approaches to both economic growth and foreign policy. A mixed economy was developed, and education's contribution to development was emphasized. The nation's Mediterranean coast was transformed into a vacation spot for European tourists.

But by the 1980s, amid economic recession and after 30 years of single-party rule, Tunisians became increasingly impatient with their aging leader's refusal to recognize opposition political parties. Strikes, demonstrations, and opposition from Muslim fundamentalists as well as underground secular movements were the context for Bourguiba's forced retirement in 1987 (he died in 2000, at age 96). He was succeeded by his prime minister, Zine al-Abidine ben Ali, whose efforts in 1988 to release jailed Muslim activists and to open political dialogue led to a period of optimism and widespread support. By the middle of that year, he had replaced most of the cabinet ministers who had served under Bourguiba. Multiparty elections were held in 1989, but they were marred by opposition charges of fraud. This pattern was repeated in 1994, when Ben Ali was reelected unopposed, while his party captured all but 19 of the 163 parliamentary seats. In October 1999, it was announced that Ben Ali had been reelected by 99.4 percent of the vote! To the BBC, outspoken Tunisian journalist Taoufik Ben Brik noted: "How can I answer the question were the elections free? when we lack freedom of expression, freedom of organisation, even freedom of movement." Ben Ali was due to retire in 2004, but in May 2002 he secured support in a referendum for changes to the Constitution that will allow him to stay on for a further two terms. Opposition groups came together to brand the move as authoritarian.

Economically, Tunisia continued to make substantial progress through an export-oriented market strategy based on manufactures, tourism, agriculture, and petroleum. In 1999, real per capita income grew by 5 percent, with the majority of Tunisians enjoying a middle-class lifestyle. The country benefits from high literacy, relatively low population growth, and wide distribution of health care.

Algeria

Algeria, wracked by the long and destructive revolution that preceded independence in 1962, was long ruled by a coalition of military and civilian leaders who rose to power as revolutionary partisans of the National Liberation Front (FNL) during the war. Although FNL leaders have differed over what policies and programs to emphasize, in the past they were able to forge a governing consensus in favor of secularism (but with respect for Islam's special status), a socialist domestic economy, and a foreign policy based on nonalignment. The country's substantial oil and gas revenues were invested in large-scale industrial projects, which were carried out by the state sector. But by the end of the 1970s, serious declines in agricultural productivity and growing urban unemployment, partially due to the country's high overall rate of population growth, sent hundreds of thousands of Algerian workers to France in search of jobs. As a result, cautious encouragement began to be given to private-sector development.

In 1988, rising bread prices led to severe rioting, which left more than 100 people dead. In the aftermath, the FNL's long period of one-party rule came to an end, with the legalization of opposition parties. In the 1990 local elections, the Islamic Salvation Front (FIS), a coalition group of Muslim fundamentalists, managed to take control of about 80 percent of the country's municipal and departmental councils. This triumph was followed by FIS success in the first round of voting for a new National Assembly in December 1991; the Front captured 187 out of 230 seats. But, just before a second round of voting could be held, in January 1992, the military seized power in a coup. A state of emergency was declared, and thousands of FIS supporters and other opponents of the new regime were detained. In response, some turned to armed resistance. In June 1992, the political temperature was raised further by the mysterious assassination of Mohamed Boudiaf, a veteran nationalist who had been installed by the military's High State Committee

(United Nations photo by Bill Graham)

Nomadic traditions, including loyalty to family and love of independence, are still integral to the cultures of North Africa.

as the president. In 1993, elements of the Islamic resistance began an increasingly effective campaign of isolating Algeria internationally by assassinating foreign expatriates residing in the country. In 1995, Liamine Zeroual was elected president in a poll boycotted by the FIS and other major opposition parties. He cited the relatively high voter turnout as a mandate for a peaceful settlement based on dialogue. But the nation has remained polarized by civil war, whose death toll is now estimated to be in excess of 80,000.

In the aftermath of the 1995 elections, armed self-styled Islamists turned to massacring large numbers of civilians of all ages, in apparent retaliation against villages and extended families who ceased providing support to them. There are also credible accusations that some mass killings have been carried out by government forces. In an effort to push forward a political settlement, in February 1999, President Zeroual stepped down. He was replaced by Abdelaziz Bouteflika, who in April received 73 percent of the vote in a poll in which leading opponents dropped out of the race. In September 1999, Bouteflika received a greater mandate when a majority of Algerians voted in favor of new peace policies aimed at bringing an end to the years of bloodshed.

There have also been tensions over the demand for cultural recognition by Algeria's large Berber-speaking minority. In October 2001, the government agreed to a series of demands, including official recognition of the Berber language, after months of unrest involving Berber youths.

Internationally, Algeria has been known for its trouble-shooting role in difficult diplomatic negotiations. In 1980, it mediated the release of the U.S. hostages held in Iran. After years of tension, largely over the war in Western Sahara, Algeria resumed diplomatic relations with its western neighbor, Morocco, in 1988.

Morocco

In August 1999, Morocco's King Hassan II died. He had ruled since 1961. He was succeeded by his son Muhammad VI, who has since encouraged a degree of political and economic liberalization in the kingdom. Under Hassan II, the political parties that developed during the struggle against French rule (led by the current king's grandfather, Muhammad V) had continued to contest elections. But Hassan rarely permitted them to have any genuine influence in policy making, preferring to reserve the role for himself and his advisers. As in Tunisia, Moroccan agricultural development has been based on technological innovations rather than on land reform. (The latter, while it could raise productivity, would also likely anger the propertied supporters of the king. Elites also oppose business-tax reforms, yet the government needs revenues to repay its multibillion-dollar debt.) Much of the country's economic development has been left to the private sector. High birth rates and unemployment have led many Moroccans to join the Algerians and Tunisians in seeking employment in Europe.

(United Nations photo by Saw Lwin)

Morocco's King Hassan II was a leader whose influence was often pivotal in North African regional planning.

By the mid-1990s, Morocco's three-decade-long war to retain control of the phosphate-rich Western Sahara—a former Spanish colony whose independence is being fought for by a local nationalist movement known as Polisario—had become an unsettled stalemate, with Morocco controlling most of the Western Sahara. During the late 1980s, Moroccan forces had become increasingly effective in frustrating the infiltration of Polisario guerrillas in the main centers of the territory by enclosing them behind a network of security walls. These walls have also effectively shut out some 120,000 refugees (the number is bitterly disputed by the contestants) living in Polisario-controlled camps in Algeria. A UN peace plan calling for a referendum over the territory's future was agreed to by both sides in 1990. But, though the two parties have generally maintained a cease-fire since 1991, other provisions of the plan have not been implemented, largely as a result of continued Moroccan intransigence. In 1995, Polisario formally renounced the agreement. But, with waning international support and the Moroccan forces now well dug in, Polisario's short-term prospects of making either a political or a military breakthrough appear unpromising.

Morocco maintains a mixed economy based largely on agriculture, fishing, light industry, phosphate mining, tourism, and remittances from citizens working abroad. The illegal cultivation of cannabis is also an important source of income for many Moroccans. Economic growth is highly dependent on agricultural output and has experienced wide fluctuations in recent years, due to drought.

REGIONAL AND CONTINENTAL LINKS

There have been many calls for greater regional integration in North Africa since the 1950s. Under Nasser, Egypt was the leader of the pan-Arab movement; it even joined Syria in a brief political union, from 1958 to 1961. Others have attempted to create a union of the countries of the *Maghrib* (Arabic for "west") region—that is, Algeria, Morocco, and Tunisia. Recently, these three countries, along with the adjacent states of Libya and Mauritania, agreed to work toward an economic community, but they continue to be politically divided. At one time or another, Qadhafi has been accused of subverting all the region's governments; Algeria and Morocco have disagreed over the disposition of the Western Sahara; and each country has closed its borders to its neighbors' citizens. Still, the logic of closer political and economic links and the example of increasing European unity on the other side of the Mediterranean will likely keep the issue of regional unity alive.

Both as members of the Organization of African Unity (OAU) and as individual states, the North African countries have had strong diplomatic and political ties to the rest of the continent. They are, however, also deeply involved in regional affairs outside Africa, particularly those of the Arab and Mediterranean worlds. There have also been some modest tensions across the Sahara. Requests by the North African nations that other OAU countries break diplomatic relations with Israel were promptly met in the aftermath of the 1974 Arab–Israeli War. Many sub-Saharan countries hoped that, in return for their solidarity, the Arab nations would extend development aid to help them, in particular to cope with rising oil prices. Although some aid was forthcoming (mostly from the Persian Gulf countries rather than the North African oil producers), it was less generous than many had expected. During the 1980s, a number of sub-Saharan countries resumed diplomatic relations with Israel.

Border disputes, ideological differences, and internal conflicts have caused additional tensions. The Polisario cause in Western Sahara, for example, badly divided the Organization of African Unity. When the OAU recognized the Polisario's exiled government in 1984, Morocco, along with Zaire, withdrew from membership in the body. However, the OAU has also had significant regional successes. In 1974, for example, its mediation led to a settlement of a long-standing border dispute between Algeria and Morocco.

In recent years, the states of North Africa have begun to seek greater economic ties with Europe as well as with one another. In 1998, Algeria, Egypt, Morocco, and Tunisia signed accords with the European Union, agreements that are intended to promote free trade across the Mediterranean over the coming decade.

Editor's Note: The countries of North Africa are covered in detail in *Global Studies: The Middle East.*

Southern Africa

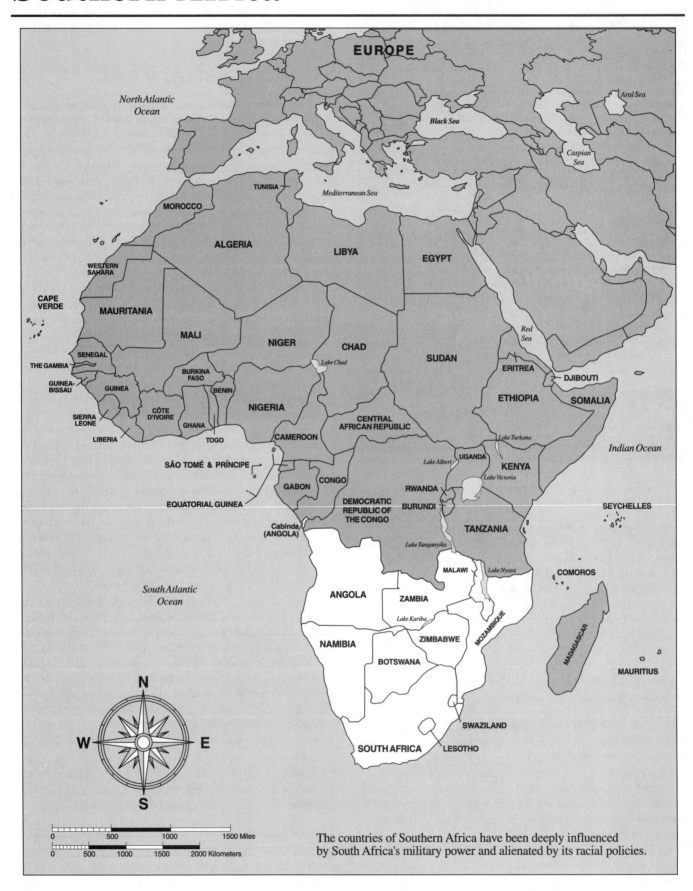

The countries of Southern Africa have been deeply influenced by South Africa's military power and alienated by its racial policies.

Southern Africa: The Continuing Struggle for Self-Determination

Southern Africa—which includes the nations of Angola, Botswana, Lesotho, Malawi, Mozambique, Namibia, South Africa, Swaziland, Zambia, and Zimbabwe—is a diverse region made up of savannas and forest, snow-topped mountains and desert, temperate Mediterranean and torrid tropical climates. Until recently much of the region was marred by armed conflict. But the wars and civil conflicts that have long plagued Angola, Mozambique, Lesotho, Namibia, and South Africa in particular have for now ended. In the case of Angola, this has been facilitated by developments in the Democratic Republic of the Congo (D.R.C., the former Zaire), which for decades had served as a source of, often Western-backed, regional destabilization. The more recent involvement of the governments of Namibia, Angola, and Zimbabwe in the D.R.C.'s internal conflicts appears to be coming to a close with a peace agreement initiated by South Africa. As a result, military complexes can be scaled down in favor of diverting more resources to the challenges of economic development while addressing the very serious challenge of the HIV/AIDS pandemic.

Southern African identity is as much defined by the region's peoples and their past and present interactions as by the region's geographic features. An appreciation of local history is crucial to understanding the forces that both divide and unite the region today. One of the primary themes in Southern Africa today is land, and the domination of the land by various groups of Europeans. Much of the history of Southern Africa has been shaped by the loss of land by indigenous communities to European settlers. The resulting imbalance in land distribution has not been resolved by the establishment of black-majority rule in the former settler-dominated states of Zimbabwe, Namibia, and South Africa. The resulting land shortage for the majority of rural dwellers remains a major constraint to socioeconomic development.

EUROPEAN MIGRATION AND DOMINANCE

A dominant theme in the modern history of Southern Africa has been the evolving struggle of the region's indigenous black African majority to free itself of the racial hegemony of white settlers from Europe and their descendants. By the eighth century A.D., but probably earlier, the southernmost regions of the continent were populated by a variety of black African ethnic groups who spoke languages belonging to the Bantu as well as the Khoisan linguistic classifications. Members of these two groupings practiced both agriculture and pastoralism; archaeological evidence indicates that livestock keeping in Southern Africa pre-dates the time of Christ. Some, such as the BaKongo of northern Angola and the Shona peoples of the Zimbabwean plateaux, had, by the fifteenth century, organized strong states; others, like most Nguni-speakers prior to the early 1800s, lived in smaller communities. Trade networks existed throughout the region, linking various local peoples not only to one another but also to the markets of the Indian Ocean and beyond. For example, porcelains from China have been unearthed in the grounds of the Great Zimbabwe, a stone-walled settlement that flourished in the fifteenth century.

In the 1500s, small numbers of Portuguese began settling along the coasts of Angola and Mozambique. A century later, in 1652, the Dutch established a settlement at Africa's southernmost tip, the Cape of Good Hope. While the Portuguese flag generally remained confined to coastal enclaves until the late 1800s, the Dutch colony expanded steadily into the interior throughout the 1700s, seizing the land of local Khoisan communities. Unlike the colonial footholds of the Portuguese and other Europeans on the continent, which prior to the 1800s were mostly centers for the export of slaves, the Dutch Cape Colony imported slaves from Asia as well as from elsewhere in Africa. Although not legally enslaved, conquered Khoisan were also reduced to servitude. In the process, a new society evolved at the Cape. Much like the American South before the U.S. Civil War, the Cape Colony was racially divided between free white settlers and subordinated peoples of mixed African and Afro-Asian descent.

During the Napoleonic Wars, Britain took over the Cape Colony. Shortly thereafter, in 1820, significant numbers of English-speaking colonists began arriving in the region. The arrival of the British coincided with a period of political realignment throughout much of Southern Africa that is commonly referred to as the "Mfecane." Until recently, the historical literature has generally attributed this upheaval to dislocations caused by the rise of the Zulu state, under the great warrior prince Shaka. However, more recent scholarship on the Mfecane has focused on the disruptive effects of increased traffic in contraband slaves from the interior to the Cape and the Portuguese stations of Mozambique, following the international ban on slave trading.

In the 1830s, the British abolished slavery throughout their empire and extended limited civil rights to nonwhites at the Cape. In response, a large number of white Dutch-descended Boers, or Afrikaners, moved into the interior, where they founded two republics that were free of British control. This migration, known as the Great Trek, did not lead the white settlers into an empty land. The territory was home to many African groups, who lost their farms and pastures to the superior firepower of the early Afrikaners, who often coerced local communities into supplying corvee labor for their farms and public works. But a few African polities, like Lesotho and the western Botswana kingdoms, were able to preserve their independence by acquiring their own firearms.

In the second half of the nineteenth century, white migration and dominance spread throughout the rest of Southern Africa. The discovery of diamonds and gold in northeastern South Africa encouraged white exploration and subsequent occupation farther north. In the 1890s, Cecil Rhodes's British South Africa Company occupied modern Zambia and Zimbabwe, which then became known as the Rhodesias. British traders, missionaries, and settlers also invaded the area now known as Malawi. Mean-

(United Nations photo by J. P. Laffont)

Angolan youths celebrated when the nation became independent in 1974.

while, the Germans seized Namibia, while the Portuguese began to expand inland from their coastal enclaves. Thus, by 1900, the entire region had fallen under white colonial control.

With the exception of Lesotho and Botswana, which were occupied as British "protectorates," all of the European colonies in Southern Africa had significant populations of white settlers, who in each case played a predominant political and economic role in their respective territories. Throughout the region, this white supremacy was fostered and maintained through racially discriminatory policies of land alienation, labor regulation, and the denial of full civil rights to nonwhites. In South Africa, where the largest and longest-settled white population resided, the Afrikaners and English-speaking settlers were granted full self-government in 1910—with a Constitution that left the country's black majority virtually powerless.

BLACK NATIONALISM AND SOUTH AFRICAN DESTABILIZATION

After World War II, new movements advocating black self-determination gained ascendancy throughout the region. However, the progress of these struggles for majority rule and independence was gradual. By 1968, the countries of Botswana, Lesotho, Malawi, Swaziland, and Zambia had gained their independence. The area was then polarized between liberated and nonliberated nations. In 1974, a military uprising in Portugal brought statehood to Angola and Mozambique, after long armed struggles by liberation forces in the two territories. Wars of liberation also led to the overthrow of white-settler rule in Zimbabwe, in 1980, and the independence of Namibia, in 1990. Finally, in 1994, South Africa completed a negotiated transition to a nonracial government.

South Africa's liberation has far-reaching implications for the entire Southern African subcontinent as well as the country's own historically oppressed masses. Since the late nineteenth century, South Africa has been the region's economic hub. Today, it accounts for about 80 percent of the total Southern African gross domestic product. Most of the subcontinent's roads and rails also run through South Africa. For generations, the country has recruited expatriate as well as local black African workers for its industries and mines. Today, it is the most economically developed country on the continent, with manufactured goods and agricultural surpluses that are in high demand elsewhere. By the late 1980s, when the imposition of economic sanctions against the then-apartheid regime was at its height, some 46 African countries were importing South African products. With sanctions now lifted, South Africa's economic role on the continent is likely to increase substantially.

A significant milestone was South Africa's admittance in 1994 as the 11th member of the Southern African Development Community (SADC). This organization's ultimate goal is to emulate the European Union (formerly called the European Community or Common Market) by promoting economic integration and political coordination among Southern Africa's states (including Tanzania). While South Africa is expected to be at the center of the Community, SADC's roots lie in past efforts by its other members to reduce their ties to that country. The organization grew out of the Southern African Development Coordination Conference (SADCC), which was created by the region's then–black-ruled states in 1980 to lessen their dependency on white-ruled South Africa. Each SADCC government assumed responsibility for research and planning in a specific developmental area: Angola for energy, Mozambique for transport and communication, Tanzania for industry, and so on.

In its first decade, SADCC succeeded in attracting considerable outside aid for building and rehabilitating its member states' infrastructure. The organization's greatest success was the Beira corridor project, which enabled the Mozambican port to serve once more as a major regional transit point. Other successes included telecommunications independence of South Africa, new regional power grids, and the upgrading of Tanzanian roads to carry Malawian goods to the port of Dar es Salaam. In 1992, with South Africa's liberation on the horizon, the potential for a more ambitious and inclusive SADC grouping became possible.

(United Nations photo by Jerry Frank)

South Africa's economic and military dominance overshadows the region's planning. Pictured above is Cape Town, South Africa's chief port and the country's legislative capital.

In 1996, South African president Nelson Mandela replaced Botswana's president, Sir Ketumile Masire, as SADC chairman, while Pretoria became the headquarters of a new SADC "Organ for Politics Defense and Security." The new South Africa's role as security coordinator within SADC was especially ironic: Before 1990, it had been the violent destabilizing policies of South Africa's military that had sabotaged efforts toward building greater regional cooperation. SADCC members, especially those that were further linked as the so-called Frontline States (Angola, Botswana, Mozambique, Tanzania, Zambia, and Zimbabwe), were then hostile to South Africa's racial policies. To varying degrees, they provided havens for those oppressed by these policies. South Africa responded by striking out against its exiled opponents through overt and covert military operations, while encouraging insurgent movements among some of its neighbors, most notably in Angola and Mozambique.

In Angola, South Africa (along with the United States) backed the rebel movement UNITA, while in Mozambique, it assisted RENAMO. Both of these movements resorted to the destruction of the railways and roads in their operational areas, a tactic that greatly increased the dependence on South African communications of the landlocked states of Botswana, Malawi, Zambia, and Zimbabwe. It is estimated that in the 1980s, the overall monetary cost to the Frontline States of South Africa's destabilization campaign was about $60 billion. (The same countries' combined annual gross national product was only

about $25 billion in 1989.) The human costs were even greater: Hundreds of thousands of people were killed; at least equal numbers were maimed; and in Mozambique alone, more than 1 million people became refugees.

In the Southern African context, South Africa remains a military superpower. Despite the imposition of a United Nations arms embargo between 1977 and 1994, the country's military establishment was able to secure both the arms and sophisticated technology needed to develop its own military/industrial complex. Now a global arms exporter, South Africa is nearly self-sufficient in basic munitions, with a vast and advanced arsenal of weapons. Whereas in 1978 it imported 75 percent of its weapons, today that figure is less than .5 percent. By the 1980s, the country had also developed a nuclear arsenal, which it now claims to have dismantled. However, the former embargo was not entirely ineffective—while South African industry produced many sophisticated weapons systems, it found it increasingly difficult to maintain its regional superiority in such high-technology fields as fighter aircraft. By 1989, the increasing edge of Angolan pilots and air-defense systems was a significant factor in the former South African regime's decision to disengage from the Angolan Civil War. The economic costs of South African militarization were also steep. In addition to draining some 20 percent of its total budget outlays, the destabilization campaign contributed to increased international economic sanctions, which between 1985 and 1990 cost its own economy at least $20 billion. Today, both South Africa and its neighbors hope to benefit from a "peace dividend." But after a generation of militarization, progress in shifting resources from lethal to peaceful pursuits will be gradual.

Throughout the 1980s, South Africa justified its acts of aggression by claiming that it was engaged in counterinsurgency operations against guerrillas of the African National Congress (ANC) and Pan Africanist Congress (PAC), which were then struggling for the regime's overthrow. In fact, the various Frontline States took a cautious attitude toward the activities of South African political refugees, generally forbidding them from launching armed attacks from within their borders. In 1984, both Angola and Mozambique formally signed agreements of mutual noninterference with South Africa. But within a year, these accords had repeatedly and blatantly been violated by South Africa.

Drought, along with continued warfare, has resulted in recurrent food shortages in much of Southern Africa in recent decades—again especially in Angola and Mozambique. The early 1980s' drought in Southern Africa neither lasted as long as nor was as widely publicized as those of West Africa and the Horn, yet it was as destructive. Although some countries, such as Botswana, Mozambique, and Zimbabwe, as well as areas of South Africa, suffered more from nature than others, the main features of the crisis were the same: water reserves were depleted; cattle and game died; and crop production declined, often by 50 percent or more.

Maize and cereal production suffered everywhere. South Africa and Zimbabwe, which are usually grain exporters, had to import food. The countries of Angola, Botswana, and Lesotho each had more than half a million people who were affected by the shortfalls, while some 2 million were malnourished in Mozambique. But in 1988, the rains returned to the region,

(United Nations photo)

This sign, once displayed in a park in Pretoria, South Africa, reflected the restrictions of apartheid, formerly the South African government's official policy of racial discrimination.

raising cereal production by 40 percent. Zimbabwe was able not only to export but also to provide food aid to other African countries. However, South African destabilization contributed to continuing food scarcities in many parts of Angola and Mozambique.

In 1991–1992, the entire region was once more pushed toward catastrophe, with the onset of the worst single drought year in at least a century. Although most of the region experienced improved rainfall in 1993–1994, many areas are still afflicted by food shortages, while the entire region remains vulnerable to famine. Up to 4.5 million people remain at risk of starvation in Mozambique, and another 3 million in Angola,

while Malawi has had to struggle to feed hundreds of thousands of Mozambican refugees along with its own population.

A NEW ERA

Recent events have given rise to hopes for a new era of peace and progress in the region. In 1988, Angola, Cuba, and South Africa reached an agreement, with U.S. and Soviet support, that led to South Africa's withdrawal from Namibia and the removal of Cuban troops from Angola, where they had been supporting government forces. In 1990, Namibia gained its independence under the elected leadership of SWAPO—the liberation movement that had fought against local South African occupation for more than a quarter of a century. In Mozambique, two decades of fighting between the FRELIMO government and South African–backed RENAMO rebels ended in a peace process that has resulted in the successful holding of two multiparty elections. In Zambia and Malawi, multiparty democracy was restored, resulting in the electoral defeat of long-serving authoritarian rulers. The most significant development in the region, however, has been South Africa's transformation. There, the 1990 release of prominent political prisoners, particularly Nelson Mandela, the unbanning of the ANC and PAC, and the lifting of internal state-of-emergency restrictions resulted in extended negotiations that led to an end to white-minority rule.

Southern Africa has also experienced some reversals in recent years. After an on-again, off-again start, direct negotiations between the Angolan government and UNITA rebels led in 1991 to a UN–supervised peace process. But this agreement collapsed in 1992, when UNITA rejected multiparty election results. Although external support for UNITA declined, the movement was able to continue to wreak havoc on much of the Angolan countryside through its trafficking of illegal diamonds and other commodities across the uncontrolled border between Angola and the D.R.C. This in turn encouraged Angola to intervene in the D.R.C. on behalf of the government against UNITA–aligned rebels. By thus effectively encircling UNITA, the Angolan government was ultimately able to achieve military victory where mediation had failed.

Having finally come to the end of its epoch of struggle against white-minority rule, Southern Africa as a whole may be on the threshold of sustained growth. Besides their now-shared commitment to nonracialism, cooperation among the states is being facilitated by their new-found, yet still tenuous, commitment to democracy. Economic thinking within the region has also converged toward a consensus favorable to the growth of market economies. While Angola, Mozambique, Tanzania, Zambia, and Zimbabwe have all moved away from past commitments to various shades of state-centered socialism, the South African economy is being freed from the statist distortions of apartheid. While reconstruction will take time, its resource base and human as well as physical infrastructure could make Southern Africa a major global nexus in the twenty-first century.

Angola (Republic of Angola)

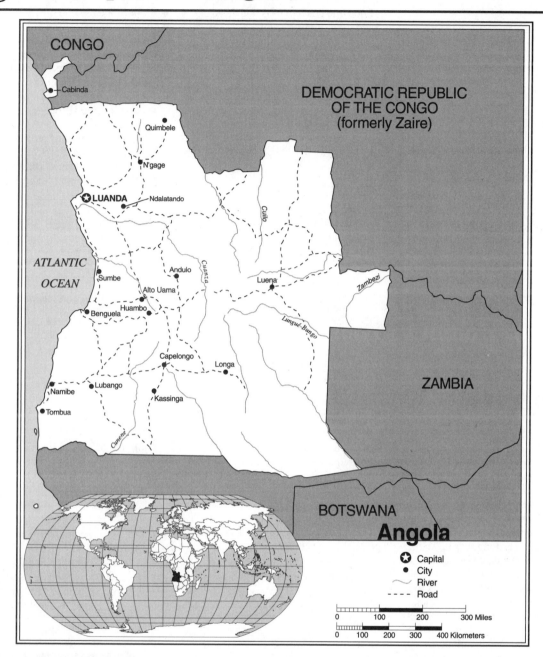

Angola Statistics

GEOGRAPHY

Area in Square Miles (Kilometers):
481,351 (1,246,699) (about twice the size of Texas)

Capital (Population): Luanda (2,819,000)

Environmental Concerns: soil erosion; desertification; deforestation; loss of habitat and biodiversity; water pollution

Geographical Features: a narrow coastal plain rises abruptly to a vast interior plateau

Climate: semiarid in south and along coast to Luanda; the north has a cool, dry season and then a hot, rainy season

PEOPLE

Population

Total: 10,594,000

Annual Growth Rate: 2.18%

Rural/Urban Population Ratio: 66/34

Major Languages: Portuguese; Bantu and other African languages

Ethnic Makeup: 37% Ovimbundu; 25% Kimbundu; 13% Bakongo; 25% others

Religions: 47% indigenous beliefs; 38% Roman Catholic; 15% Protestant

Health

Life Expectancy at Birth: 37 years (male); 40 years (female)

Infant Mortality: 192/1,000 live births

123

Physicians Available: 1/15,136 people
HIV/AIDS Rate in Adults: 2.78%

Education

Adult Literacy Rate: 42%
Compulsory (Ages): 7–15; free

COMMUNICATION

Telephones: 70,000 main lines
Daily Newspaper Circulation: 11/1,000
  people
Televisions: 48/1,000 people
Internet Users: 30,000 (2001)

TRANSPORTATION

Highways in Miles (Kilometers): 47,508
  (76,626)
Railroads in Miles (Kilometers): 1,982
  (3,189)
Usable Airfields: 244
Motor Vehicles in Use: 223,000

GOVERNMENT

Type: republic

Independence Date: November 11, 1975
  (from Portugal)
Head of State/Government: President José
  Edouardo dos Santos is both head of state
  and head of government
Political Parties: Popular Movement for
  the Liberation of Angola; National Front
  for the Liberation of Angola; others
Suffrage: universal at 18

MILITARY

Military Expenditures (% of GDP): 22%
Current Disputes: apparent end of civil
  war

ECONOMY

Currency ($ U.S. Equivalent): 58.1 new
  kwanzas = $1
Per Capita Income/GDP: $1,330/$13.3
  billion
GDP Growth Rate: 5.4%
Inflation Rate: 110%
Unemployment Rate: unemployment and
  underemployment affect more than half
  the population

Labor Force by Occupation: 85%
  agriculture; 15% industry and services
Natural Resources: petroleum; diamonds;
  phosphates; iron ore; copper; feldspar;
  gold; bauxite; uranium
Agriculture: coffee; sisal; cotton;
  sugarcane; tobacco; vegetables; bananas;
  plantains; livestock; forest products; fish
Industry: petroleum; minerals; fish
  processing; brewing; tobacco products;
  textiles; food processing; construction;
  others
Exports: $7 billion (primary partners
  United States, European Union, China)
Imports: $2.7 billion (primary partners
  European Union, South Korea, South
  Africa)

SUGGESTED WEB SITES

```
http://www.angola.org
http://www.angolapress-angop.as/
  index-e.asp
http://www.angolanews.com
http://www.sas.upenn.edu/
  African_Studies/
  Country_Specific/Angola.html
```

Angola Country Report

In April 2002, after the death of Jonas Savimbi, the leader of the rebel Union for Total Independence of Angola (UNITA) forces, and the concurrent decimation of his army, a cease-fire was signed in Luanda between UNITA and the governing Popular Movement for the Liberation of Angola (MPLA). The 27-year-long war in Angola seems to be over. The death of Savimbi in a gunfight with government forces in February 2002 rendered UNITA ineffective. UNITA is now in the process of surrendering its arms and moving into demobilization camps to return to civilian life. Scattered remnants may continue to engage in sporadic acts of violence, but UNITA's ability to wage war against the government of Angola has been fundamentally eliminated.

By early 2002 UNITA had been driven out of the Democratic Republic of the Congo (D.R.C.), its rear base, by the combined forces of Angola, Zimbabwe, Namibia, and the D.R.C. Without its rear base of support, funds derived from diamond sales after the Blood diamonds campaign, and leader Savimbi, UNITA was largely a spent force.

A peace accord had been signed in 1994, after which the United Nations sent in peacekeepers. But the fighting steadily

worsened again, and the MPLA government requested in 1999 that the UN withdraw its peacekeeping forces. The peacekeepers withdrew, and sanctions were instituted against UNITA that froze the European bank accounts for UNITA's leaders and isolated funds used to trade in gems.

DEVELOPMENT

Most of Angola's export revenues currently come from oil. There are important diamond and iron mines, but their output has suffered due to the war, which also prevented the exploitation of the country's considerable reserves of other minerals. Angola has enormous agricultural potential, but only about 2% of its arable land is under cultivation.

Angola, however, remains with large quantities of land mines spread throughout the countryside, and the physical infrastructure has been left in tatters. It will take years to recover from the effects of the civil war, but the country's mineral resources will provide the foundation for development in the future.

In 1996, hopes for a cease-fire had been raised when UNITA agreed in principle to

join the MPLA in forming a "Government of National Unity," but a settlement was never implemented. Instead, both sides maintained their military capacities while supporting proxy forces in neighboring states.

FREEDOM

Despite new constitutional guarantees, pessimists note that neither UNITA nor MPLA has demonstrated a strong commitment to democracy and human rights in the past. Within UNITA, Jonas Savimbi's word was law; he was known to have critics within his movement burned as "witches." For a time UNITA had its own Internet address, perhaps the first armed faction to do so.

Since 1975, more than half a million Angolans perished as a result of fighting between the two movements, including many passive victims of land mines. Up to 1 million others fled the country, while another 1 million or so were internally displaced. According to a report by the human-rights organization Africa Watch, tens of thousands of Angolans have lost their limbs "because of the indiscriminate use of landmines by both sides of the conflict." Angola's small and impoverished

(United Nations photo by J. P. Laffont)

Angola's war for independence from Portugal led to the creation of a one-party state.

population could not have perpetuated such carnage were it not for decades of external interference in the nation's internal affairs. The United States, the Soviet Union, South Africa, Cuba, Zaire, and many others helped to create and sustain this tragedy.

After the end of the Cold War and the demise of South Africa's apartheid regime, there was an almost complete cutoff of outside support for the conflict. An agreement in April 1991 between the MPLA and UNITA to participate in United Nations–sponsored elections led to a dramatic decline in violence during 16 months of "phony peace." The successful holding of elections in September 1992 further raised hopes of a new beginning for Angola. While the MPLA appeared to have topped the poll, UNITA and the smaller National Front for the Liberation of Angola (FNLA) secured a considerable vote. But hopes for a new beginning under an all-party government of national unity were quickly dashed by UNITA's rejection of the election result. As a result, the country was plunged into renewed civil war.

THE COLONIAL LEGACY

The roots of Angola's long suffering lie in the area's colonial underdevelopment. The Portuguese first made contact with the peoples of the region in 1483. They initially established peaceful trading contact with the powerful Kongo kingdom and other coastal peoples, some of whom were converted to Catholicism by Jesuit missionaries. But from then to the mid-1800s, the outsiders primarily saw the area as a source of slaves.

Angola has been called the "mother of Brazil" because up to 4 million Angolans were carried away from its shores to that land, chained in the holds of slave ships. With the possible exception of Nigeria, no African territory lost more of its people to the trans-Atlantic slave trade.

Following the nineteenth-century suppression of the slave trade, the Portuguese introduced internal systems of exploitation that very often amounted to slavery in all but name. Large numbers of Angolans were pressed into working on coffee plantations owned by a growing community of white settlers. Others were forced to labor in other sectors, such as diamond mines or public-works projects.

HEALTH/WELFARE

Civil war caused a serious deterioration of Angola's health services, resulting in lower life expectancy and one of the highest infant mortality rates in the world.

Although the Portuguese claimed that they encouraged Angolans to learn Portuguese and practice Catholicism, thus becoming "assimilated" into the world of the colonizers, they actually made little effort to provide education. No more than 2 percent of the population ever achieved the legal status of *assimilado*. The assimilados, many of whom were of mixed race, were concentrated in the coastal towns. Of the few interior Angolans who became literate, a large proportion were the products of Protestant, non-Portuguese, mission schools. Because each mission tended to

operate in a particular region and teach from its own syllabus, usually in the local language, an unfortunate by-product of these schools was the reinforcement (the creation, some would argue) of ethnic rivalries among the territory's educated elite.

In the late colonial period, the FNLA, MPLA, and UNITA emerged as the three major liberation movements challenging Portuguese rule. Although all three sought a national following, each built up an ethnoregional core of support by 1975. The FNLA grew out of a movement whose original focus was limited to the northern Kongo-speaking population, while UNITA's principal stronghold was the largely Ovimbundu-speaking south-central plateaux. The MPLA had its strongest following among assimilados and Kimbundu-speakers, who are predominant in Luanda, the capital, and the interior to the west of the city. From the beginning, all three movements cultivated separate sources of external support.

ACHIEVEMENTS

Between 1975 and 1980, the Angolan government claimed that it had tripled the nation's primary-school enrollment, to 76%. That figure subsequently dropped as a result of war.

The armed struggle against the Portuguese began in 1961, with a massive FNLA–inspired uprising in the north and MPLA–led unrest in Luanda. To counter the northern rebellion, the Portuguese resorted to the saturation bombing of villages. In the first year of fighting, this left an estimated 50,000 dead (about half the total number killed throughout the anticolonial struggle). The liberation forces were as much hampered by their own disunity as by the brutality of Portugal's counterinsurgency tactics. Undisciplined rebels associated with the FNLA, for example, were known to massacre not only Portuguese plantation owners but many of their southern workers as well. Such incidents contributed to UNITA's split from the FNLA in 1966. There is also evidence of UNITA forces cooperating with the Portuguese in attacks on the MPLA. Besides competition with its two rivals, the MPLA also encountered some difficulty in keeping its urban and rural factions united.

CIVIL WAR

The overthrow of Portugal's Fascist government in 1974 led to Angola's rapid decolonization. Attempts to create a transitional government of national unity among the three major nationalist movements failed. The MPLA succeeded in seizing Luanda, which led to a loose alliance between the FNLA and UNITA. As fighting between the groups escalated, so did the involvement of their foreign backers. Meanwhile, most of Angola's 300,000 or more white settlers fled the country, triggering the collapse of much of the local economy. With the notable exception of Angola's offshore oil industry, most economic sectors have since failed to recover their preindependence output as a result of the war.

While the chronology of outside intervention in the Angolan conflict is a matter of dispute, it is nonetheless clear that, by October 1975, up to 2,000 South African troops were assisting the FNLA–UNITA forces in the south. In response, Cuba dispatched a force of 18,000 to 20,000 to assist the MPLA, which earlier had gained control of Luanda. These events proved decisive during the war's first phase. On the one hand, collaboration with South Africa led to the withdrawal of Chinese and much of the

African support for the FNLA–UNITA cause. It also contributed to the U.S. Congress' passage of the Clarke Amendment, which abruptly terminated the United States' direct involvement. On the other hand, the arrival of the Cubans allowed the MPLA quickly to gain the upper hand on the battlefield. Not wishing to fight a conventional war against the Cubans by themselves, the South Africans withdrew their conventional forces in March 1976.

By 1977, the MPLA's "People's Republic" had established itself in all of Angola's provinces. It was recognized by the United Nations and most of its membership as the nation's sole legitimate government, the United States numbering among the few that continued to withhold recognition. However, the MPLA's apparent victory did not bring an end to the hostilities. Although the remaining pockets of FNLA resistance were overcome following an Angola–Zaire rapprochement in 1978, UNITA maintained its largely guerrilla struggle. Until 1989, UNITA's major supporter was South Africa, whose destabilization of the Luanda government was motivated by its desire to keep the Benguela railway closed (thus diverting traffic to its own system) and harass Angola-based SWAPO forces. Besides supplying UNITA with logistical support, the South Africans repeatedly invaded southern Angola and on occasion infiltrated sabotage units into other areas of the country. South African aggression in turn justified Cuba's maintenance (by 1988) of some 50,000 troops in support of the government. In 1986, the U.S. Congress approved the resumption of "covert" U.S. material assistance to UNITA via Zaire.

An escalation of the fighting in 1987 and 1988 was accompanied by a revival of negotiations for a peace settlement among representatives of the Angolan government, Cuba, South Africa, and the United States. In 1988, South African forces were checked in a battle at the Angolan town of

Cuito Cuanavale. South Africa agreed to withdraw from Namibia and end its involvement in the Angolan conflict. It was further agreed that Cuba would complete a phased withdrawal of its forces from the region by mid-1991.

Timeline: PAST

1400s
The Kongo state develops

1483
The Portuguese make contact with the Kongo state

1640
Queen Nzinga defends the Mbundu kingdom against the Portuguese

1956
The MPLA is founded in Luanda

1961
The national war of liberation begins

1975
Angola gains independence from Portugal

1976
South African–initiated air and ground incursions into Angola

1979
President Agostinho Neto dies; José dos Santos becomes president

1986
Jonas Savimbi visits the United States; U.S. "material and moral" support for UNITA resumes

1990s
Talks for national reconciliation break down; multiparty elections are held

PRESENT

2000s
Savimbi is killed in a gun battle with government forces

Civil war appears to be over

Botswana (Republic of Botswana)

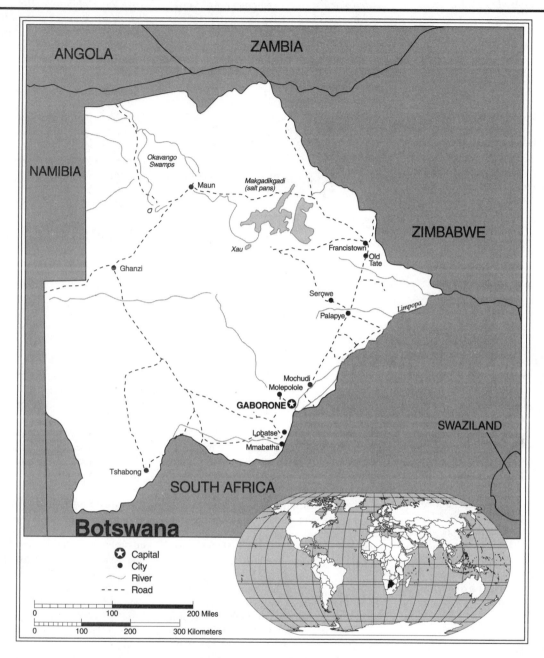

Botswana Statistics

GEOGRAPHY

Area in Square Miles (Kilometers): 231,804 (600,372) (about the size of Texas)

Capital (Population): Gaborone (248,000)

Environmental Concerns: overgrazing; desertification; limited freshwater resources

Geographical Features: mainly flat to gently rolling tableland; Kalahari sandvelt covers 50% of the country; wetlands in the north

Climate: semiarid; warm winters and hot summers

PEOPLE

Population

Total: 1,681,000

Annual Growth Rate: 2.1%

Rural/Urban Population Ratio: 52/48

Major Languages: English; Setswana

Ethnic Makeup: 80% Tswana; 12% Kalanga; 18% others

Religions: 85% indigenous beliefs; 15% Christian

Health

Life Expectancy at Birth: 37 years (male); 40 years (female)

Infant Mortality: 64.7/1,000 live births
Physicians Available: 1/4,395 people
HIV/AIDS Rate in Adults: 35.8%

Education

Adult Literacy Rate: 87%

COMMUNICATION

Telephones: 150,000 main lines
Daily Newspaper Circulation: 29/1,000 people
Televisions: 24/1,000 people
Internet Users: 33,000 (2001)

TRANSPORTATION

Highways in Miles (Kilometers): 6,130 (10,217)
Railroads in Miles (Kilometers): 583 (940)
Usable Airfields: 92
Motor Vehicles in Use: 150,000

GOVERNMENT

Type: parliamentary republic
Independence Date: September 30, 1966 (from the United Kingdom)
Head of State/Government: President Festus Mogae is both head of state and head of government
Political Parties: Botswana Democratic Party; Botswana Alliance Movement; Botswana National Front; Botswana Congress Party; others
Suffrage: universal at 18

MILITARY

Military Expenditures (% of GDP): 3.5%
Current Disputes: none

ECONOMY

Currency ($ U.S. Equivalent): 5.25 pulas = $1
Per Capita Income/GDP: $7,800/$12.4 billion

GDP Growth Rate: 4.7%
Inflation Rate: 6.6%
Unemployment Rate: 20%
Population Below Poverty Line: 47%
Natural Resources: diamonds; copper; nickel; salt; soda ash; potash; coal; iron ore; silver
Agriculture: sorghum; maize; pulses; peanuts; cowpeas; beans; sunflower seeds; livestock
Industry: diamonds; copper; nickel; salt; soda ash; potash; tourism; livestock processing
Exports: $2.5 billion (primary partners Europe, Southern Africa)
Imports: $2.1 billion (primary partners Southern Africa, Europe)

SUGGESTED WEB SITES

http://www.gov.bw

http://www.info.bw

Botswana Country Report

Since its independence in 1966, Botswana has enjoyed one of the highest average economic growth rates in the world. This has resulted in a better life for many of its citizens. But serious challenges remain if the country is to realize the ambitious goals contained within its "Vision 2016" program of national development. Economic growth has not been accompanied by equity. More than 40 percent of households remain below the official poverty line. Another concern is the feeling of many Botswana that their country's economy continues to be dominated by noncitizens, while many locals remain unemployed or underemployed.

DEVELOPMENT

The United Nations' 1990 Human Development Report singles out Botswana among the nations of Africa for significantly improving the living conditions of its people. In 1989, President Masire was awarded The Hunger Project's leadership prize, based on Botswana's record of improving rural nutritional levels during the 1980s despite 7 years of severe drought.

Perhaps the greatest challenge currently facing Botswana is its sad distinction of having the world's highest recorded rate of HIV/AIDS infection. About one third of all Batswana between the ages of 15 and 40

are believed to be living with HIV/AIDS. This has caused a dramatic drop in the country's overall life expectancy, from 69 years in 1995 to just under 40 years today. Based on current trends, UNAIDS estimates that overall life expectancy could drop further to as low as 27 years by 2010.

The man currently leading Botswana in the face of the above challenges is the country's third president, Festus Mogae, who was inaugurated in April 1998 following the retirement of the long-serving Sir Ketumile Masire (1980–1998). In October 1999, Mogae and his Botswana Democratic Party won an easy victory in national elections, against a divided opposition. But his government was subsequently rocked by conflict between several veteran members of the cabinet and the youthful, charismatic vice-president, Ian Khama, a former army commander and the son of the country's first president, Sir Seretse Khama (1966–1980). In August 2000, it was announced that Ian Khama would be given a new coordinating role to assure adequate performance by various ministries, leading many local commentators to proclaim him as Botswana's first "prime minister."

Over the past four decades, Botswana has been hailed as a model of postcolonial development in Africa. When the country emerged from 81 years of British colonial occupation, it was ranked as one of the 10 poorest countries in the world, with an an-

nual per capita income of just $69. In the years since, the nation's economy has grown at an average annual rate of 9 percent. Social services have been steadily expanded, and infrastructure has been created. Whereas at independence the country had no significant paved roads, most major towns and villages are now interlinked by ribbons of asphalt and tarmac. The nation's capital, Gaborone, has emerged as a vibrant city. Schools, hospitals, and businesses dot the landscape, while the country's all-digital telecommunications network is among the most advanced in the world. But the failure of such growth to promote greater equity has also led to a rise in social tensions.

FREEDOM

Democratic pluralism has been strengthened by the growth of a strong civil society and an independent press. Concern has been voiced about social and economic discrimination against Khoisan-speaking communities living in remote areas of the Kalahari, who are known to many outsiders as "Bushmen."

Botswana's economic success has come in the context of its unbroken postindependence commitment to political pluralism, respect for human rights, and racial and ethnic tolerance. Freedoms of speech and

(United Nations photo by E. Darroch)

Botswana, like many other African nations, is subject to periodic drought. The country, however, has a good supply of underground water and the governmental competence to utilize this resource. In this photograph, antelopes drink from a hole dug to allow water seepage.

association have been upheld. In October 1999, the nation held its eighth successive multiparty election. The Botswana Democratic Party, which has ruled since independence, increased its majority in part due to a split in the main opposition party, the left-leaning opposition Botswana National Front.

HEALTH/WELFARE

After years of being praised as a model of primary health-care delivery, Botswana's public-health service has come under increased criticism for a perceived decline in quality. Botswana's high HIV-positive rate has placed the system under serious stress.

Most of Botswana's people share Setswana as their first language, which is understood by about 95% of the population. There also exist a number of sizable minority communities—Kalanga, Herero, Kalagari, Khoisan groups, and others—but contemporary ethnic conflict is relatively modest. In 2002, Parliament agreed to amend the Constitution to remove all references to "tribes" while widening the representation within the Ntlo ya Dikgosi, an

advisory house of traditional leaders. In the nineteenth century, most of Botswana was incorporated into five Tswana states, each centering around a large settlement of 10,000 people or more. These states, which incorporated non-Tswana communities, survived through agropastoralism, hunting, and their control of trade routes linking Southern and Central Africa.

Lucrative dealing in ivory and ostrich feathers allowed local rulers to build up their arsenals and thus deter the aggressive designs of South African whites. An attempt by white settlers to seize control of southeastern Botswana was defeated in an 1852–1853 war. However, European missionaries and traders were welcomed, leading to a growth of Christian education and the consumption of industrial goods.

A radical transformation took place after the imposition of British overrule in 1885. Colonial taxes and economic decline stimulated the growth of migrant labor to the mines and industries of South Africa. (In a few regions, migrant earnings remain the major source of income to this day.) Although colonial rule brought much hardship and little benefit, the twentieth-century relationship between the people of Botswana and the British was complicated by local

fears of being incorporated into South Africa. For many decades, leading nationalists championed continued rule from London as a shield against their powerful, racially oppressive neighbor.

ACHIEVEMENTS

In 1999, Botswana's Mpule Kwelagobe was crowned as Miss Universe. In July 2000, Botswana launched its first national television service. A UN report ranked Botswana first in Africa in its percentage of women holding middle- and senior-level managerial positions.

ECONOMIC DEVELOPMENT

Economic growth since independence has been largely fueled by the rapid expansion of mining activity. Botswana has become one of the world's leading producers of diamonds, which typically account for 80 percent of its export earnings. Local production is managed by Debswana Corporation, an even partnership between the Botswana government and DeBeers, a South Africa–based global corporation; DeBeers' Central Selling Organization has a near monopoly on diamond sales worldwide. The Botswana government has a good record of maximizing the local benefits of Debswana's production.

The nickel/copper/cobalt mining complex at Selibi-Pikwe is the largest nongovernment employer in Botswana. Falling metal prices and high development costs have reduced the mine's profitability. Despite high operating efficiency, it is expected to close by the end of the decade. Given that mining can make only a modest contribution to local employment, and because of the potential vulnerability of the diamond market, Botswana is seeking to expand its small manufacturing and service sectors. Meat processing is currently the largest industrial activity outside minerals, but efforts are under way to attract overseas investment in both private and parastatal production. Botswana already has a liberal foreign-exchange policy and has established Bedia as an agency to promote foreign direct investment.

Although it has been negatively affected by events in neighboring Zimbabwe, tourism is of growing importance. Northern Botswana is particularly noted for its bountiful wildlife and stunning scenery. The region includes the Okavango Delta, a vast and uniquely beautiful swamp area, and the Chobe National Park, home of the world's largest elephant herds.

Agriculture is still a leading economic activity for many Batswana. The standard Tswana greeting, *Pula* ("Rain"), reflects the significance attached to water in a society prone to its periodic scarcity. Botswana suffered severe droughts between 1980 and 1987 and again in 1991 and 1992, which—despite the availability of underground water supplies—had a devastating effect on both crops and livestock. Up to 1 million cattle are believed to have perished. Small-scale agropastoralists, who make up the largest segment of the population, were

particularly hard hit. However, government relief measures prevented famine. The government provides generous subsidies to farmers, but environmental constraints hamper efforts to achieve food self-sufficiency even in nondrought years.

Commercial agriculture is dominated by livestock. The Lobatse abbatoir, opened in 1954, stimulated the growth of the cattle industry. Despite periodic challenges from disease and drought, beef exports have become relatively stable. Much of the output of the Botswana Meat Commission has preferential access to the European Union. There is some concern about the potential for future reductions in the European quota. Because most of Botswana's herds are grazed in communal lands, questions about the allocation of pasture are a source of local debate. There is also a growing, but largely misinformed, international concern that wildlife are being threatened by overgrazing livestock.

SOUTH AFRICA

Before 1990, Botswana's progress took place against a backdrop of political hostility on the part of its powerful neighbor, South Africa. Since the nineteenth century, Botswana has sheltered many South African refugees fleeing racist oppression. This led to periodic acts of aggression against the country, especially during the 1980s, when Botswana became the repeated victim of overt military raids and covert terrorist operations. The establishment of a nonracial democracy in South Africa has led to a normalization of relations.

Gaborone is the headquarters of the Southern African Development Community, which was originally conceived to reduce the economic dependence of its 10 member nations on the South African apartheid state. With South Africa now a member, SADC now plans to transform itself into a common market. Despite the countries' past political differences, Botswana has maintained a customs union with South Africa that dates back to the colonial era.

Timeline: PAST

700s
Emergence of the Tswana trading center at Toutswemogala

1820s
Kololo and Ndebele invaders devastate the countryside

1830s
Tswana begin to acquire guns through trade in ivory and other game products

1852–1853
Batswana defeat Boer invaders

1885
The British establish colonial rule over Botswana

1966
Botswana regains its independence

1980s
Elections in 1984 and 1989 result in landslide victories for the Democratic Party; the National Front is the major opposition party

1990s
The ruling Democratic Party wins the 1994 and 1999 elections; the opposition National Front makes gains in 1994 but loses seats in 1999

PRESENT

2000s
Despite astounding national economic growth, many Batswana remain poor

HIV/AIDS crisis

Lesotho (Kingdom of Lesotho)

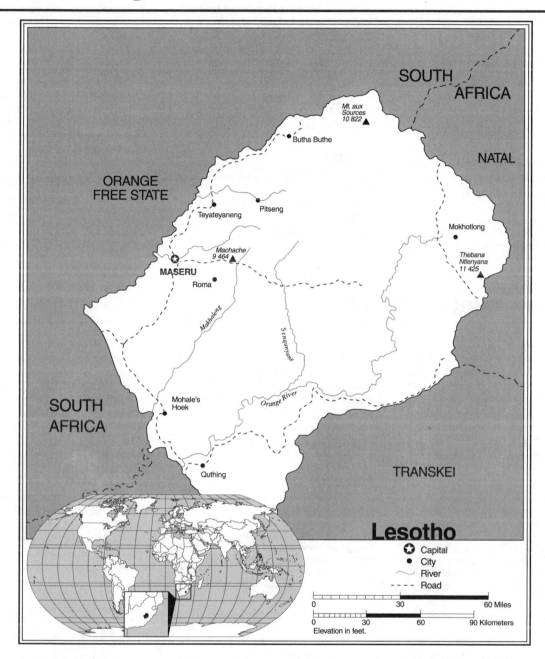

Lesotho Statistics

GEOGRAPHY

Area in Square Miles (Kilometers):
11,716 (30,344) (about the size of Maryland)

Capital (Population): Maseru (271,000)

Environmental Concerns: overgrazing; soil erosion; soil exhaustion; desertification; water pressures

Geographical Features: mostly highland, with plateaus, hills, and mountains; landlocked

Climate: temperate

PEOPLE

Population

Total: 2,208,000

Annual Growth Rate: 1.33%

Rural/Urban Population Ratio: 73/27

Major Languages: English; Sesotho

Ethnic Makeup: 99.7% Sotho

Religions: 80% Christian; 20% indigenous beliefs

Health

Life Expectancy at Birth: 46 years (male); 48 years (female)

Infant Mortality: 83/1,000 live births

Physicians Available: 1/14,306 people

HIV/AIDS Rate in Adults: 23.5%

Education

Adult Literacy Rate: 71.3%

131

Compulsory (Ages): 6–13; free

COMMUNICATION

Telephones: 23,000 main lines
Daily Newspaper Circulation: 7/1,000 people
Televisions: 7/1,000 people
Internet Users: 4,000 (2000)

TRANSPORTATION

Highways in Miles (Kilometers): 2,973 (4,955)
Railroads in Miles (Kilometers): 1.6 (2.6)
Usable Airfields: 28
Motor Vehicles in Use: 23,000

GOVERNMENT

Type: parliamentary constitutional monarchy

Independence Date: October 4, 1966 (from the United Kingdom)
Head of State/Government: King Letsie III; Prime Minister Pakalitha Mosisili
Political Parties: Lesotho Congress for Democracy; Basotho National Party; Basutoland Congress Party; Marematlou Freedom Party; others
Suffrage: universal at 18

MILITARY

Current Disputes: none

ECONOMY

Currency ($ U.S. Equivalent): 8.28 malotis = $1
Per Capita Income/GDP: $2,450/$5.3 billion
GDP Growth Rate: 2.6%
Inflation Rate: 7%
Unemployment Rate: 45%

Labor Force by Occupation: 86% subsistence agriculture; 35% of male wage earners work in South Africa
Natural Resources: water; agricultural and grazing land; some diamonds and other minerals
Agriculture: corn; wheat; sorghum; pulses; barley; livestock
Industry: food and beverages; textiles; handicrafts; construction; tourism
Exports: $250 million (primary partners Southern Africa, North America)
Imports: $720 million (primary partners Southern Africa, Asia)

SUGGESTED WEB SITES

http://www.lesotho.gov/ls
http://www.mbendi.co.za/cylecy.htm
http://www.sas.upenn.edu/African_Studies/Country_Specific/Lesotho.html

Lesotho Country Report

In 2002, the Lesotho Congress for Democracy (LCD) party returned to power after securing a majority of parliamentary seats in elections. The elections, which were endorsed as free and fair by international election observers, were held under a revised Constitution designed to give smaller parties a voice in Parliament.

DEVELOPMENT

Despite an infusion of international aid, Lesotho's economic dependence on South Africa has not decreased since independence; indeed, it has been calculated that the majority of outside funds have actually ended up paying for South African services.

Since October 1988, Lesotho has been ruled by a transitional executive, put into place following the intervention of Botswana and South African troops in support of the previous LCD–led government, which was being threatened by a military coup. Large segments of Lesotho's defense force resisted the intervention, causing scores of deaths on both sides. In the process, many businesses in the capital city, Maseru, and other centers were heavily looted by rioters, who directed much of their anger against Asians.

Listed by the United Nations as one of the world's least-developed nations, each year the lack of opportunity at home causes up to half of Lesotho's adult males to seek employment in neighboring South Africa, where jobs are becoming increasingly scarce. The retrenchment of Basotho mineworkers has led to a local unemployment rate of nearly 50 percent. This dire economic situation is aggravated by Lesotho's chronic political instability.

FREEDOM

In Lesotho, basic freedoms and rights are compromised by continuing political instability. Basotho journalists have come out against proposed measures that they say will gag Lesotho's vigorous independent press.

Lesotho is one of the most ethnically homogeneous nations in Africa; almost all of its citizens are Sotho. The country's emergence and survival were largely the product of the diplomatic and military prowess of its nineteenth-century rulers, especially its great founder, King Moshoeshoe I. In the 1860s, warfare with South African whites led to the loss of land and people as well as an acceptance of British overrule. For nearly a century, the British preserved the country but also taxed the inhabitants and generally neglected their interests. Consequently, Lesotho remained dependent on South Africa. However, despite South African attempts to incorporate the country politically as well as economically, Lesotho's independence was restored by the British in 1966.

Lesotho's politicians were bitterly divided at independence. The conservative Basotho National Party (BNP) had won an upset victory in preindependence elections, with strong backing from the South African government and the local Roman Catholic Church—Lesotho's largest Christian denomination. The opposition, which walked out of the independence talks, was polarized between a pro-royalist faction, the Marematlou Freedom Party (whose regional sympathies largely lay with the African National Congress of South Africa), and the Basotho Congress Party (or BCP, which was allied with the rival Pan-Africanist Congress).

HEALTH/WELFARE

With many of Lesotho's young men working in the mines of South Africa, much of the resident population relies on subsistence agriculture. Despite efforts to boost production, malnutrition, aggravated by drought, is a serious problem.

Soon after independence, the BNP prime minister, Leabua Jonathan, placed the king, Moshoeshoe II, under house arrest. (Later, the king was temporarily exiled.) The BCP won the 1970 elections, but Jonathan, possibly at the behest of South Africa, declared a state of emergency and nullified the results.

In the early 1980s, armed resistance to Jonathan's dictatorship was carried out by the Lesotho Liberation Army (LLA), an armed faction of the BCP. The Lesotho government maintained that the LLA was aided and abetted by South Africa as part of that country's regional destabilization efforts. By 1983, both the South African government and the Catholic Church hierarchy were becoming nervous about Jonathan's establishment of diplomatic ties with various Communist-ruled countries and the growing sympathy within the BNP, in particular its increasingly radical youth wing, for the ANC. South African military raids and terrorist attacks targeting anti-apartheid refugees in Lesotho became increasingly common. Finally, a South African blockade of Lesotho in 1986 led directly to Jonathan's ouster by his military.

ACHIEVEMENTS

Lesotho has long been known for the high quality of its schools, which for more than a century and a half have trained many of the leading citizens of Southern Africa.

Lesotho's new ruling Military Council, initially led by Major General Justinus Lekhanya, was closely linked to South Africa. In 1990, Lekhanya had Moshoeshoe II exiled (for the second time), after he refused to agree to the dismissals of several senior officers. Moshoeshoe's son Letsie was installed in his place. In 1991, Lekhanya was himself toppled by the army. The new leader, General Elias Rameama, promised to hold multiparty elections. In 1992, Moshoeshoe returned, to a hero's welcome, but he was prevented from resuming his role as monarch. His status was uncertain after elections in March 1993 brought the BCP back to power.

Under its aging leader, Ntsu Mokhele, the BCP faced military opposition to its rule. An outbreak of internal fighting within the Royal Lesotho Defense Force (RLDF) culminated, in April 1994, in the assassination of the deputy prime minister and the kidnapping of cabinet members by mutinous soldiers. In August, King Letsie tried to dismiss the government and suspend the constitution. In the face of growing unrest, Botswana, South Africa, and Zimbabwe acted on behalf of the Southern African Development Community as mediators—and subsequently as guarantors—of constitutional rule. The BCP and Moshoeshoe were both restored to power; the latter was killed in January 1996 in an auto accident. In June 1997, a schism in the BCP's ranks led Mokhele to establish the new Lesotho Congress for Democracy party, taking most of the BCP with him. The LCD won a landslide victory in May 1998 elections, but the remnants of the BCP and other opposition parties refused to accept the result, inciting King Letsie and the military to intervene. The resulting mutiny within the RLDF led to the South African/Botswana intervention to restore order.

Timeline: PAST

1820s
Lesotho emerges as a leading state in Southern Africa

1866
Afrikaners annex half of Lesotho

1870–1881
The Sotho successfully fight to preserve local autonomy under the British

1966
Independence is restored

1970
The elections and Constitution are declared void by Leabua Jonathan

1974
An uprising against the government fails

1979
The Lesotho Liberation Army begins a sabotage campaign

1986
South African destabilization leads to the overthrow of Jonathan by the military

1990s
Troops from South Africa and Botswana intervene in Lesotho to avert a coup

PRESENT

2000s
The Lesotho Congress for Democracy is returned to power in parliamentary elections

Malawi (Republic of Malawi)

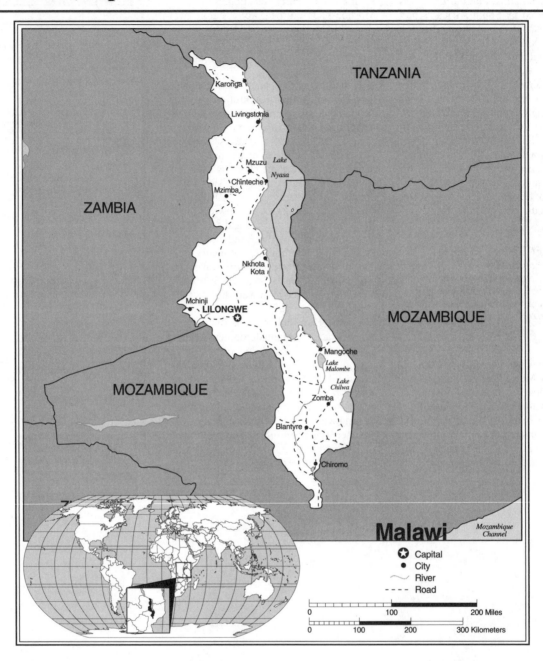

Malawi Statistics

GEOGRAPHY

Area in Square Miles (Kilometers):
45,747 (118,484) (about the size of Pennsylvania)

Capital (Population): Lilongwe (523,000)

Environmental Concerns: deforestation; land degradation; water pollution; siltation of fish spawning grounds

Geographical Features: narrow, elongated plateau with rolling plains, rounded hills, some mountains; landlocked

Climate: subtropical

PEOPLE

Population

Total: 10,702,000

Annual Growth Rate: 1.39%

Rural/Urban Population Ratio: 76/24

Major Languages: Chichewa; English; regional languages

Ethnic Makeup: 90% Chewa; 10% Nyanja, Lomwe, other Bantu groups

Religions: 75% Christian; 20% Muslim; 5% indigenous beliefs

Health

Life Expectancy at Birth: 36 years (male); 37 years (female)

Infant Mortality: 120/1,000 live births

Physicians Available: 1/47,634 people
HIV/AIDS Rate in Adults: 15.96%

Education

Adult Literacy Rate: 58%
Compulsory (Ages): 6–14

COMMUNICATION

Telephones: 38,000 main lines
Internet Users: 15,000 (2000)

TRANSPORTATION

Highways in Miles (Kilometers): 8,756 (14,594)
Railroads in Miles (Kilometers): 489 (789)
Usable Airfields: 44
Motor Vehicles in Use: 55,000

GOVERNMENT

Type: multiparty democracy

Independence Date: July 6, 1964 (from the United Kingdom)
Head of State/Government: President Bakili Muluzi is both head of state and head of government
Political Parties: United Democratic Front; Malawi Congress Party; Alliance for Democracy; Malawi Democratic Party
Suffrage: universal at 18

MILITARY

Military Expenditures (% of GDP): 0.76%
Current Disputes: boundary dispute with Tanzania

ECONOMY

Currency ($ U.S. Equivalent): 91.36 kwachas = $1
Per Capita Income/GDP: $660/$7 billion
GDP Growth Rate: 1.7%
Inflation Rate: 28.6%

Labor Force by Occupation: 86% agriculture
Population Below Poverty Line: more than half
Natural Resources: limestone; uranium; coal; bauxite; arable land; hydropower
Agriculture: tobacco; tea; sugarcane; cotton; potatoes; cassava; sorghum; pulses; livestock
Industry: tobacco; sugar; tea; sawmill products; cement; consumer goods
Exports: $415.5 million (primary partners South Africa, United States, Germany)
Imports: $463.6 million (primary partners South Africa, United Kingdom, Zimbabwe)

SUGGESTED WEB SITES

```
http://www.malawi.net
http://www.maform.malawi.net
http://www.sas.upenn.edu/
   African_Studies/
   Country_Specific/Malawi.html
```

Malawi Country Report

Malawi is facing its worst famine in more than 50 years. In a land where most of the population engages in farming, the government is struggling to feed the people after the lack of rains in 2001–2002. The harvest in 2002 was 25 percent below that of the last five years. Some 6 million Malawians live below the poverty line, of whom an estimated 70 percent are at serious risk of starvation unless help is found quickly. The situation has been aggravated by Malawi's sale of strategic grain reserves, allegedly on the advice of IMF/World Bank experts. Very little of the money has been collected from the sales. As $40 million worth of maize grain (the staple food) went missing and unpaid for, apparently to local speculators as well as international buyers, people have began asking questions. The scandal could not have come at a worse time for President Bakili Muluzi, who was trying to get Parliament to approve a bill that would allow him to serve a third term in office instead of the normal two terms (which he has already completed). In his drive to remain in power, there have been increasing reports of systematic intimidation directed against Muluzi's opponents.

Malawi's current crisis stands in contrast to the optimism that was generated in 1994 when Muluzi and his United Democratic Front (UDF) ended 30 years of dictatorial rule under Malawi's first president,

Dr. Hastings Banda. Muluzi's victory over the then–96-year-old Banda brought an end to what had been one of Africa's most repressive regimes. But since Banda's death in 1997, a degree of nostalgia has emerged regarding his era, perhaps reflecting the failure of political liberalization to bring improvement to the country's weak economy, which is largely dependent on tobacco production. The spread of the HIV/AIDS pandemic and a rising crime rate have also shaken public confidence. While the old political order has been swept aside, the new order has yet to deliver better conditions for most Malawians.

DEVELOPMENT

As in other parts of Africa, there is increasing recognition in Malawi that rural development must be addressed from a perspective that recognizes the key role of women, especially in arable agriculture. Securing property rights for women has become an important development as well as human-rights issue.

Since the early months of independence in 1964, when he purged his cabinet and ruling Malawi Congress Party (MCP) of most of the young politicians who had promoted him to leadership in the nationalist struggle, Banda ruthlessly used his secret police and MCP's militia, the Malawi

Young Pioneers (MYP), to eliminate potential alternatives to his highly personalized dictatorship. Generations of Malawians, including those living abroad, grew up with the knowledge that voicing critical thoughts about the self-proclaimed "Life President," or *Ngwazi* ("Great Lion"), could prove fatal. Only senior army officers, Banda's long-time "official state hostess" Tamanda Kadzamira, and her uncle John Tembo, the powerful minister of state, survived Ngwazi's jealous exercise of power.

FREEDOM

Although greatly improved since the end of the Banda era, serious human-rights problems remain, including the abuse and death of detainees by police. Prison conditions are poor. Lengthy pretrial detention, an inefficient judicial system, and limited resources have called into question the ability of defendants to receive timely and fair trials. High levels of crime have prompted a rise in vigilante justice.

In 1992, Banda's grip began to weaken. Unprecedented antigovernment unrest gave rise to an internal opposition, spearheaded by clergy, underground trade unionists, and a new generation of dissident politicians. By 1993, this opposition had coalesced into two major movements: the southern-based UDF, and the northern-based Alliance for Democracy (AFORD). The detention of

AFORD's leader, Chakufwa Chihana, and others failed to stem the tide of opposition. A referendum in June showed two-to-one support for a return to multiparty politics. In November, while Banda was hospitalized in South Africa, young army officers seized the initiative by launching a crackdown against the MYP while purging a number of senior officers from their own ranks. Thereafter, the army played a neutral role in assuring the success of the election.

HEALTH/WELFARE

Malawi's health service is considered exceptionally poor even for an impoverished country. The country has one of the highest child mortality rates in the world, and more than half of its children under age 5 are stunted by malnutrition.

While the ruthless efficiency of its security apparatus contributed to past perceptions of Malawi's stability, Banda did not survive by mere repression. A few greatly benefited from the regime. Until 1979, the country enjoyed an economic growth rate averaging 6 percent per year. Almost all this growth came from increased agricultural production. The postindependence government favored large estates specializing in exported cash crops. While in the past the estates were almost exclusively the preserve of a few hundred white settlers, today many are controlled by either the state or local individuals.

In the 1970s, the prosperity of the estates helped to fuel a boom in industries involved in agricultural processing. Malawi's limited economic success prior to the 1980s came at the expense of the vast majority of its citizens, who survive as small landholders growing food crops. By 1985, some 86 percent of rural households farmed on less than five acres. Overcrowding has contributed to serious soil depletion while marginalizing most farmers to the point where they can have little hope of generating a significant surplus. In addition to land shortage, peasant production has suffered from low official produce prices and lack of other inputs. The northern half of the country, which has almost no estate production, has been relatively neglected in terms of government expenditure on transport and other forms of infrastructure. Many Malawian peasants have for generations turned to migrant labor as a means of coping with their poverty, but there have been far fewer opportunities for them in South Africa and Zimbabwe in recent years.

ACHIEVEMENTS

Although it is the poorest, most overcrowded country in the region, Malawi's response to the influx of Mozambican refugees was described by the U.S. Committee for Refugees as "no less than heroic."

Under pressure from the World Bank, the Malawian government has since 1981 modestly increased its incentives to the small landholders. Yet these reforms have been insufficient to overcome the continuing impoverishment of rural households, which has been aggravated in recent decades by a decline in migrant-labor remittances. The maldistribution of land in many areas remains a major challenge. On a more positive note, peace in Mozambique has reopened landlocked Malawi's access to the Indian Ocean ports of Beira and Nacala while reducing the burden of dealing with what once numbered 600,000 refugees. Communications infrastructure to the ports, damaged by war, is now being repaired.

Timeline: PAST

1500s
Malawi trading kingdoms develop

1859
Explorer David Livingstone arrives along Lake Malawi; missionaries follow

1891
The British protectorate of Nyasaland (present-day Malawi) is declared

1915
Reverend John Chilembwe and followers rise against settlers and are suppressed

1944
The Nyasaland African Congress, the first nationalist movement, is formed

1964
Independence, under the leadership of Hastings Banda

1967
Diplomatic ties are established with South Africa

1971
"Ngwazi" Hastings Kamuzu Banda becomes president-for-life

1990s
Bakili Muluzi is elected president, ending Banda's 30-year dictatorship; Banda dies in 1997

PRESENT

2000s
Political liberalization fails to improve the economy

Famine threatens millions of Malawians

Mozambique (Republic of Mozambique)

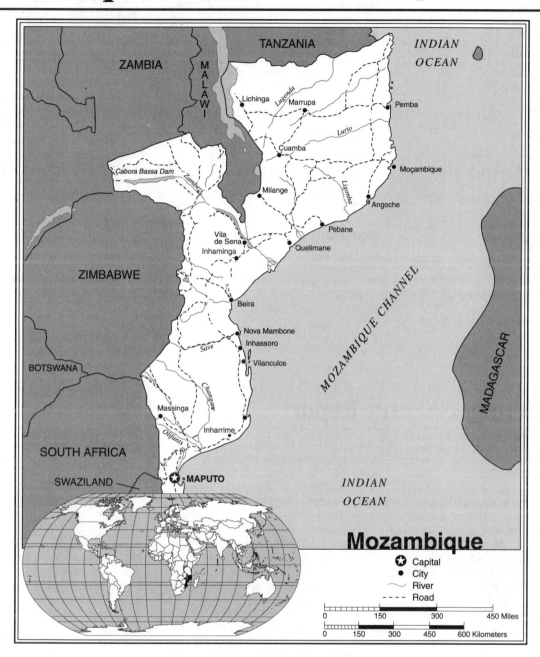

Mozambique Statistics

GEOGRAPHY

Area in Square Miles (Kilometers):
309,494 (801,590) (about twice the size of California)

Capital (Population): Maputo (1,134,000)

Environmental Concerns: civil war and drought have had adverse consequences on the environment; water pollution; desertification

Geographical Features: mostly coastal lowlands; uplands in center; high plateaus in northwest; mountains in west

Climate: tropical to subtropical

PEOPLE

Population

Total: 19,608,000

Annual Growth Rate: 1.13%

Rural/Urban Population Ratio: 61/39

Major Languages: Portuguese; indigenous dialects

Ethnic Makeup: nearly 100% indigenous groups

Religions: 50% indigenous beliefs; 30% Christian; 20% Muslim

Health

Life Expectancy at Birth: 38 years (male); 37 years (female)
Infant Mortality: 138.5/1,000 live births
Physicians Available: 1/131,991 people
HIV/AIDS Rate in Adults: 12.6%–16.4%

Education

Adult Literacy Rate: 40.1%
Compulsory (Ages): 7–14

COMMUNICATION

Telephones: 90,000 main lines
Daily Newspaper Circulation: 8/1,000 people
Televisions: 3.5/1,000 people
Internet Users: 22,500 (2000)

TRANSPORTATION

Highways in Miles (Kilometers): 17,886 (29,810)
Railroads in Miles (Kilometers): 1,879 (3,131)
Usable Airfields: 166
Motor Vehicles in Use: 89,000

GOVERNMENT

Type: republic
Independence Date: June 25, 1975 (from Portugal)
Head of State/Government: President Joaquim Alberto Chissano; Prime Minister Pascoal Mocumbi
Political Parties: Front for the Liberation of Mozambique (Frelimo); Mozambique National Resistance—Electoral Union (Renamo)
Suffrage: universal at 18

MILITARY

Military Expenditures (% of GDP): 1%
Current Disputes: none; cease-fire since 1992

ECONOMY

Currency ($ U.S. Equivalent): 23,467 meticais = $1
Per Capita Income/GDP: $900/$17.5 billion
GDP Growth Rate: 9.2%
Inflation Rate: 10%
Unemployment Rate: 21%

Labor Force by Occupation: 81% agriculture; 13% services; 6% industry
Population Below Poverty Line: 70%
Natural Resources: coal; titanium; natural gas; hydropower
Agriculture: cotton; cassava; cashews; sugarcane; tea; corn; rice; fruits; livestock
Industry: processed foods; textiles; beverages; chemicals; tobacco; cement; glass; asbestos; petroleum products
Exports: $746 million (primary partners South Africa, Zimbabwe, Spain)
Imports: $1.25 billion (primary partners South Africa, Portugal, United States)

SUGGESTED WEB SITES

```
http://www.
  mozambique.mz.eindex.htm
http://poptel.org.uk/mozambique-
  news
http://www.mozambiquenews.com
http://www.africaindex.
  africainfo.no/africaindex1/
  countries/mozambique.html
http://www.sas.upenn.edu/
  African_Studies/
  Country_Specific/Mozambique.html
```

Mozambique Country Report

Mozambique has made steady economic and political progress over the past decade in the face of grinding poverty, natural disasters, and the immense burden of overcoming the legacy of a bitter civil war. In June 2002, the ruling Mozambique Liberation Front (Frelimo) party chose Armando Guebuza, an independence-struggle veteran, as its presidential candidate for the 2004 presidential elections, after its long-serving incumbent, Joaquim Chissano, declined to run again.

DEVELOPMENT

To maintain minimum services and to recover from wartime and flood destruction, Mozambique relies on the commitment of its citizens and international assistance. Western churches have sent relief supplies, food aid, and vehicles.

In February 2000, the eyes of the world focused on devastating floods in Mozambique. In some places the country's two main rivers, the Limpopo and Save, expanded miles beyond their normal banks, engulfing hundreds of villages and destroying property and infrastructure. Many

Mozambicans were left homeless. The disaster was a serious setback for the nation, which had been making steady economic progress after three decades of civil war. Even before the floods, Mozambique (which remains one of the world's poorest countries) faced immense economic, political, and social challenges.

A 1992 cease-fire agreement, followed by the holding of multiparty elections in 1994 and 1999, has seemingly brought peace to the country. However, the opposition Mozambique National Resistance Movement (Renamo), which narrowly lost both polls, rejected the 1999 electoral outcome.

Since the 1994 elections Frelimo, which has governed the country since independence in 1975, has faced a large opposition bloc in Parliament from Renamo, its old civil-war adversary, which currently controls 117 out of 250 seats. A now-peaceful Renamo is, arguably, less of a challenge to the government than the dictates of international donors, upon whose funding it now depends. While the government must be concerned about improving the dismal living conditions endured by the majority of its people, the donors have insisted on fiscal austerity and a privatization program that has led to retrenchments as well as a

FREEDOM

While the status of political and civil liberties has improved, the government's overall human-rights record continues to be marred by security-force abuses (including extra-judicial killings, excessive use of force, torture, and arbitrary detention) and an ineffective and only nominally independent judicial system.

Frelimo originally came to power as a result of a liberation war. Between 1964 and 1974, it struggled against Portuguese colonial rule. At a cost of some 30,000 lives, Mozambique gained its independence in 1975 under Frelimo's leadership. Although the new nation was one of the least-developed countries in the world, many were optimistic that the lessons learned in the struggle could be applied to the task of building a dynamic new society based on Marxist-Leninist principles.

Unfortunately, hopes for any sort of postindependence progress were quickly dashed by Renamo, which was originally established as a counterrevolutionary fifth column by Rhodesia's (Zimbabwe) Central Intelligence Organization. More than 1 million people died due to the rebellion,

(United Nations photo by Kate Truscott)

The drain on natural resources resulting from civil war, the displacement of approximately one fifth of the population, persistent drought, and, most recently, devastating floods, have led to the need for Mozambique to import food to stave off famine.

a large proportion murdered in cold blood by Renamo forces. It is further estimated that, out of a total population of 17 million, some 5 million people were internally displaced, and about 2 million others fled to neighboring states. No African nation paid a higher price in its resistance to white supremacy.

HEALTH/WELFARE

Civil strife, widespread Renamo attacks on health units, and food shortages drastically curtailed health-care goals and led to Mozambique's astronomical infant mortality rate.

Although some parts of Mozambique were occupied by the Portuguese for more than 400 years, most of the country came under colonial control only in the early twentieth century. The territory was developed as a dependency of neighboring colonial economies rather than that of Portugal itself. Mozambican ports were linked by rail to South Africa and the interior colonies of British Central Africa—that is, modern Malawi, Zambia, and Zimbabwe. In the southern provinces, most men, and many women, spent time as migrant labor-

ers in South Africa. The majority of the males worked in the gold mines.

Most of northern Mozambique was granted to three predominantly British concessions companies, whose abusive policies led many to flee the colony. For decades, the colonial state and many local enterprises also relied on forced labor. After World War II, new demands were put on Mozambicans by a growing influx of Portuguese settlers, whose numbers swelled during the 1960s, from 90,000 to more than 200,000. Meanwhile, even by the dismal standards of European colonialism in Africa, there continued to be a notable lack of concern for human development. At independence, 93 percent of the African population in Mozambique were illiterate. Furthermore, most of those who had acquired literacy or other skills had done so despite the Portuguese presence.

Although a welcome event in itself, the sudden nature of the Portuguese empire's collapse contributed to the destabilization of postindependence Mozambique. Because Frelimo had succeeded in establishing itself as a unified nationalist front, Mozambique was spared an immediate descent into civil conflict, such as that which engulfed Angola, Portugal's other major African possession. However, the economy

was already bankrupt due to the Portuguese policy of running Mozambique on a nonconvertible local currency. The rapid transition to independence compounded this problem by encouraging the sudden exodus of almost all the Portuguese settlers.

ACHIEVEMENTS

Between 1975 and 1980, the illiteracy rate in Mozambique declined from 93% to 72% while classroom attendance more than doubled. Progress slowed during the 1980s due to civil war. Today, the overall literacy rate stands at about 40%.

Perhaps even more costly to Mozambique in the long term was the polarization between Frelimo and African supporters of the former regime, who included about 100,000 who had been active in its security forces. The rapid Portuguese withdrawal was not conducive to the difficult task of reconciliation. While Frelimo did not subject the "compromised ones" to bloody reprisals, their rights were circumscribed, and many were sent, along with prostitutes and other "antisocial" elements, to "reeducation camps." While the historically pro-Portuguese stance of the local Catholic hierarchy would have complicated its relations with the new state under any circumstance, Frelimo's Marxist rejection of religion initially alienated it from broader numbers of believers.

A TROUBLED INDEPENDENCE

Frelimo assumed power without the benefit or burden of a strong sense of administrative continuity. While it had begun to create alternative social structures in its "liberated zones" during the anticolonial struggle, these areas had encompassed only a small percentage of Mozambique's population and infrastructure. But Frelimo was initially able to fill the vacuum and launch aggressive development efforts. Health care and education were expanded, worker committees successfully ran many of the enterprises abandoned by the settlers, and communal villages coordinated rural development. However, efforts to promote agricultural collectivization as the foundation of a command economy generally led to peasant resistance and economic failure. Frelimo's ability to adapt and implement many of its programs under trying conditions was due largely to its disciplined mass base (the party's 1990 membership stood at about 200,000).

No sooner had Mozambique begun to stabilize itself from the immediate dislocations of its decolonization process than it became embroiled in the Rhodesian conflict. Mozambique was the only neighboring state to impose fully the "mandatory" United Nations economic sanctions against Rhodesia (present-day Zimbabwe). Between 1976 and 1980, this decision led to the direct loss of half a billion dollars in rail and port revenues. Furthermore, Frelimo's decision to provide bases for the fighters of the Patriotic Front led to a state of undeclared war with Rhodesia as well as its Re-

Timeline: PAST

1497
Portuguese explorers land in Mozambique

1820s
The Northern Nguni of Shosagaane invade southern Mozambique, establishing the Gaza kingdom

1962
The Frelimo liberation movement is officially launched

1975
The liberation struggle is successful

1980s
Increased Renamo attacks on civilian and military targets; President Samora Machel is killed in a mysterious airplane crash; Joaquim Chissano becomes president

1990s
Renamo agrees to end fighting, participate in multiparty elections

PRESENT

2000s
Floods lead to enormous losses in life, property, and infrastructure

Chissano decides not to run again; elections are scheduled for 2004

Unfortunately, the fall of Rhodesia did not bring an end to externally sponsored destabilization. Renamo had the support of South Africa. By continuing Renamo's campaign of destabilization, the South African regime gained leverage over its hostile neighbors, for the continued closure of Mozambique's ports meant that most of their traffic had to pass through South Africa. In 1984, Mozambique signed a nonag-

gression pact with South Africa, which should have put an end to the latter's support of Renamo. However, captured documents and other evidence indicate that official South African support for Renamo continued at least until 1989, while South African supplies were still reaching the rebels under mysterious circumstances. In response, Zimbabwe, and to a lesser extent Malawi and Tanzania, contributed troops to assist in the defense of Mozambique.

In its 1989 Congress, Frelimo formally abandoned its commitment to the primacy of Marxist-Leninist ideology and opened the door to further political and economic reforms. Multipartyism was formally embraced in 1991. With the help of the Catholic Church and international mediators, the government opened talks with Renamo. In October 1992, Renamo's leader, Alfonso Dlakama, signed a peace accord that called for UN–supervised elections. The cease-fire finally came into actual effect in the early months of 1993, by which time the UN personnel on the ground reported that some 3 million Mozambicans were suffering from famine.

Besides their mutual distrust, reconciliation between Renamo and Frelimo was troubled by their leaderships' inability to control their armed supporters. With neither movement able to pay its troops, apolitical banditry by former fighters for both sides increased. International financial and military support, mobilized through the United Nations, was inadequate to meet this challenge. In June and July 1994, a number of UN personnel, along with foreign-aid workers, were seized as hostages. The near-complete collapse of the country, however, has so far encouraged Mozambique's political leaders to sustain the peace drive.

Namibia (Republic of Namibia)

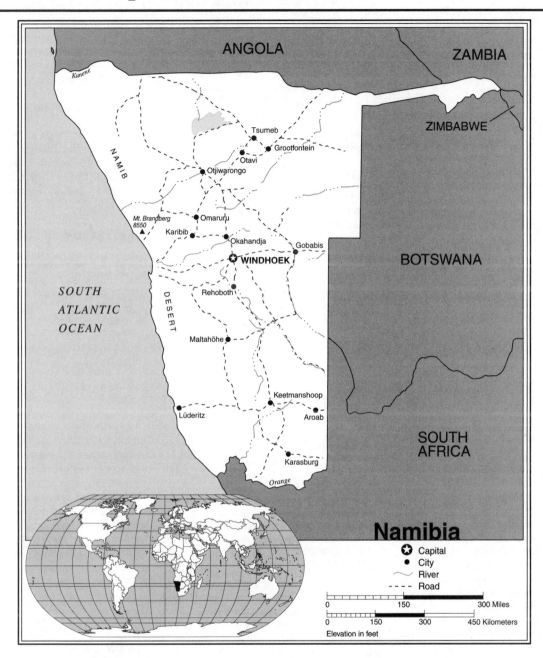

Namibia Statistics

GEOGRAPHY

Area in Square Miles (Kilometers):
318,261 (824,292) (about half the size of Alaska)

Capital (Population): Windhoek (216,000)

Environmental Concerns: very limited natural freshwater resources; desertification

Geographical Features: mostly high plateau; desert along coast and in east

Climate: desert

PEOPLE

Population

Total: 1,821,000
Annual Growth Rate: 1.19%

Rural/Urban Population Ratio: 70/30

Major Languages: English; Ovambo; Kavango; Nama/Damara; Herero; Khoisan; German; Afrikaans

Ethnic Makeup: 50% Ovambo; 9% Kavango; 7% Herero; 7% Damara; 27% others

Religions: 80%–90% Christian; 10%–20% indigenous beliefs

Health

Life Expectancy at Birth: 44 years (male); 41 years (female)
Infant Mortality: 72.4/1,000 live births
Physicians Available: 1/4,594 people
HIV/AIDS Rate in Adults: 19.54%

Education

Adult Literacy Rate: 38%
Compulsory (Ages): 6–16

COMMUNICATION

Telephones: 111,000 main lines
Daily Newspaper Circulation: 27/1,000 people
Televisions: 28/1,000 people
Internet Users: 30,000 (2001)

TRANSPORTATION

Highways in Miles (Kilometers): 39,220 (63,258)
Railroads in Miles (Kilometers): 1,429 (2,382)
Usable Airfields: 137
Motor Vehicles in Use: 129,000

GOVERNMENT

Type: republic
Independence Date: March 21, 1990 (from South African mandate)
Head of State/Government: President Samuel Shafishuna Nujoma is both head of state and head of government
Political Parties: South West Africa People's Organization; Democratic Turnhalle Alliance of Namibia; United Democratic Front; Monitor Action Group; others
Suffrage: universal at 18

MILITARY

Military Expenditures (% of GDP): 2.6%
Current Disputes: border dispute over at least one island in the Linyanti River

ECONOMY

Currency ($ U.S. equivalent): 8.33 dollars = $1
Per Capita Income/GDP: $4,500/$8.1 billion
GDP Growth Rate: 4%
Inflation Rate: 8.8%
Unemployment Rate: 30%–40%
Labor Force by Occupation: 47% agriculture; 33% services; 20% industry
Natural Resources: diamonds; gold; tin; copper; lead; zinc; uranium; salt; cadmium; lithium; natural gas; possible oil, coal reserves; fish; vanadium; hydropower
Agriculture: millet; sorghum; peanuts; livestock; fish
Industry: meat packing; dairy products; fish processing; mining
Exports: $1.58 billion (primary partners United Kingdom, South Africa, Spain)
Imports: $1.7 billion (primary partners South Africa, United States, Germany)

SUGGESTED WEB SITES

http://www.govnet.gov.na/
   intro.htm
http://www.sas.upenn.edu/
   African_Studies/
   Country_Specific/Namibia.html
http://www.republicofnamibia.com

Namibia Country Report

The collapse of the UNITA rebel movement in neighboring Angola has brought to a close what had been a perennial source of insecurity along Namibia's northern border. In August 1999, Namibians were shocked when a small group of self-proclaimed separatists launched an armed attack on the town of Katima Mulilo. (Ultimately defeated, the separatists sought refuge in Botswana; most were involuntarily repatriated back to Namibia in 2001–2002.) Subsequent attacks along the border with war-torn Angola and the involvement of Namibian forces in the war in the Democratic Republic of the Congo (Congo-Kinshasa, or D.R.C.) further shook the country's peaceful international image. The incidents, however, did not fundamentally disturb Namibia's steady progress since its liberation from South African rule in 1990.

With the country now fully at peace for the first time in 30 years, the Namibian government is free to concentrate on domestic development. The material resources that went into the war effort can now be put toward peaceful use.

One area earmarked for urgent attention is land reform. Local observers were little surprised in September 2002 when President Samuel Nujoma voiced his support for

DEVELOPMENT

The Nujoma government has instituted English as the medium of instruction in all schools. (Before independence, English was discouraged for African schoolchildren, a means of controlling their access to skills necessary to compete in the modern world.) This effort requires new curricula and textbooks for the entire country.

the Zimbabwean government's seizure of white-held farms. In fact, Nujomo's own ruling party, the South West Africa People's Organization (SWAPO), has committed itself to redistributing to the African farmers half of the land currently owned by white absentee landlords. It is unlikely, however, that land reform in Namibia will be accompanied by the lawless violence that has marred recent developments in Zimbabwe.

Namibia became independent in 1990, after a long liberation struggle. Its transition from the continent's last colony to a developing nation-state marked the end of a century of often brutal colonization, first by Germany and later South Africa. The German colonial period (1884–1917) was marked by the annihilation of more than 60 percent of the African population in the

southern two thirds of the country, during the uprising of 1904–1907. The South African period (1917–1990) saw the imposition of apartheid as well as a bitter 26-year war for independence between the South African Army (SADF) and SWAPO. During that war, countless civilians, especially in the northern areas of the country, were harassed, detained, and abused by South African–created death squads, such as the *Koevoet* (the Afrikaans word for "crowbar").

FREEDOM

In line with Namibia's liberal Constitution, human rights are generally respected. There are some problem areas: Demonstrations that do not have prior police approval are banned. The president and other high officials have repeatedly attacked the independent press. There has not been a full accounting of missing detainees who were in SWAPO camps before independence. Security forces have admitted to cases of extra-judicial killing along the Angolan border.

Namibia's final liberation was the result of South African military misadventures and U.S.–Soviet cooperation in reducing tensions in the region. In 1987, as it had done many times before, South Africa in-

(United Nations photo by J. Isaac)

Developing agricultural production in Namibia is key to the country's economic future. The international economic sanctions that applied before independence have been lifted, and Namibia is now free to enter the potentially profitable markets of Europe and North America. This man working in a cornfield near Grootfontein is part of Namibia's crucial agricultural economy.

vaded Angola to assist Jonas Savimbi's UNITA movement. Its objective was Cuito Cuanavale, a small town in southeastern Angola where the Luanda government had set up an air-defense installation to keep South African aircraft from supplying UNITA troops. The SADF met with fierce resistance from the Angolan Army and eventually committed thousands of its own troops to the battle. In addition, black Namibian soldiers were recruited and given UNITA uniforms to fight on the side of the SADF. Many of these proxy UNITA troops later mutinied because of their poor treatment at the hands of white South African soldiers.

South Africa failed to capture Cuito Cuanavale, and its forces were eventually surrounded. Faced with military disaster, the Pretoria government bowed to decades of international pressure and agreed to withdraw from its illegal occupation of Namibia. In return, Angola and its ally Cuba agreed to send home troops sent by Havana in 1974 after South Africa invaded Angola for the first time. Key brokers of the cease-fire, negotiations, and implementation of this

agreement were the United States and the Soviet Union. This was the first instance of their post–Cold War cooperation.

HEALTH/WELFARE

The social-service delivery system of Namibia must be rebuilt to eliminate the structural inequities of apartheid. Health care for the black majority, especially those in remote rural areas, will require significant improvements. Public-health programs for blacks, nonexistent prior to independence, must be created.

A plebiscite was held in Namibia in 1989. Under United Nations supervision, more than 97 percent of eligible voters cast their ballots—a remarkable achievement given the vast distances that many had to travel to reach polling stations. SWAPO emerged as the clear winner, with 57 percent of the votes cast. The party's share of the vote increased to 73 percent in the subsequent 1995 elections; support for its main political rival, the Democratic Turnhalle Alliance, dropped to 15 percent.

CHALLENGES AND PROSPECTS

Namibia is a sparsely populated land. More than half of its more than 1.8 million residents live in the northern region known as Ovamboland. Rich in minerals, Namibia is a major producer of diamonds, uranium, copper, silver, tin, and lithium. A large gold mine recently began production, and the end of hostilities has opened up northern parts of the country to further mineral explorations.

ACHIEVEMENTS

The government of President Sam Nujoma has received high praise for its efforts at racial and political reconciliation after a bitter 26-year war for independence. Nujoma has led these efforts and has impressed many observers with his exceptional political and consensus-building skills.

Much of Namibia is arid. Until recently, pastoral farming was the primary agricultural activity, with beef, mutton, and goat meat the main products. Independence brought an end to international sanctions applied when South Africa ruled the country, giving Namibian agricultural goods access to the world market. Although some new investment has been attracted to the relatively well-watered but historically neglected northern border regions, most of Namibia's rural majority are barely able to eke out a living, even in nondrought years.

Despite the economic promise, the fledgling government of Namibia faces severe problems. It inherits an economy structurally perverted by apartheid to favor the tiny white minority. With a glaring division between fabulously wealthy whites and the oppressively poor black majority, the government is faced with the daunting problem of promoting economic development while encouraging the redistribution of wealth. Apartheid ensured that managerial positions were filled by whites, leaving a dearth of qualified and experienced nonwhite executives in the country. This past pattern of discrimination has contributed to high levels of black unemployment today.

The demobilization of 53,000 former SWAPO and South African combatants and the return of 44,000 exiles aggravated this problem. A few former soldiers—notably the Botsotsos, made up of former Koevoet members—turned to organized crime. Having already inherited a civil service bloated by too many white sinecures, the SWAPO administration resisted the temptation of trying to hire its way out of the problem. In 1991–1992, several thousand ex-combatants received vocational training in Development Brigades, modeled after similar initiatives in Botswana and Zimbabwe, but inadequate funding and preparation limited the program's success.

Another major problem lies in Namibia's economic dependence on South Africa. Before independence, Namibia had been developed as a captive market for South African goods, while its resources had been depleted by overexploitation on the part of South African firms. In 1990, all rail and most road links between Namibia and the rest of the world ran through South Africa. But South Africa's March 1, 1994, return of Walvis Bay, Namibia's only port,

has greatly reduced this dependence. The port has now been declared a free-trade area. Namibia has also been linked to South Africa through a Common Monetary Area. In 1994, a new Namibian dollar was introduced, replacing the South African rand. But, at least for the time being, the currency's value remains tied to the rand.

The Nujoma government has taken a hard look at these and other economic problems and embarked on programs to solve them. SWAPO surprised everyone during the election campaign by modifying its previously strident socialist rhetoric and calling for a market-oriented economy. Since taking power, it has joined the International Monetary Fund and proposed a code for foreign investors that includes protection against undue nationalizations. Since independence, the Ministry of Finance has pursued conservative policies, which have calmed the country's largely white business community but have been criticized as insufficient to transform the economy for the greater benefit of the impoverished masses.

The government recognizes the need to attract significant foreign investment to overcome the colonial legacy of underdevelopment. In 1993, a generous package of manufacturing incentives was introduced by the Ministry of Trade and Industry. In the same year, the Namibia National Development Corporation was established to channel public investment into the economy. It is too early to assess the success of these initiatives.

NAMIBIA'S FISHING INDUSTRY

Namibia's fishing sector has made an impressive recovery after years of decline. The country's coastal waters had long supported exceptionally high concentrations of sea life due to the upwelling of nutrients by the cold offshore current. But in the years before independence, overfishing, mostly by foreign vessels, nearly wiped out many species. Since then, the government has established a 200-nautical-mile Exclusive Economic Zone along Namibia's coast and passed a Sea Fisheries Act designed to promote the conservation and

controlled exploitation of the country's marine resources. These measures have been backed up by effective monitoring on the part of the new Ministry of Fisheries and Marine Resources and the creation of a National Fisheries Research and Information Centre. A rapid recovery in fish stocks has led to an annual growth of 35 percent in the sector's value.

Timeline: PAST

1884–1885
Germany is given rights to colonize Namibia at the Conference of Berlin

1904–1907
Herero, Nama, and Damara rebellions against German rule

1966
The UN General Assembly revokes a 1920 South African mandate; SWAPO begins war for independence

1968
Bantustans, or "homelands," are created by South Africa

1971
A massive strike paralyzes the economy

1978
An internal government is formed by South Africa

1980s
Defeat at Cuito Cuanavale leads to a South African agreement to withdraw from Namibia; SWAPO wins UN-supervised elections; a new Constitution is approved

1990s
More than 1,000 refugees flee to Botswana from Namibia's Caprivi regions

PRESENT

2000s
The International Court of Justice awards the disputed Kasikili/Sedudu Island to Botswana

Namibian involvement in the Congo-Kinshasa war

Unrest continues

South Africa (Republic of South Africa)

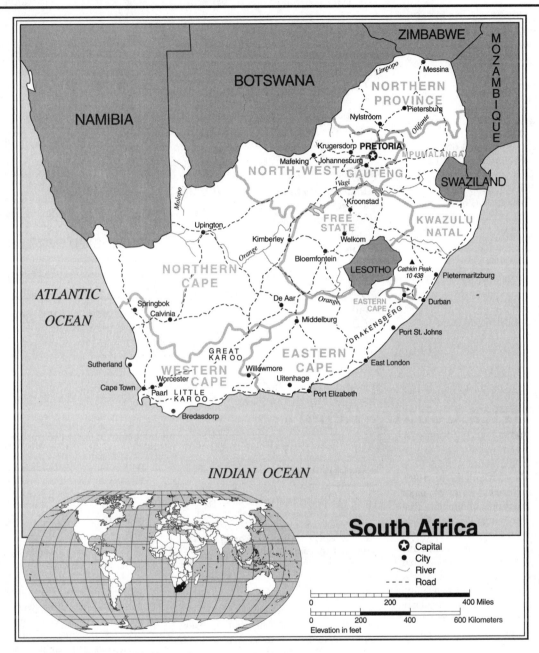

South Africa Statistics*

GEOGRAPHY

Area in Square Miles (Kilometers):
437,872 (1,222,480) (about twice the size of Texas)

Capital (Population): Pretoria (administrative) (1,590,000); Cape Town (legislative) (2,993,000); Bloemfontein (judicial) (364,000)

Environmental Concerns: water and air pollution; acid rain; soil erosion; desertification; lack of fresh water

Geographical Features: vast interior plateau rimmed by rugged hills and a narrow coastal plain

Climate: mostly semiarid; subtropical along the east coast

PEOPLE

Population

Total: 43,648,000

Annual Growth Rate: 0.02%

Rural/Urban Population Ratio: 50/50

Major Languages: Afrikaans; English; Ndebele; Pedi; Sotho; Swati; Tsonga; Tswana; Venda; Xhosa; Zulu

Ethnic Makeup: 75% black; 14% white; 9% Colored; 2% Indian

Religions: 68% Christian; 28.5% indigenous beliefs and animist; 2% Muslim; 1.5% Hindu

Health

Life Expectancy at Birth: 45 years (male); 46 years (female)
Infant Mortality: 61.7/1,000 live births
Physicians Available: 1/1,529 people
HIV/AIDS Rate in Adults: 19.94%

Education

Adult Literacy Rate: 85%
Compulsory (Ages): 7–16

COMMUNICATION

Telephones: 5,000,000+ main lines
Daily Newspaper Circulation: 29/1,000 people
Televisions: 128/1,000 people
Internet Users: 2,400,000 (2001)

TRANSPORTATION

Highways in Miles (Kilometers): 215,157 (358,596)
Railroads in Miles (Kilometers): 12,859 (21,431)
Usable Airfields: 740
Motor Vehicles in Use: 6,000,000

GOVERNMENT

Type: republic
Independence Date: May 31, 1910 (from the United Kingdom)
Head of State/Government: President Thabo Mbeki is both head of state and head of government
Political Parties: African National Congress; New National Party; Inkatha Freedom Party; African Christian Democratic Party; Freedom Front; Pan-Africanist Congress; others
Suffrage: universal at 18

MILITARY

Military Expenditures (% of GDP): 1.5%
Current Disputes: civil unrest; territorial issues with Swaziland

ECONOMY

Currency ($ U.S. Equivalent): 8.15 rands = $1
Per Capita Income/GDP: $9,400/$412 billion
GDP Growth Rate: 2.6%

Inflation Rate: 5.8%
Unemployment Rate: 37%
Labor Force by Occupation: 45% services; 30% agriculture; 25% industry
Population Below Poverty Line: 50%
Natural Resources: gold; chromium; coal; antimony; iron ore; manganese; nickel; phosphates; tin; diamonds; others
Agriculture: corn; wheat; sugarcane; fruits; vegetables; livestock products
Industry: mining; automobile assembly; metalworking; machinery; textiles; iron and steel; chemicals; fertilizer; foodstuffs
Exports: $32.3 billion (primary partners Europe, United States, Japan)
Imports: $28.1 billion (primary partners Europe, United States, Saudi Arabia)

SUGGESTED WEB SITES

http://www.gov.za
http://www.southafrica.co.za

*Note: When separated, figures for blacks and whites vary greatly.

South Africa Country Report

Since taking over the South African presidency from Nelson Mandela in 1999, Thabo Mbeki has faced a series of major domestic and international challenges. Rising crime, the burgeoning HIV/AIDS pandemic, and the falling value of the country's currency (the rand) against the euro and the U.S. dollar have complicated the immense task of post-apartheid transformation. The economy remains healthy, with the tourist industry doing particularly well despite the international fallout from the 9/11 terrorist attacks in the United States. However, despite the growth of the black middle class, the distribution of wealth in South Africa remains highly skewed, largely reflecting the racial divisions of the old apartheid order. The Mbeki government's basic commitment to market-driven growth, coupled with affirmative-action policies for the previously disadvantaged, has been challenged from both the left and the right. The labor movement, led by the Congress of South African Trade Unions (COSATU), held a general strike against the government's privatization and general economic policies during 2002, notwithstanding COSATU's alliance with Mbeki's political party, the African National Congress (ANC).

DEVELOPMENT

The Government of National Unity's major priority was the implementation of the comprehensive Reconstruction and Development Plan. A major aspect of the plan was a government commitment to build 1 million low-cost houses each year for 5 years.

In December 2000 local-government elections, the ANC took most of the 237 local councils (59 percent), but the Democratic Alliance—created five months previously from a merger of the Democratic Party (DP), the new National Party (NP), and the Federal Alliance—captured nearly a quarter of the votes. The Inkatha Freedom Party (IFP) won 9 percent. By 2002, however, many of the members from the NP had abandoned the Democratic Alliance in favor of an alliance with the ANC. The ANC also benefited in 2002 from the government being cleared of allegations of official corruption surrounding a large 1999 arms deal.

In the area of HIV/AIDS, Mbeki, along with many of his key associates, courted controversy by expressing skepticism about commonly accepted scientific beliefs about the causal link between the HIV virus and full-blown AIDS, as well as the value of antiretroviral therapies. In December 2001, the country's High Court ruled that pregnant women must be given antiretroviral drugs to help prevent transmission of HIV to their babies. In July 2002, the Constitutional Court ordered the government to provide key anti-AIDS drugs at all public hospitals. The government had argued that such drugs were too costly; but earlier, in April 2001, a group of 39 multinational pharmaceutical companies had withdrawn its legal battle to stop South Africa from importing generic (cheaper) AIDS drugs. The decision to drop the landmark court case was hailed as a major victory for the world's poorest countries.

In hosting the inaugural meeting of the African Union and the World Summit of Sustainable Development in 2002, after the 2001 United Nations Race Conference, South Africa enhanced its diplomatic standing while proving to be an ideal setting for such gatherings. Mbeki's leading efforts to build a "New Economic Partnership for African Development" (NEPAD) were, however, undermined by a lack of consensus with other key African leaders on its blueprint, as well as European and American unease at his inability or unwillingness to take a strong stand against the ruinous policies of Robert Mugabe in neighboring Zimbabwe.

Mbeki's predecessor, Mandela, remains as a unifying father figure to most South Africans. His retirement ended an extraordinary period in which South Africa struggled to come to terms with its new status as a nonracial democracy, after a long history of white-minority rule.

In April 1994, millions of South Africans turned out to vote in their country's first nonracial elections. Most waited patiently for hours to cast their ballots for the first time. The result was a landslide victory for the African National Congress, which, under the new interim Constitution, would nonetheless cooperate with two of its long-standing rivals, the National Party and the Inkatha Freedom Party, in a "Government of National Unity" (GNU). On May 10, 1994, the ANC's leader was sworn in as South Africa's first black president. Despite the history of often violent animosity between its components, the GNU survived for two years, facilitating national reconciliation. In July 1996, the NP pulled out of the GNU, giving the ANC a freer hand to pursue its ambitious but largely unrealized program of "Reconstruction and Development."

FREEDOM

The government is committed to upholding human rights, which are generally respected. Members of the security forces have committed abuses, however, including torture and excessive use of force. Action has been taken to punish some of those involved. The Truth and Reconciliation Commission, created to investigate apartheid-era human-rights abuses, completed its investigations in 1998. It made recommendations for reparations for victims and granted amnesty for full disclosure of politically motivated crimes.

South Africa has decisively turned away from its long, tragic history of racism. For nearly 3 1/2 centuries, the territory's white minority expanded and entrenched its racial hegemony over the nonwhite majority. After 1948, successive NP governments consolidated white supremacy into a governing system known as *apartheid* ("separatehood"). But in a dramatic political about-face, the NP government, under the new leadership of F. W. de Klerk, committed itself in 1990 to a negotiated end to apartheid. Political restrictions inside the country were relaxed through the unbanning of anti-apartheid resistance

organizations, most notably the ANC, the Pan-Africanist Congress (PAC), and the South African Communist Party (SACP). Thereafter, three years of on-again, off-again negotiations, incorporating virtually all sections of public opinion, resulted in a 1993 consensus in favor of a five-year, nonracial, interim Constitution.

Notwithstanding its remarkable political progress in recent years, South Africa remains a deeply divided country. In general, whites continue to enjoy relatively affluent, comfortable lives, while the vast majority of nonwhites survive in a state of impoverished deprivation. The boundary between these two worlds remains deep. Under the pre-1990 apartheid system, nonwhites were legally divided as members of three officially subordinate race classifications: "Bantu" (black Africans), "Coloureds" (people of mixed race), or "Asians." (*Note:* Many members of these three groups prefer the common label of "black," which the government now commonly uses in place of Bantu as an exclusive term for black Africans, hereafter referred to in this text as *blacks*.)

THE ROOTS OF APARTHEID

White supremacy in South Africa began with the Dutch settlement at Cape Town in 1652. For 1 1/2 centuries, the domestic economy of the Dutch Cape Colony, which gradually expanded to include the southern third of modern South Africa, rested on a foundation of slavery and servitude. Much like the American South before the Civil War, Cape Colonial society was racially divided between free white settlers and nonwhite slaves and servants. Most of the slaves were Africans imported from outside the local region, although a minority were taken from Asia. The local blacks, who spoke various Khiosan languages, were not enslaved. However, they were robbed by the Europeans of their land and herds. Many were also killed either by European bullets or diseases. As a result, most of the Cape's Khiosan were reduced to a status of servitude. Gradually, the servant and slave populations, with a considerable admixture of European blood, merged to form the core of the so-called Coloured group.

At the beginning of the nineteenth century, the Cape Colony reverted to British control. In the 1830s, the British abolished slavery and extended legal rights to servants. But, as with the American South, emancipation did not end racial barriers to the political and economic advancement of nonwhites. Nonetheless, even the limited reforms that were introduced upset many of the white "Cape Dutch" (or "Boers"), whose society was evolving its own "Afri-

HEALTH/WELFARE

Public-health and educational facilities are being desegregated. In its first 100 days, the new government introduced free child health-care and AIDS-prevention programs. A 10-year program of schooling is to be free to all children. Students have returned to school in large numbers. Crime remains a major problem, with a recent study concluding that South Africa is the most murderous country in the world. HIV/AIDS has reached pandemic proportions in South Africa.

kaner" identity. (Today, some 60 percent of the whites and 90 percent of the Coloureds in South Africa speak the Dutch-derived Afrikaans language.) In the mid-nineteenth century, thousands of Afrikaners, accompanied by their Coloured clients, escaped British rule by migrating into the interior. They established two independent republics, the Transvaal and the Orange Free State, whose Constitutions recognized only whites as having any civil rights.

The Afrikaners, and the British who followed them, did not settle an empty land. Then, as now, most of the people living in the area beyond the borders of the old Dutch Cape Colony were blacks who spoke languages linguistically classified as Bantu. While there are nine officially recognized Bantu languages in South Africa, all but two (Tsonga and Venda) belong to either the Sotho-Tswana (Pedi, Sotho, Tswana) or Nguni (Ndebele, Swati, Xhosa, and Zulu) subgroupings of mutually intelligible dialects.

Throughout the 1700s and 1800s, the indigenous populations of the interior and the eastern coast offered strong resistance to the white invaders. Unlike the Khiosan of the Cape, these communities were able to preserve their ethnolinguistic identities. However, the settlers eventually robbed them of most of their land as well as their independence. Black subjugation served the economic interests of white farmers, and later industrialists, who were able to coerce the conquered communities into providing cheap and forced labor. After 1860, many Asians, mostly from what was then British-ruled India, were also brought into South Africa to work for next to nothing on sugar plantations. As with the blacks and Coloureds, the Asians were denied civil rights.

The lines of racial stratification were already well entrenched at the turn of the twentieth century, when the British waged a war of conquest against the Afrikaner republics. During this South African, or Boer, War, tens of thousands of Afrikaners, blacks, and Coloureds died while interned

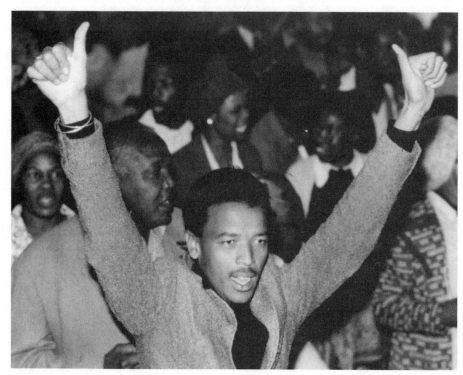

(United Nations photo)
Resistance groups gained international recognition in their struggle against the South African apartheid regime.

in British concentration camps. The camps helped to defeat the Afrikaner resistance but left bitter divisions between the resistance and pro-British English-speaking whites. However, it was the nonwhites who were the war's greatest losers. A compromise peace between the Afrikaners and the British Empire paved the way for the emergence, in 1910, of a self-governing "Union of South Africa," made up of the former British colonies and Afrikaner republics. In this new state, political power remained in the hands of the white minority.

"GRAND APARTHEID"

In 1948, the Afrikaner-dominated Nationalist Party was voted into office by the white electorate on a platform promising apartheid. Under this system, existing patterns of segregation were reinforced by a vast array of new laws. "Pass laws," which had long limited the movement of blacks in many areas, were extended to apply throughout the country. Black men and women were required to carry "passbooks" at all times to prove their right to be residing in a particular area. Under the Group Areas Act, more than 80 percent of South Africa was reserved for whites (who now make up no more than 14 percent of the population). In this area, blacks were confined to townships or white-owned farms, where, until recently, they were considered to be temporary residents. If they lacked a properly registered job, they were subject to deportation to one of the 10 "homelands."

ACHIEVEMENTS

 With the end of international cultural and sporting boycotts, South African artists and athletes have become increasingly prominent. In 1993, Nelson Mandela and F. W. de Klerk were awarded the Nobel Peace Prize, following in the footsteps of their countrymen Albert Luthuli and Desmond Tutu.

Under apartheid, the homelands—poor, noncontiguous rural territories that together account for less than 13 percent of South Africa's land—were the designated "nations" of South Africa's blacks, who made up more than 70 percent of the population. Each black was assigned membership in a particular homeland, in accordance with ethnolinguistic criteria invented by the white government. Thus, in apartheid theory, there was no majority in South Africa but, rather, a single white nation—which in reality remained divided among speakers of Afrikaans, English, and other languages, and 10 separate black nations. The Coloureds and the Asians were consigned a never clearly defined intermediate position as powerless communities associated with, but segregated from, white South Africa. The apartheid "ideal" was that each black homeland would eventually become "independent," leaving white South Africa without the "burden" of a black majority. Of course, black "immigrants" could still work for the "white economy," which would remain reliant on

black labor. To assure that racial stratification was maintained at the workplace, a system of job classification was created that reserved the best positions for whites, certain middle-level jobs for Asians and Coloureds, and unskilled labor for blacks.

Before 1990, the NP ruthlessly pursued its ultimate goal of legislating away South Africa's black majority. Four homelands—Bophutatswana, Ciskei, Transkei, and Venda—were declared independent. The 9 million blacks who were assigned as citizens of these pseudo-states (which were not recognized by any outside country) did not appear in the 1989 South African Census, even though most lived outside of the homelands. Indeed, despite generations of forced removals and influx control, today there is not a single magistrate's district (the equivalent of a U.S. county) that has a white majority.

While for whites apartheid was an ideology of mass delusion, for blacks it meant continuous suffering. In the 1970s alone, some 3.5 million blacks were forcibly relocated because they were living in "black spots" within white areas. Many more at some point in their lives fell victim to the pass laws. Within the townships and squatter camps that ringed the white cities, families survived from day to day not knowing when the police might burst into their homes to discover that their passbooks were not in order.

Under apartheid, blacks were as much divided by their residential status as by their assigned ethnicity. In a relative sense, the most privileged were those who had established their right to reside legally within a township like Soweto. Township dwellers had the advantage of being able to live with their families and seek work in a nearby white urban center. Many of their coworkers lived much farther away, in the peri-urban areas of the homelands. Some in this less fortunate category spent as much as one third of their lives on Putco buses, traveling to and from their places of employment. Still, the peri-urban homeland workers were in many ways better off than their colleagues who were confined to crowded worker hostels for months at a time while their families remained in distant rural homelands. There were also millions of female domestic workers who generally earned next to nothing while living away from their families in the servant quarters of white households. Many of these conditions still persist in South Africa.

Further down the black social ladder were those living in the illegal squatter camps that existed outside the urban areas. Without secure homes or steady jobs, the squatters were frequent victims of nighttime police raids. When caught, they were generally transported back to their homelands, from whence they would usually try once more to escape. The relaxation of influx-control regulations eased the tribulations of many squatters, but their lives remained insecure.

Yet even the violent destruction of squatter settlements by the state did not stem their explosive growth. For many blacks, living without permanent employment in a cardboard house was preferable to the hardships of the rural homelands. Nearly half of all blacks live in these areas today. Unemployment there tops 80 percent, and agricultural production is limited by marginal, overcrowded environments.

Economic changes in the 1970s and 1980s tended further to accentuate the importance of these residential patterns. Although their wages on average were only a fraction of those enjoyed by whites, many township dwellers saw their wages rise over several decades, partially due to their own success in organizing strong labor federations. At the same time, however, life in the homelands became more desperate as their populations mushroomed.

Apartheid was a totalitarian system. Before 1994, an array of security legislation gave the state vast powers over individual citizens, even in the absence of a state of emergency, such as existed throughout much of the country between 1985 and 1990. Control was more subtly exercised through the schools and other public institutions. An important element of apartheid was "Bantu Education." Beyond being

segregated and unequal, black educational curricula were specifically designed to assure underachievement, by preparing most students for only semiskilled and unskilled occupations. The schools were also divided by language and ethnicity. A student who was classified as Zulu was taught in the Zulu language to be loyal to the Zulu nation, while his or her playmates might be receiving similar instruction in Tsonga or Sotho. Ethnic divisions were also often encouraged at the workplace.

LIMITED REFORMS

In 1982 and 1983, there was much official publicity about reforming apartheid. Yet the Nationalist Party's moves to liberalize the system were limited and were accompanied by increased repression. Some changes were simply semantic. In official publications, the term "apartheid" was replaced by "separate development," which was subsequently dropped in favor of "plural democracy."

A bill passed in the white Parliament in 1983 brought Asian and Coloured representatives into the South African government—but only in their own separate chambers, which remained completely subordinate to the white chamber. The bill also concentrated power in the office of the presidency, which eroded the oversight prerogatives of white parliamentarians. Significantly, the new dispensation completely excluded blacks. Seeing the new Constitution as another transparent attempt at divide-and-rule while offering them nothing in the way of genuine empowerment, most Asians and Coloureds refused to participate in the new political order. Instead, many joined together with blacks and a handful of progressive whites in creating a new organization, the United Democratic Front (UDF), which opposed the Constitution.

In other moves, the NP gradually did away with many examples of "petty" apartheid. In many areas, signs announcing separate facilities were removed from public places; but very often, new, more subtle signs were put up to assure continued segregation. Many gas stations in the Transvaal, for example, marked their bathroom facilities with blue and white figures to assure that everyone continued to know his or her place. Another example of purely cosmetic reform was the legalization of interracial marriage—although it was no longer a crime for a man and a woman belonging to different racial classifications to be wed, until 1992 it remained an offense for such a couple to live in the same house. In 1986, the hated passbooks were replaced with new "identity cards." Unions were legalized in the 1980s, but in the Orwellian

world of apartheid, their leaders were regularly arrested. The UDF was not banned but, rather, was forbidden from holding meetings. Although such reforms were meaningless to most nonwhites living within South Africa, some outsiders, including the Reagan administration, were impressed by the "progress."

BLACK RESISTANCE

Resistance to white domination dates back to 1659, when the Khiosan first attempted to counter Dutch encroachments on their pastures. In the first half of the twentieth century, the African National Congress (founded in 1912 to unify what until then had been regionally based black associations) and other political and labor organizations attempted to wage a peaceful civil-rights struggle. An early leader within the Asian community was Mohandas (the Mahatma) Gandhi, who pioneered his strategy of passive resistance in South Africa while resisting the pass laws. In the 1950s, the ANC and associated organizations adopted Gandhian tactics on a massive scale, in a vain attempt to block the enactment of apartheid legislation. Although ANC president Albert Luthuli was awarded the Nobel Peace Prize, the NP regime remained unmoved.

The year 1960 was a turning point. Police massacred more than 60 persons when they fired on a passbook-burning demonstration at Sharpeville. Thereafter, the government assumed emergency powers, banning the ANC and the more recently formed Pan-Africanist Congress. As underground movements, both turned to armed struggle. The ANC's guerrilla organization, *Umkonto we Sizwe* ("Spear of the Nation"), attempted to avoid taking human lives in its attacks. *Poqo* ("Ourselves Alone"), the PAC's armed wing, was less constrained in its choice of targets but proved less able to sustain its struggle. By the mid-1960s, with the capture of such figures as Umkonto leader Nelson Mandela, active resistance had been all but fully suppressed.

A new generation of resistance emerged in the 1970s. Many nonwhite youths were attracted to the teachings of the Black Consciousness Movement (BMC), led by Steve Biko. The BMC and like-minded organizations rejected the racial and ethnic classifications of apartheid by insisting on the fundamental unity of all oppressed black peoples (that is, all nonwhites) against the white power structure. Black consciousness also rejected all forms of collaboration with the apartheid state, which brought the movement into direct opposition with homeland leaders like Gatsha Buthelezi, whom they looked upon as sellouts. In the aftermath of student demonstrations in Soweto, which sparked months of unrest across the country, the government suppressed the BMC. Biko was subsequently murdered while in detention. During the crackdown, thousands of young people fled South Africa. Many joined the exiled ANC, helping to reinvigorate its ranks.

Despite the government's heavy-handed repression, internal resistance to apartheid continued to grow. Hundreds of new and revitalized organizations—community groups, labor unions, and religious bodies—emerged to contribute to the struggle. Many became affiliated through coordinating bodies such as the UDF, the Congress of South African Trade Unions (COSATU), and the South African Council of Churches (SACC). SACC leader Archbishop Desmond Tutu became the second black South African to be awarded the Nobel Peace Prize for his nonviolent efforts to bring about change. But in the face of continued oppression, black youths, in particular, became increasingly willing to use whatever means necessary to overthrow the oppressors.

The year 1985 was another turning point. Arrests and bannings of black leaders led to calls to make the townships "ungovernable." A state of emergency was proclaimed by the government in July, which allowed for the increased use of detention without trial. By March 1990, some 53,000 people, including an estimated 10,000 children, had been arrested. Many detainees were tortured while in custody. Stone-throwing youths nonetheless continued to challenge the heavily armed security forces sent into the townships to restore order. By 1993, more than 10,000 people had died during the unrest.

TOWARD A NEW SOUTH AFRICA

Despite the Nationalist Party's ability to marshall the resources of a sophisticated military–industrial complex to maintain its totalitarian control, it was forced to abandon apartheid along with its four-decade-long monopoly of power. Throughout the 1980s, South Africa's advanced economy was in a state of crisis due to the effects of unrest and, to a lesser extent, of sanctions and other forms of international pressure. Under President P. W. Botha, the NP regime stubbornly refused to offer any openings to genuine reform. However, Botha's replacement in 1989 by F. W. de Klerk opened up new possibilities. The unbanning of the ANC, PAC, and SACP was accompanied by the release of many political prisoners. As many had anticipated, after gaining his freedom in March 1990, ANC leader Nelson Mandela emerged as the leading advocate for change. More surprising was the de Klerk government's willingness to engage in serious negotiations with Mandela and others. By August 1990, the ANC felt that the progress being made justified the formal suspension of its armed struggle.

Many obstacles blocked the transition to a postapartheid state. The NP initially advocated a form of power sharing that fell short of the concept of one person, one vote in a unified state. The ANC, UDF (disbanded in 1991), COSATU, and SACP, which were associated as the Mass Democratic Movement (MDM), however, remained steadfast in their loyalty to the nonracial principles of the 1955 Freedom Charter. Many members of the PAC and other radical critics of the ANC initially feared that the apartheid regime was not prepared to agree to its dismantlement and that the ongoing talks could only serve to weaken black resistance. On the opposite side of the spectrum were still-powerful elements of the white community who remained openly committed to continued white supremacy. In addition to the Conservative Party, the principal opposition in the old white Parliament, there were a number of militant racist organizations, which resorted to terrorism in an attempt to block reforms. Some within the South African security establishment also sought to sabotage the prospects of peace. In March 1992, these far-right elements suffered a setback when nearly 70 percent of white voters approved continued negotiation for democratic reform.

Another troublesome factor was Buthelezi's Inkatha Freedom Party and other, smaller black groups that had aligned themselves in the past with the South African state. Prior to the elections, thousands were killed in clashes between Inkatha and ANC/MDM supporters, especially in the Natal/Kwazulu region. As the positions of the ANC and NP began to converge in 1993, the IFP delegation walked out of the negotiations and formed a "Freedom Alliance" with white conservatives and the leaders of the Bophutatswana and Ciskei homelands. It collapsed in March 1994, following the violent overthrow of the Bophutatswana regime and the defeat of groups of armed white right-wingers that rallied to its defense. Following this debacle, the IFP and more moderate white conservatives—the "Freedom Front"—agreed to participate in national elections. Attempts by more extreme right-wingers to disrupt the elections through a terrorist bombing campaign were crushed in a belated security crackdown.

The elections and the subsequent installation of the GNU were remarkably peaceful, despite organizational difficulties and instances of voting irregularities. In the end, all major parties accepted the result in which the ANC (incorporating MDM) attracted 63 percent of the vote, the NP 20 percent, IFP 10 percent, the Freedom Front 2.2 percent, and the PAC a disappointing 1.2 percent.

Timeline: PAST

1000–1500
Migration of Bantu-speakers into Southern Africa

1652
The first settlement of Dutch people in the Cape of Good Hope area

1659
The first Khiosan attempt to resist white encroachment

1815
The British gain possession of the Cape Colony

1820s
Shaka develops the Zulu nation and sets in motion the wars and migrations known as the Mfecane

1899–1902
The Boer War: the British fight the Afrikaners (Boers)

1910
The Union of South Africa gives rights of self-government to whites

1912
The African National Congress is founded

1948
The Nationalist Party comes to power on an apartheid platform

1960
The Sharpeville Massacre: police fire on demonstration; more than 60 deaths result

1976
Soweto riots are sparked off by student protests

1980s
Unrest in the black townships leads to the declaration of a state of emergency; thousands are detained while violence escalates; F. W. de Klerk replaces P. W. Botha as president; anti-apartheid movements are unbanned; political prisoners are released

1990s
Negotiations begin for a nonracial interim Constitution; nonracial elections in May 1994 result in an ANC-led Government of National Unity; Nelson Mandela becomes president

PRESENT

2000s
Thabo Mbeki is inaugurated as president

Mbeki's responses to the HIV/AIDS pandemic draw intense criticism

Swaziland (Kingdom of Swaziland)

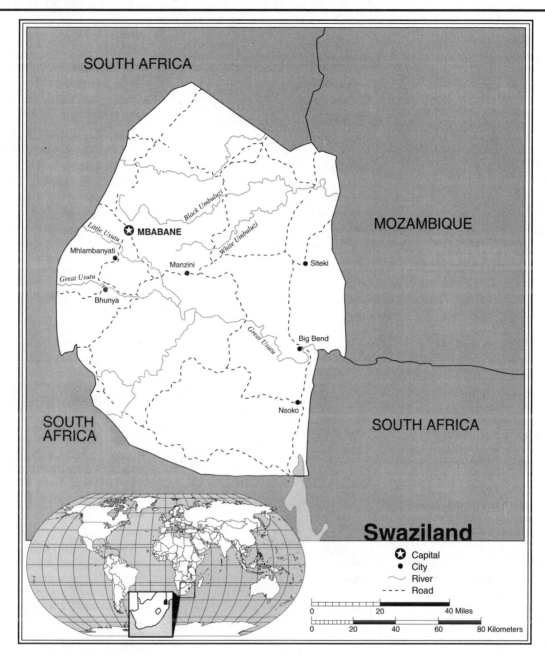

Swaziland Statistics

GEOGRAPHY

Area in Square Miles (Kilometers): 6,704 (17,366) (about the size of New Jersey)

Capital (Population): Mbabane (administrative) (80,000); Lobamba (legislative) (na)

Environmental Concerns: depletion of wildlife populations; overgrazing; soil degradation; soil erosion; limited potable water

Geographical Features: mostly mountains and hills; some sloping plains; landlocked

Climate: from tropical to temperate

PEOPLE

Population

Total: 1,124,000
Annual Growth Rate: 1.63%
Rural/Urban Population Ratio: 74/26

Major Languages: English; SiSwati
Ethnic Makeup: 97% African; 3% European
Religions: 60% Christian; 40% indigenous beliefs

Health

Life Expectancy at Birth: 36 years (male); 38 years (female)
Infant Mortality: 109/1,000 live births
Physicians Available: 1/9,265 people

151

HIV/AIDS Rate in Adults: 35.6%

Education

Adult Literacy Rate: 78%

COMMUNICATION

Telephones: 38,500 main lines
Televisions: 96/1,000 people
Internet Users: 6,000 (2001)

TRANSPORTATION

Highways in Miles (Kilometers): 1,769
  (2,853)
Railroads in Miles (Kilometers): 184
  (297)
Usable Airfields: 18
Motor Vehicles in Use: 37,000

GOVERNMENT

Type: monarchy; independent member of
  the British Commonwealth

Independence Date: September 6, 1968
Head of State/Government: King Mswati
  III; Prime Minister Sibusiso Barnabas
  Dlamini
Political Parties: banned by the
  Constitution
Suffrage: n/a

MILITARY

Military Expenditures (% of GDP): 4.7%
Current Disputes: territorial issues with
  South Africa

ECONOMY

Currency ($ U.S. Equivalent): 8.2
  lilangenis = $1
Per Capita Income/GDP: $4,200/$4.6
  billion
GDP Growth Rate: 2.5%
Inflation Rate: 7.5%
Unemployment Rate: 34%

Natural Resources: iron ore; asbestos;
  coal; clay; hydropower; forests; gold;
  diamonds; quarry stone; talc

Agriculture: corn; livestock; sugarcane;
  fruits; cotton; rice; sorghum; tobacco;
  peanuts

Industry: sugar processing; mining; wood
  pulp; beverages

Exports: $702 million (primary partners
  South Africa, Europe, Mozambique)

Imports: $850 million (primary partners
  South Africa, Europe, Japan)

SUGGESTED WEB SITES

```
http://www.swazi.com/government
http://www.sas.upenn.edu/
  African_Studies/
  Country_Specific/Swaziland.html
http://www.realnet.co.sz
```

Swaziland Country Report

The alleged abduction of a female high school student in October 2002 focused international attention on the struggle between modernist and royal traditionalist forces in Swaziland. The young woman had been removed from her school to be considered for the honor of becoming the 11th wife of King Mswati III. In an unprecedented move, the potential bride's mother accused two of the Swazi king's close associates with kidnapping.

DEVELOPMENT

Much of Swaziland's economy is managed by the Tibiyo TakaNgwana, a royally controlled institution established in 1968 by Sobuza. It is responsible for the financial assets of the communal lands (upon which most Swazis farm) and mining operations.

A small, landlocked kingdom sandwiched between the much larger states of Mozambique and South Africa, casual observers have tended to look upon Swaziland as a peaceful island of traditional Africa that has been immune to the continent's contemporary conflicts. This image, now being increasingly challenged from within is a product of the country's status as the only precolonial monarchy in sub-Saharan Africa to have survived into the modern era. Swazi sociopolitical organization is os-

tensibly governed in accordance with age-old structures and norms. But below this veneer of timelessness lies a dynamic society that has been subject to internal and external pressures. The holding of restricted, nonparty elections in 1993 and 1998 has not quelled the debate over the country's political future between defenders of the status quo and those who advocate a restoration of multiparty democracy.

FREEDOM

The current political order restricts many forms of opposition, although its defenders claim that local councils, *Tikhudlas,* allow for popular participation in decision making. The leading opposition group is the People's United Democratic Movement.

During the 1993 elections, a "stay-away" campaign in favor of reform, accompanied by quiet diplomacy by neighboring states, helped push the Swazi government toward dialogue on the issue. In 1996, King Mswati announced the appointment of a committee to draw up a new constitution. But progress has since been stalled. The officially banned People's United Democratic Movement (PU-DEMO) and other civil-society groups have long called for a repeal of the 1973 royal decree that abolished constitutional rule, including the guarantee of basic freedoms.

HEALTH/WELFARE

Swaziland's low life expectancy and high infant mortality rates have resulted in greater public-health budget allocations. A greater emphasis has also been placed on preventive medicine. However, the extremely high rate of HIV/AIDS poses severe and long-term threats to the nation.

From 1903 until the restoration of independence in 1968, the country remained a British colonial protectorate, despite sustained pressure for its incorporation into South Africa. Throughout the colonial period, the ruling Dlamini dynasty, which was led by the energetic Sobuza II after 1921, served successfully as a rallying point for national self-assertion on the key issues of regaining control of alienated land and opposing union with South Africa. Sobuza's personal leadership in both struggles contributed to the overwhelming popularity of his royalist Imbokodvo Party in the elections of 1964, 1967, and 1972. In 1973, faced with a modest but articulate opposition party, the Ngwane National Liberatory Congress, Sobuza dissolved Parliament and repealed the Westminster-style Constitution, characterizing it as "un-Swazi." In 1979, a new, nonpartisan Parliament was chosen; but authority remained with the king, assisted by his advisory council, the Liqoqo.

Sobuza's death in 1982 left many wondering if Swaziland's unique monarchist institutions would survive. A prolonged power struggle increased tensions within the ruling order. Members of the Liqoqo seized effective power and appointed a new "Queen Regent," Ntombi. However, palace intrigue continued until Prince Makhosetive, Ntombi's teenage son, was installed as King Mswati III in 1986, at age 18. The new king approved the demotion of the Liqoqo back to its advisory status and has ruled through his appointed prime minister and cabinet.

ACHIEVEMENTS

The University of Swaziland was established in the 1970s and now offers a full range of degree and diploma programs.

One of the major challenges facing any Swazi government is its relationship with South Africa. Under Sobuza, Swaziland managed to maintain its political autonomy while accepting its economic dependence on its powerful neighbor. The king also maintained a delicate balance between the apartheid state and the forces opposing it. In the 1980s, this balance became tilted, with a greater degree of cooperation between the two countries' security forces in curbing suspected African National Congress (ANC) activists. In an abrupt reversal of fortunes, Swaziland's prodemocracy activists now look to the new ANC–led government in South Africa for support.

Swaziland's economy, like its politics, is the product of both internal and external initiatives. Since independence, the nation has enjoyed a high rate of economic growth, led by the expansion and diversification of its agriculture. Success in agriculture has promoted the development of secondary industries, such as a sugar refinery and a paper mill. There has also been increased exploitation of coal and asbestos. Another important source of revenue is tourism, which depends on weekend traffic from South Africa.

Swazi development has relied on capital-intensive, rather than labor-intensive, projects. As a result, disparities in local wealth and dependence on South African investment have increased. Only 16 percent of the Swazi population, including migrant workers in South Africa, were in formal-sector employment by 1989. Until recently the economy was boosted by international investors looking for a politically preferable window to the South African market. An example is Coca Cola's decision to move its regional headquarters and concentrate plant from South Africa to Swaziland; the plant employs only about 100 workers but accounts for 20 percent of all foreign-exchange earnings. The current reform process in South Africa, however, is reducing Swaziland's attraction as a center for corporate relocation and sanctions-busting. It has also increased pressure for greater democracy.

Timeline: PAST

1800s
Zulu and South African whites encroach on Swazi territory

1900
A protectorate is established by the British

1903
Britain assumes control over Swaziland

1968
Independence is restored

1973
Parliament is dissolved and political parties are banned

1982
King Sobuza dies

1986
King Mswati III is crowned, ending the regency period marked by political instability

1990s
Swaziland's relationship with South Africa shifts; pressures mount for increased democracy

PRESENT

2000s
AIDS is recognized as a formidable threat to the Swazi people

Pressure continues for multipartyism

An alleged kidnapping case draws international attention

Zambia (Republic of Zambia)

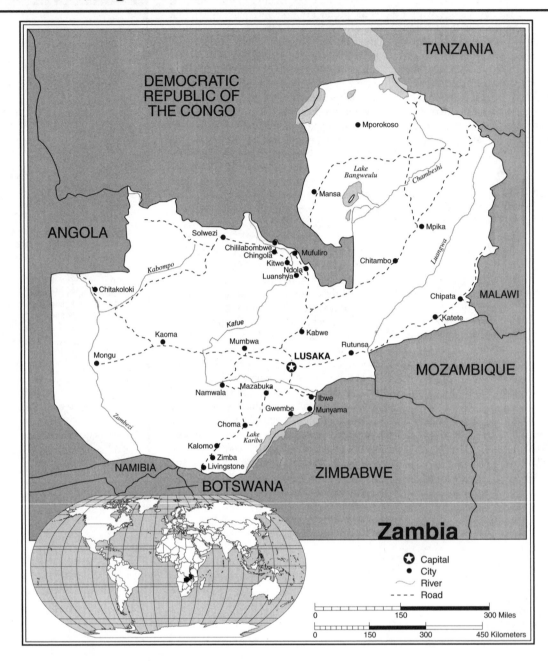

Zambia Statistics

GEOGRAPHY

Area in Square Miles (Kilometers):
290,724 (752,972) (about the size of Texas)

Capital (Population): Lusaka (1,718,000)

Environmental Concerns: air pollution; acid rain; poaching; deforestation; soil erosion; desertification; lack of adequate water treatment

Geographical Features: mostly high plateau with some hills and mountains; landlocked

Climate: tropical; modified by altitude

PEOPLE

Population

Total: 9,960,000
Annual Growth Rate: 1.9%

Rural/Urban Population Ratio: 60/40

Major Languages: English; Bemba; Nyanja; Ila-Tonga; Lozi; others

Ethnic Makeup: 99% African

Religions: 50% Christian; 48% indigenous beliefs; 2% others

Health

Life Expectancy at Birth: 37 years (male); 37 years (female)

154

Infant Mortality: 89.3/1,000 live births
Physicians Available: 1/10,917 people
HIV/AIDS Rate in Adults: 19.95%

Education

Adult Literacy Rate: 79%
Compulsory (Ages): 7–14

COMMUNICATION

Telephones: 130,000 main lines
Daily Newspaper Circulation: 13/1,000
  people
Televisions: 32/1,000 people
Internet Users: 15,000 (2000)

TRANSPORTATION

Highways in Miles (Kilometers): 41,404
  (66,781)
Railroads in Miles (Kilometers): 1,294
  (2,087)
Usable Airfields: 111
Motor Vehicles in Use: 215,000

GOVERNMENT

Type: republic
Independence Date: October 24, 1964
  (from the United Kingdom)
Head of State/Government: President
  Levy Mwanawasa is both head of state
  and head of government
Political Parties: Movement for
  Multiparty Democracy; United National
  Independence Party; others
Suffrage: universal at 18

MILITARY

Military Expenditures (% of GDP): 0.9%
Current Disputes: none

ECONOMY

Currency ($ U.S. Equivalent): 5,000
  kwachas = $1
Per Capita Income/GDP: $870/$8.5
  billion
GDP Growth Rate: 3.9%
Inflation Rate: 21.5%

Unemployment Rate: 50%
Labor Force by Occupation: 85%
  agriculture; 9% services; 6% industry
Natural Resources: copper; zinc; lead;
  cobalt; coal; emeralds; gold; silver;
  uranium; hydropower
Agriculture: corn; sorghum; rice; tobacco;
  cotton; seeds; cassava; peanuts;
  sugarcane
Industry: livestock; mining; foodstuffs;
  beverages; chemicals; textiles; fertilizer
Exports: $876 million (primary partners
  United Kingdom, South Africa,
  Switzerland)
Imports: $1.2 billion (primary partners
  South Africa, United Kingdom,
  Zimbabwe)

SUGGESTED WEB SITES

http://www.zambia.co.zm
http://www.zamnet.zm
http://www.state.gov/www/
  background_notes/
  zambia_0997_bgn.html

Zambia Country Report

After nearly four decades of independence, some three quarters of Zambia's population continue to live below the poverty line, earning less than $1 a day. Up to one in five has also been afflicted by the HIV virus, while many people continue to be struck down by malaria. Floods and drought since 2001 have aggravated Zambia's already poor agricultural output, placing some 2 million people at risk from starvation. The end of the civil war in Angola has, however, brought a welcome relief from border incursions.

DEVELOPMENT

Higher producer prices for agriculture, technical assistance, and rural-resettlement schemes are part of government efforts to raise Zambia's agricultural production. The agricultural sector has shown growth.

The man leading the country in the face of the above challenges is Zambia's third president, Levy Mwanawasa. In January 2002, he was sworn in, replacing his mentor, Frederick Chiluba. Chiluba had tried to alter the Constitution to allow him to run for a third term in office, but members of his own party in Parliament defeated the move. After obtaining the presidency, Mwanawasa stunned the nation by calling upon

Parliament to nullify Chiluba's immunity from prosecution, declaring that the former president was implicated in a series of corruption cases. In July, Parliament voted unanimously to lift the immunity. Although members of the opposition concurred with the ruling party, they further called for a commission of inquiry to also investigate Mwanawasa, who they alleged was implicated in Chiluba-era misdeeds.

FREEDOM

Under the MMD, police have continued to commit extra-judicial killings and other abuses. The government has continued to try to limit press freedom, while failing to honor its 1991 promise to privatize the public media.

For many, Chiluba's presidency was a disappointment. They had hoped that his election in 1991 as the leader of the Movement for Multiparty Democracy (MMD) was the beginning of a new era of democracy and development, after decades of decline under the one-party rule of his predecessor Kenneth Kaunda. But the Chiluba government largely failed to overcome the effects of high inflation and a shrinking gross domestic product, as well

as the mounting challenge of the HIV/AIDS pandemic. Many educated Zambians have left their country in search of opportunities elsewhere.

Zambian politics and society became polarized following the controversial elections of 1996. In that poll, Chiluba and the MMD were reelected amid a boycott by opposition groups, including the former ruling party, the United National Independence Party (UNIP). There were reports of voter-registration irregularities and the enactment of a law barring UNIP's leader, Kaunda, and others from running on account of their being "foreigners." (Kaunda's father had been born before colonial boundaries in what is today Malawi.) The fairness of the elections was also compromised by MMD's use of state resources, especially the public media, in its campaign, and by the charging of Kaunda and nine other UNIP members with treason following a brief bombing campaign by an otherwise still shadowy group calling itself the Black Mambas (after an especially poisonous snake).

In 1997, a Zambian Army captain took control of the national radio station and announced a coup. The plot had little support and was quickly crushed. In its aftermath, however, Parliament approved a 90-day state of emergency, during which time prominent critics of the government were

HEALTH/WELFARE

Life expectancy rates have increased in Zambia since independence, as a result of improved health-care facilities. However, AIDS increasingly looms as a critical problem in Zambia.

ACHIEVEMENTS

Zambia has long played a major role in the fight against white supremacy in Southern Africa. From 1964 until 1980, it was a major base for Zimbabwe nationalists.

detained on allegations of involvement in the coup. Domestic and international human-rights groups reported serious abuses throughout the period.

The deeper roots of Zambia's woes lie in Kenneth Kaunda's 27-year rule. During much of that period, the nation's economy steadily declined. Kaunda consistently blamed his country's setbacks on external forces rather than on his government's failings. There was some justification for his position. The high rate of return on exported copper made the nation one of the most prosperous in Africa until 1975. Since then, fluctuating, but generally depressed, prices for the metal—and the disruption of landlocked Zambia's traditional sea outlets as a result of strife in neighboring states—have had disastrous economic consequences.

Nonetheless, internal factors have also contributed to Zambia's decay. From the early years of Zambia's independence, Kaunda and UNIP showed little tolerance for political opposition. In 1972, the country was legally transformed into a one-party state in which power was concentrated in the hands of Kaunda and his fellow members of UNIP's Central Committee. After 1976, the government ruled with emer-

gency powers. Although Zambia was never as repressive as such neighboring states as Malawi and Zaire/Democratic Republic of the Congo, torture and political detention without trial were common.

Timeline: PAST

A.D. 1889
Rhodes' South African Company is chartered by the British government

1924–1934
Development of the Copperbelt

1953–1963
Federation of Northern Rhodesia, Southern Rhodesia, and Nyasaland is formed; still part of British Empire

1964
Zambia gains independence

1972
Zambia becomes a one-party state under Kenneth Kaunda's United National Independence Party

1980s
South African military raids on Zambia

1990s
Kaunda is defeated in multiparty elections; Frederick Chiluba becomes the nation's second president; a coup attempt is thwarted; the government imposes a state of emergency

PRESENT

2000s
The country grapples with the HIV/AIDS pandemic

Levy Mwanawasa gains the presidency

In its rule, UNIP was supposedly guided by the philosophy of "humanism," a term that became synonymous with the "thoughts of Kaunda." The party also claimed adherence to socialism. But though it was once a mass party that spear-

headed Zambia's struggle for majority rule and independence, UNIP came to stand for little other than the perpetuation of its own power.

An underlying economic problem has remained the decline of rural production, despite Zambia's considerable agricultural potential. The underdevelopment of agriculture is rooted in the colonial policies that favored mining to the exclusion of other sectors. Since independence, the rural areas have continued to be neglected in terms of infrastructural investment. Until recently Zambian farmers were paid little for their produce, with government subsidization of imported food. The result has been a continuous influx of individuals into urban areas, despite a lack of jobs for them, and falling food production.

Zambia's rural decline has severely constrained the government's ability to meet the challenge imposed by depressed export earnings from tourism, copper, and other industries. This has led to severe shortages of foreign exchange and chronic indebtedness. After years of relative inertia, the government, during the 1980s, devoted greater attention to rural development. Agricultural production rose modestly in response to increased incentives. But the size and desperate condition of the urban population discouraged the government from decontrolling prices altogether; rising maize prices in 1986 set off riots that left at least 30 people dead. The new MMD government ended the subsidies, but there has since been insufficient investment to revive the collapsed rural economy. Despite such continued neglect, in 2002 the government was adamant in its opposition to allowing the import of genetically modified grains as food aid, supposedly in order to protect domestic crops from the threat of contamination. Meanwhile, famine threatens millions of Zambians.

Zimbabwe (Republic of Zimbabwe)

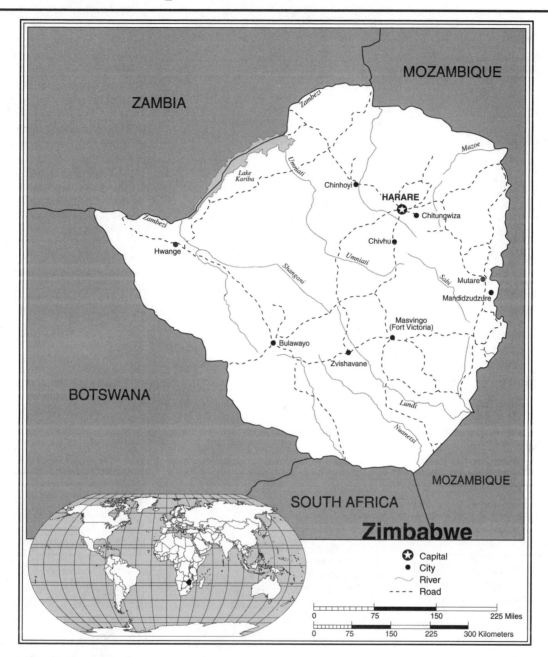

Zimbabwe Statistics

GEOGRAPHY

Area in Square Miles (Kilometers):
150,873 (390,759) (about the size of Montana)

Capital (Population): Harare (1,868,000)

Environmental Concerns: deforestation; soil erosion; land degradation; air and water pollution; poaching

Geographical Features: high plateau with higher central plateau (high veld); mountains in the east; landlocked

Climate: tropical; moderated by altitude

PEOPLE

Population

Total: 11,377,000

Annual Growth Rate: 0.05%

Rural/Urban Population Ratio: 65/35

Major Languages: English; Shona; Ndebele; Sidebele

Ethnic Makeup: 71% Shona; 16% Ndebele; 11% other African; 2% others

Religions: 50% syncretic (part Christian, part indigenous beliefs); 25% Christian; 24% indigenous beliefs; 1% Muslim

Health

Life Expectancy at Birth: 38 years (male); 35 years (female)

Infant Mortality: 62.2/1,000 live births

Physicians Available: 1/6,909 people

HIV/AIDS Rate in Adults: 25.06%

157

Education

Adult Literacy Rate: 85%
Compulsory (Ages): 6–13

COMMUNICATION

Telephones: 222,000 main lines
Daily Newspaper Circulation: 17/1,000 people
Televisions: 12/1,000 people
Internet Users: 30,000 (1999)

TRANSPORTATION

Highways in Miles (Kilometers): 11,369 (18,338)
Railroads in Miles (Kilometers): 1,655 (2,759)
Usable Airfields: 454
Motor Vehicles in Use: 358,000

GOVERNMENT

Type: parliamentary democracy
Independence Date: April 18, 1980 (from the United Kingdom)

Head of State/Government: Executive President Robert Mugabe is both head of state and head of government
Political Parties: Zimbabwe African National Union–Patriotic Front; Zimbabwe African National Union–Ndonga; Movement for Democratic Change; others
Suffrage: universal at 18

MILITARY

Military Expenditures (% of GDP): 3.8%
Current Disputes: involvement in the war in the Democratic Republic of the Congo

ECONOMY

Currency ($ U.S. Equivalent): 56.86 dollars = $1
Per Capita Income/GDP: $2,500/$28 billion
GDP Growth Rate: -6.5%
Inflation Rate: 100%
Unemployment Rate: 60%

Labor Force by Occupation: 66% agriculture; 24% services; 10% industry
Natural Resources: coal; minerals and metals
Agriculture: coffee; tobacco; corn; sugarcane; peanuts; wheat; cotton; livestock
Industry: mining; steel; textiles
Exports: $2.1 billion (primary partners South Africa, United Kingdom, Japan)
Imports: $1.5 billion (primary partners South Africa, United Kingdom, United States)

SUGGESTED WEB SITES

```
http://www.zimembassy-usa.org
http://www.zcbc.co.zw
http://www.dailynews.co.zw
http://www.sas.upenn.edu/African-
   Studies/Country-Specific/
   Zimbabwe.html
http://www.zimweb.com/
   Dzimbabwe.html
```

Zimbabwe Country Report

Zimbabwe has been in the international spotlight in recent years. The emergence of a strong opposition party, the Movement for Democratic Change (MDC), appears to have been the spark that ignited a series of major initiatives aimed at winning the support of the masses of the people by the country's aging president, Robert Mugabe. In 2000, Mugabe gave his blessing to the occupation of white-owned commercial farms by supposed veterans of Zimbabwe's liberation war, which had ended two decades earlier. The takeover of the farms, and its accompanying campaign of intimidation, played a role in a narrow, disputed victory by Mugabe's Zimbabwean African National Union–Patriotic Front (ZANU–PF) party over the MDC in June 2000 parliamentary elections.

DEVELOPMENT

Peasant production has increased dramatically since independence, creating grain reserves and providing exports for the region. The contribution of communal farmers has been recognized both within Zimbabwe and internationally.

With the approach of the 2002 presidential elections, the land-seizure program was expanded. In the process, thousands of African farm workers as well as white farm owners were evicted and disenfranchised.

At the same time, there were growing reports of violence and intimidation directed against suspected MDC supporters, especially in rural areas, as well as attacks on independent journalists. During the presidential campaign, President Mugabe and his supporters blamed western countries, particularly Britain, for being behind a campaign to "recolonize" Zimbabwe. Mugabe described British prime minister Tony Blair as a liar, a scoundrel, and a thief. In the face of credible allegations of massive vote rigging, Mugabe was proclaimed the victor in the elections. Among the international election observers present (a number of European observers were barred) there was a division of opinion as to whether the outcome sufficiently reflected the will of the people. The British Commonwealth ultimately opted to suspend Zimbabwe's membership, while the European Union imposed travel and economic sanctions against Mugabe and his close associates.

The chaos of the last few years has been accompanied by severe economic deprivation for the masses of Zimbabweans. Those relocated on former white-owned farms have suffered along with others due to the lack of such essential inputs as fertilizers,

fuel, and water. To make matters worse, the country has been hit by severe drought. An estimated 1 million people now face the prospect of hunger and starvation in Zimbabwe, and the currency has lost a great deal of its value. Zimbabwe has moved from one of Southern Africa's "breadbaskets" to a dependent regional basket case. Most observers agree that things are going to worsen in the immediate future.

FREEDOM

Since the 1990 lifting of the state of emergency that had been in effect since the days of the Federation, Zimbabwe's human-rights record has generally improved. Some government institutions, however, especially the Central Intelligence Organization, are still accused of extra-judicial abuses.

The Mugabe-led ZANU-PF party has dominated Zimbabwe's politics since the country's independence from Britain in 1980. The emergence in 2000 of MDC, led by the former trade unionist Morgan Tsangarai, has posed a formidable foe to ZANU. As a coalition movement, the MDS's primary program has been to replace Mugabe. Having initially failed in this effort, cleavages between various factions of the movement may become more noticeable. In many rural areas, MDC supporters have been driven underground by official repression, but MDC remains strong in the western Matebeleland region as well as major urban areas.

Ideologically, Mugabe belongs to the African liberationist tradition of the 1960s—strong and ruthless leadership, anti-Western, suspicious of capitalism, and deeply intolerant of dissent and opposition. With more than a third of the total land and up to 80 percent of the most productive farming areas in the hands of a few thousand whites before the recent seizure, land redistribution has been a key issue for Zimbabweans. Offering land to landless African farmers is seen by many as an attractive option. In this sense, Robert Mugabe does not stand alone. By adopting mob tactics toward the emotive issue of land—where there was already a widely accepted need for redistribution reform—Mugabe may have bought his regime some additional time. But Zimbabwe's reputation as a law-abiding society now lies in tatters, along with its economy. Many have chosen to leave the country, usually illegally, in search for survival in South Africa or Botswana.

Zimbabwe achieved its formal independence in April 1980, after a 14-year armed struggle by its disenfranchised black African majority. Before 1980, the country was called Southern Rhodesia—a name that honored Cecil Rhodes, the British imperialist who had masterminded the colonial occupation of the territory in the late nineteenth century. For its black African majority, Rhodesia's name was thus an expression of their subordination to a small minority of privileged white settlers whose racial hegemony was the product of Rhodes's conquest. The new name of Zimbabwe was symbolic of the greatness of the nation's precolonial roots.

THE PRECOLONIAL PAST

By the fifteenth century, Zimbabwe had become the center of a series of states that prospered through their trade in gold and other goods with Indian Ocean merchants. These civilizations left as their architectural legacy the remains of stone settlements known as *zimbabwes*. The largest of these, the so-called Great Zimbabwe, lies near the modern town of Masvingo. Within its massive walls are dozens of stella, topped with distinctive carved birds whose likeness has become a symbol of the modern state. Unfortunately, early European fortuneseekers and archaeologists destroyed much of the archaeological evidence of this site, but what survives confirms that the state had trading contacts as far afield as China.

From the sixteenth century, the Zimbabwean civilizations seem to have declined, possibly as a result of the disruption of the East African trading networks by the Portuguese. Nevertheless, the states themselves survived until the nineteenth century. And their cultural legacy is very much alive today, especially among the 71 percent of Zimbabwe's population who speak Shona. Zimbabwe's other major ethnolinguistic community are the Ndebele-speakers, who today account for about 16 percent of the population. This group traces its local origin to the mid-nineteenth-century conquest of much of modern Zimbabwe by invaders from the south under the leadership of Umzilagazi, who established a militarily strong Ndebele kingdom, which subsequently was ruled by his son.

HEALTH/WELFARE

Public expenditure on health and education has risen dramatically since independence. Most Zimbabweans now enjoy access to medical facilities, while primary-school enrollment has quadrupled. Higher education has also been greatly expanded. But the advances are threatened by downturns in the economy, and school fees have been reintroduced.

WHITE RULE

Zimbabwe's colonial history is unique in that it was never under the direct rule of a European power. In 1890, the lands of the Ndebele and Shona were invaded by agents of Cecil Rhodes's British South Africa Company (BSACO). In the 1890s, both groups put up stiff resistance to the encroachments of the BSACO settlers, but eventually they succumbed to the invaders. In 1924, the BSACO administration was dissolved and Southern Rhodesia became a self-governing British Crown colony. "Self-government" was, in fact, confined to the white-settler community, which grew rapidly but never numbered more than 5 percent of the population.

In 1953, Southern Rhodesia was federated with the British colonial territories of Northern Rhodesia (Zambia) and Nyasaland (Malawi). This Central African Federation was supposed to evolve into a "multiracial" dominion; but from the beginning, it was perceived by the black majority in all three territories as a vehicle for continued white domination. As the federation's first prime minister put it, the partnership of blacks and whites in building the new state would be analogous to a horse and its rider—no one had any illusions as to which race group would continue to be the beast of burden.

In 1963, the federation collapsed as a result of local resistance. Black nationalists established the independent "nonracial" states of Malawi and Zambia. For a while, it appeared that majority rule would also come to Southern Rhodesia. The local black community was increasingly well organized and militant in demanding full citizenship rights. However, in 1962, the white electorate responded to this challenge by voting into office the Rhodesia Front (RF), a party determined to uphold white supremacy at any cost. Using already-existing emergency powers, the new government moved to suppress the two major black nationalist movements: the Zimbabwe African People's Union (ZAPU) and the Zimbabwe African National Union (ZANU).

RHODESIA DECLARES INDEPENDENCE

In a bid to consolidate white power along the lines of the neighboring apartheid regime of South Africa, the RF, now led by Ian Smith, made its 1965 Unilateral Declaration of Independence (UDI) of any ties to the British Crown. Great Britain, along with the United Nations, refused to recognize this move. In 1967, the United Nations imposed mandatory economic sanctions against the "illegal" RF regime. But the sanctions were not fully effective, largely due to the fact that they were flouted by South Africa and the Portuguese authorities who controlled most of Mozambique until 1974. The United States continued openly to purchase Rhodesian chrome for a number of years, while many states and individuals engaged in more covert forms of sanctions-busting. The Rhodesian economy initially benefited from the porous blockade, which encouraged the development of a wide range of import-substitution industries.

With the sanctions having only a limited effect and Britain and the rest of the international community unwilling to engage in more active measures, it soon became clear that the burden of overthrowing the RF regime would be borne by the local population. ZANU and ZAPU, as underground movements, began to engage in armed struggle beginning in 1966. The success of their attacks initially was limited; but from 1972, the Rhodesian Security Forces were increasingly besieged by the nationalists' guerrilla campaign. The 1974 liberation of Mozambique from the Portuguese greatly increased the effectiveness of the ZANU forces, who were allowed to infiltrate into

Rhodesia from Mozambican territory. Meanwhile, their ZAPU comrades launched attacks from bases in Zambia. In 1976, the two groups became loosely affiliated as the Patriotic Front.

ACHIEVEMENTS

Zimbabwe's capital city of Harare has become an arts and communications center for Southern Africa. Many regional as well as local filmmakers, musicians, and writers based in the city enjoy international reputations. And the distinctive malachite carvings of Zimbabwean sculptors are highly valued in the international art market.

Unable to stop the military advance of the Patriotic Front, which was resulting in a massive white exodus, the RF attempted to forge a power-sharing arrangement that preserved major elements of settler privilege. Although rejected by ZANU or ZAPU, this "internal settlement" was implemented in 1978–1979. A predominantly black government took office, but real power remained in white hands, and the fighting only intensified. Finally, in 1979, all the belligerent parties, meeting at Lancaster House in London, agreed to a compromise peace, which opened the door to majority rule while containing a number of constitutional provisions designed to reassure the white minority. In the subsequent elections, held in 1980, ZANU captured 57 and ZAPU 20 out of the 80 seats elected by the "common roll." Another 20 seats, which were reserved for whites for seven years as a result of the Lancaster House agreement, were captured by the Conservative Alliance (the new name for the RF). ZANU leader Robert Mugabe became independent Zimbabwe's first prime minister.

THE RHODESIAN LEGACY

The political, economic, and social problems inherited by the Mugabe government were formidable. Rhodesia had essentially been divided into two "nations": one black, the other white. Segregation prevailed in virtually all areas of life, with those facilities open to blacks being vastly inferior to those open to whites. The better half of the national territory had also been reserved for white ownership. Large commercial farms prospered in this white area, growing maize and tobacco for export as well as a diversified mix of crops for domestic consumption. In contrast, the black areas, formally known as Tribal Trust Lands, suffered from inferior soil and rainfall, overcrowding, and poor infrastructure. Most black adults had little choice but to obtain seasonal work in the white areas.

Black unskilled workers on white plantations, like the large number of domestic servants, were particularly impoverished. But until the 1970s, there were also few opportunities for blacks with higher skills as a result of a de facto "color bar," which reserved the best jobs for whites.

Despite its stated commitment to revolutionary socialist objectives, since 1980, the Mugabe government has taken an evolutionary approach in dismantling the socioeconomic structures of old Rhodesia. This cautious policy is, in part, based on an appreciation that these same structures support what, by regional standards, is a relatively prosperous and self-sufficient economy. Until 1990, the government's hands were also partially tied by the Lancaster House accords, wherein private property, including the large settler estates, could not be confiscated without compensation. In its first years, the government nevertheless made impressive progress in improving the livelihoods of the Zimbabwean majority by redistributing some of the surplus of the still white-dominated private sector. With the lifting of sanctions, mineral, maize, and tobacco exports expanded and import restrictions eased. Workers' incomes rose, and a minimum wage, which notably covered farm employees, was introduced. Rising consumer purchasing power initially benefited local manufacturers. Health and educational facilities were expanded, while a growing number of blacks began to occupy management positions in the civil service and, to a lesser extent, in businesses.

Zimbabwe had hoped that foreign investment and aid would pay for an ambitious scheme to buy out many white farmers and to settle black peasants on their land. However, funding shortfalls resulted in only modest resettlement. Approximately 4,000 white farmers owned more than one third of the land. In 1992, the government passed a bill that allowed for the involuntary purchase of up to 50 percent of this land at an officially set price. While enjoying overwhelming domestic support, this land-redistribution measure came under considerable external criticism for violating the private-property and judicial rights of the large-scale farmers. Others pointed out that, besides producing large surpluses of food in nondrought years, many jobs were tied to the commercial estates. Revelations in 1993–1994 that some confiscated properties had been turned over to leading ZANU politicians gave rise to further controversy.

While gradually abandoning its professed desire to build a socialist society, the Zimbabwean government has continued to face a classic dilemma of all industrializing

Timeline: PAST

1400s–1500s
Heyday of the gold trade and Great Zimbabwe

1840s
The Ndebele state emerges in Zimbabwe

1890
The Pioneer Column: arrival of the white settlers

1895–1897
Chimurenga: rising against the white intruders, ending in repression by whites

1924
Local government in Southern Rhodesia is placed in the hands of white settlers

1965
Unilateral Declaration of Independence

1966
Armed struggle begins

1980
ZANU leader Robert Mugabe becomes Zimbabwe's first prime minister

1990s
ZANU and ZAPU merge and win the 1990 elections; elections in 1995 result in a landslide victory for the ruling ZANU-PF

PRESENT

2000s
Mugabe pushes redistribution of white-owned commercial farms

societies: whether to continue to use tight import controls to protect its existing manufacturing base or to open up its economy in the hopes of enjoying a takeoff based on export-oriented growth. While many Zimbabwean manufacturers would be vulnerable to greater foreign competition, there is now a widespread consensus that limits of the local market have contributed to stagnating output and physical depreciation of local industry in recent years.

POLITICAL DEVELOPMENT

The Mugabe government has promoted reconciliation across the racial divide. Although the reserved seats for whites were abolished in 1987, the white minority (whites now make up less than 2 percent of the population) is well represented within government as well as business. Mugabe's ZANU administration has shown less tolerance of its political opponents, especially ZAPU. ZANU was originally a breakaway faction of ZAPU. At the time of this split, in 1963, the differences between the two movements had largely been over tactics. But elections in 1980 and 1985 confirmed that the followings of both movements have become ethnically based, with most Shona supporting ZANU and Ndebele supporting ZAPU.

Initially, ZANU agreed to share power with ZAPU. However, in 1982, the alleged discoveries of secret arms caches, which ZANU claimed ZAPU was stockpiling for a coup, led to the dismissal of the ZAPU ministers. Some leading ZAPU figures were also detained. The confrontation led to violence that very nearly degenerated into a full-scale civil war. From 1982 to 1984, the Zimbabwean Army, dominated by former ZANU and Rhodesian units, carried out a brutal counterinsurgency campaign against supposed ZAPU dissidents in the largely Ndebele areas of western Zimbabwe. Thousands of civilians were killed—especially by the notorious Fifth Brigade, which operated outside the normal military command structure. Many more fled to Botswana, including, for a period, the ZAPU leader, Joshua Nkomo.

Until 1991, Mugabe's stated intention was to create a one-party state in Zimbabwe. With his other black and white opponents compromised by their past association with the RF and its internal settlement, this largely meant coercing ZAPU into dissolving itself into ZANU. However, the increased support for ZAPU in its core Ndebele constituencies during the 1985 elections led to a renewed emphasis on the carrot over the stick in bringing about the union. In 1987, ZAPU formally merged into ZANU, but their shotgun wedding made for an uneasy marriage.

With the demise of ZAPU, new forces have emerged in opposition to Mugabe and the drive for a one-party state. Principal among these is the Zimbabwe Unity Move-ment (ZUM), led by former ZANU member Edger Tekere. In the 1990 elections, ZUM received about 20 percent of the vote, in a poll that saw a sharp drop in voter participation. The elections were also marred by serious restrictions on opposition activity and blatant voter intimidation. The deaths of ZUM supporters in the period before the elections reinforced the message of the government-controlled media that a vote for the opposition was an act of suicide. A senior member of the Central Intelligence Organization and a ZANU activist were subsequently convicted of the murder of ZUM organizing secretary Patrick Kombayi. However, they were pardoned by Mugabe.

Mugabe initially claimed that his 1990 victory was a mandate to establish a one-party state. But in 1991, the changing international climate, the continuing strength of the opposition, and growing opposition within ZANU itself caused him to shelve the project. Under 1992 election law, however, ZANU alone was made eligible for state funding.

The survival of political pluralism in Zimbabwe reflects the emergence of a civil society that is increasingly resistant to the concentration of power. Independent non-governmental organizations have successfully taken up many social human-rights issues. Less successful have been attempts to promote an independent press, which has remained almost entirely in government/ZANU hands.

In 1992, the Forum Party, a new opposition movement, was launched, under the leadership of former chief justice Enoch Dumbutshena. But it failed to break the mold of Zimbabwean politics due to its own internal splits and failure to unite with other groups. As a result, Mugabe was easily reelected in March 1996 in a poll with low voter turnout (it was ultimately boycotted by the entire opposition).

Notwithstanding its continuing electoral success, public confidence in the ZANU government has been greatly eroded by its relative failure in handling the 1992 drought crisis. Despite warning signs of the impending catastrophe, little attempt was made to stockpile food. This failure resulted in widespread hunger and dependence on expensive food imports. Long-neglected waterworks, especially those serving Bulawayo, the country's second-largest city, proved to be inadequate. The government also lost support due to its seeming insensitivity to the plight of ordinary Zimbabweans suffering from high rates of unemployment and inflation. With inflation at 22 percent, a civil servants strike was sparked in August 1996 by an across-the-board 6 percent raise for ordinary workers as compared to a 130 percent raise for members of Parliament.

While the welfare of ordinary Zimbabweans may have improved since 1980, popular frustration with the status quo is increasing. In 1998, resentment against the government, resulting from continued economic decline, was aggravated in some quarters by Mugabe's decision to dispatch nearly 3,000 troops to the Democratic Republic of the Congo (D.R.C., the former Zaire) to defend the embattled regime of Laurent Kabila. With formal-sector unemployment in excess of 50 percent and inflation approaching 200 percent, it was a foreign adventure that the country could ill afford.

West Africa

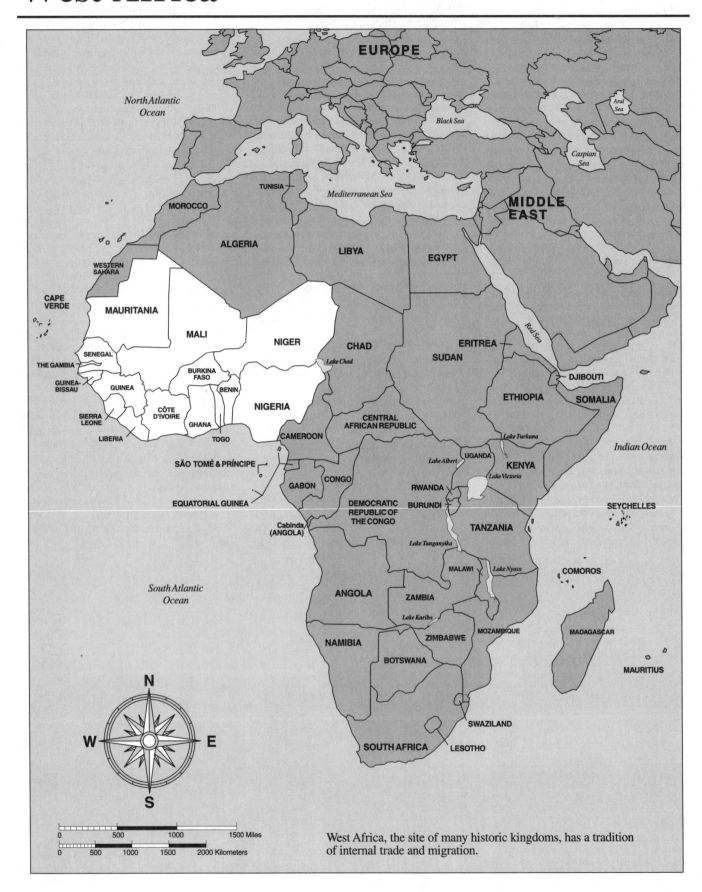

West Africa, the site of many historic kingdoms, has a tradition of internal trade and migration.

West Africa: Seeking Unity in Diversity

Anyone looking at a map of Africa can identify West Africa as the great bulge on the western coast of the continent. It is a region bound by the Sahara Desert to the north, the Atlantic Ocean to the south and west, and, in part, by the Cameroonian Mountains to the east. Each of these boundaries has historically been a bridge rather than a barrier, in that the region has long been linked through trade to the rest of the world.

At first glance, what is more striking is West Africa's great variety, rather than any of its unifying features. It contains the environmental extremes of desert and rain forest. While most of its people rely on agriculture, every type of occupation can be found, from herders to factory workers. Hundreds of languages are spoken; some are as different from one another as English is from Arabic or Japanese. Local cultural traditions and the societies that practice them are also myriad.

Yet the more closely one examines West Africa, the more one is impressed with the features that give the nations of the region a degree of coherence and unity. Some of the common characteristics and features of West Africa as a whole include the vegetation belts that stretch across the region from west to east, creating a similar environmental mix among the region's polities; the constant movement of peoples across local and national boundaries; and efforts being made by West African governments toward greater integration in the region, primarily through economic organizations. West Africans also share elements of a common history.

With the exception of Liberia, all the contemporary states of West Africa were the creations of competing European colonial powers—France, Germany, Great Britain, and Portugal—that divided most of the area during the late 1800s. Before this partition, however, much of the region was linked by the spread of Islam and patterns of trade, including the legacy of intensive involvement between the sixteenth and nineteenth centuries in the trans-Atlantic slave trade. From ancient times, great kingdoms expanded and contracted across the West African savanna and forest, giving rise to sophisticated civilizations.

WEST AFRICAN VEGETATION AND CLIMATE ZONES

Traveling north from the coastlines of such states as Nigeria, Ghana, and Côte d'Ivoire, one encounters tropical rain forests, which give way first to woodland savanna and then to more arid, more open plains. In Mali, Niger, and other landlocked areas to the north, the savanna gives way to the still drier Sahel environment, and finally to the Sahara Desert itself.

Whatever their ethnicity or nationality, the peoples living within each of these vegetation zones generally share the benefits and problems of similar livelihoods. For instance, cocoa, coffee, yams, and cassava are among the cash and food crops planted in the cleared forest and woodland zones, which stretch from Guinea to Nigeria. Groundnuts, sorghum, and millet are commonly harvested in the savanna belt that runs from Senegal to northern Nigeria. Herders in the Sahel, who historically could not go too far south with their cattle because of the presence of the deadly tsetse fly in the forest, continue to cross state boundaries in search of pasture.

People throughout West Africa have periodically suffered from drought. The effects of drought have often been aggravated in recent years by population pressures on the land. These factors have contributed to environmental changes and degradation. The condition of the Sahel in particular has deteriorated through a process of desertification, leading to large-scale relocations among many of its inhabitants. The eight Sahelian countries—Cape Verde, The Gambia, Burkina Faso, Mali, Senegal, Niger, Chad (in Central Africa), and Mauritania—have consequently formed the Committee for Struggle Against Drought in the Sahel (CILSS).

Farther to the south, large areas of woodland savanna have turned into grasslands as their forests have been cut down by land-hungry farmers. Drought has also periodically resulted in widespread brushfires in Ghana, Côte d'Ivoire, Togo, and Benin, fires that have transformed forests into savannas and savannas into deserts. Due to the depletion of forest, the Harmattan (a dry wind that blows in from the Sahara during January and February) now reaches many parts of the coast that in the recent past did not feel its breath. Its dust and haze have become a sign of the new year—and of new agricultural problems—throughout much of West Africa.

The great rivers of West Africa, such as The Gambia, Niger, Senegal, and Volta, along with their tributaries, have become increasingly important both as avenues of travel and trade and for the water they provide. Countries have joined together in large-scale projects designed to harness their waters for irrigation and hydroelectric power through regional organizations, like the Mano River grouping of Guinea, Liberia, and Sierra Leone and the Organization for the Development of the Senegal River, composed of Mali, Mauritania, and Senegal.

THE LINKS OF HISTORY AND TRADE

The peoples of West Africa have never been united as members of a single political unit. Yet some of the precolonial kingdoms that expanded across the region have great symbolic importance for those seeking to enhance interstate cooperation. The Mali empire of the thirteenth to fifteenth centuries, the Songhai empire of the sixteenth century, and the nineteenth-century Fulani caliphate of Sokoto, all based in the savanna, are widely remembered as examples of past supranational glory. The kingdoms of the southern forests, such as the Asante Confederation, the Dahomey kingdom, and the Yoruba city-states, were smaller than the great savanna empires to their north. Although generally later in origin and different in character from the northern states, the forest kingdoms are, nonetheless, sources of greater regional identity.

The precolonial states of West Africa gave rise to great urban centers, interlinked through extensive trade networks. This development was probably the result of the area's agricultural productivity, which supported a relatively high population

(IFC/World Bank photo by Ray Witkin)

A worker cuts cloth at a textile mill in Côte d'Ivoire. The textile designs seen here are similiar to indigenous regional patterns. Most of the cloth made in Côte d'Ivoire is exported.

density from early times. Many modern settlements have long histories. Present-day Timbuctu and Gao, in Mali, were important centers of learning and commerce in medieval times. Some other examples include Ouagadougou, Ibadan, Benin, and Kumasi, all in the forest zone. These southern centers prospered in the past by sending gold, kola, leather goods, cloth—and slaves—to the northern savanna and southern coast.

The cities of the savannas linked West Africa to North Africa. Beginning in the eleventh century, the ruling groups of the savanna increasingly turned to the universal vision of Islam. While Islam also spread to the forests, the southernmost areas were ultimately more strongly influenced by Christianity, which was introduced by Europeans, who became active along the West African coast in the fifteenth century. For centuries, the major commercial link among Europe, the Americas, and West Africa was the trans-Atlantic slave trade; during the 1800s, however, legitimate commerce in palm oil and other tropical products replaced it. New centers such as Dakar, Accra, and Freetown emerged—resulting either from the slave trade or from its suppression.

THE MOVEMENT OF PEOPLES

Despite the (incorrect) view of many who see Africa as being a continent made up of isolated groups, one constant characteristic of West Africa has been the transregional migration of its people. Herders have moved east and west across the savanna and south into the forests. Since colonial times, many professionals as well as laborers have sought employment outside their home areas.

Some of the peoples of West Africa, such as the Malinke, Fulani, Hausa, and Mossi, have developed especially well-established heritages of mobility. In the past, the Malinke journeyed from Mali to the coastal areas in Guinea, Senegal, and The Gambia. Other Malinke traders made their way to Burkina Faso, Liberia, and Sierra Leone, where they came to be known as Mandingoes.

The Fulani have developed their own patterns of seasonal movement. They herd their cattle south across the savanna during the dry season and return to the north during the rainy season. Urbanized Fulani groups have historically journeyed from west to east, often serving as agents of Islamization as well as promoters of trade. More recently, many Fulani have been forced to move southward as a result of the deterioration of their grazing lands. The Hausa, who live mostly in northern Nigeria and Niger, are found throughout much of West Africa. Indeed, their trading presence is so widespread that some have suggested that the Hausa language be promoted as a lingua franca, or common language, for West Africa.

Millions of migrant laborers are regularly attracted to Côte d'Ivoire and Ghana from the poorer inland states of Burkina Faso, Mali, and Niger, thus promoting continuing economic interdependence among these states. Similar large-scale migrations also occur elsewhere. The drastic expulsion of aliens by the Nigerian government in 1983 was startling to the outside world, in part because few had realized that so many Ghanaians, Nigeriens, Togolese, Beninois, and Cameroonians had taken up residence in Nigeria. Such immigration is not new, though its scale into Nigeria was greatly increased by that country's oil boom. Peoples such as the Yoruba, Ewe, and Vai, who were divided by colonialism, have often ignored modern state boundaries in order to maintain their ethnic ties. Other migrations also have roots in the colonial past. Sierra Leonians worked as clerks and craftspeople throughout the coastal areas of British West Africa, while Igbo were recruited to serve in northern Nigeria. Similarly, Beninois became the assistants of French administrators in other parts of French West Africa, while Cape Verdians occupied intermediate positions in Portugal's mainland colonies.

WEST AFRICAN INTEGRATION

Many West Africans recognize the weaknesses inherent in the region's national divisions. The peoples of the region would benefit from greater multilateral political cooperation and economic integration. Yet there are many obstacles blocking the growth of pan-regional development. National identity is probably even stronger today than it was in the days when Kwame Nkrumah, the charismatic Ghanaian leader, pushed for African unity but was frustrated by parochial interests. The larger and more prosperous states, such as Nigeria and Côte d'Ivoire, are reluctant to share their relative wealth with smaller countries, which, in turn, fear being swallowed.

One-party rule and more overt forms of dictatorship have recently been abandoned throughout West Africa. However, for the moment, the region is still politically divided between those states that have made the transition to multiparty constitutional systems of government and those that are still under effective military control. Overlapping ethnicity is also sometimes more a source of suspicion rather than a source of unity between states. Because the countries were under the rule of different colonial powers, French, English, and Portuguese serve today as official languages of the different nations, which also inherited different administrative traditions. Moreover, during colonial times, independent infrastructures were developed in each country; these continue to orient economic activities toward the coast and Europe rather than encouraging links among West African countries.

Political changes also affect regional cooperation and domestic development. Senegambia, the now-defunct confederation of Senegal and The Gambia, was dominated by Senegal and resented by many Gambians. The Liberian Civil War has also led to division between the supporters and opponents of the intervention of a multinational peacekeeping force.

Despite the many roadblocks to unity, a number of multinational organizations have developed in West Africa, stimulated in large part by the severity of the common problems that the countries face. The West African countries have a good record of cooperating to avoid armed conflict and to settle their occasional border disputes. In addition to the multilateral agencies that are coordinating the struggle against drought and the development of various river basins, there are also various regional commodity cartels, such as the five-member Groundnut Council. The West African Examinations Council standardizes secondary-school examinations in most of the countries where English is an official language, and most of the Francophonic states have the same currency.

The most ambitious and broad organization in the region is the Economic Organization of West African States (ECOWAS), which includes all the states incorporated in the West African section of this text. Established in 1975 by the Treaty of Lagos, ECOWAS aims to promote trade, cooperation, and self-reliance. The progress of the organization in these areas has thus far been limited. But ECOWAS can point to some significant achievements. Several joint ventures have been developed; steps toward tariff reduction are being taken; and ECOWAS members have agreed in principle to eventually establish a common currency.

The ECOWAS states have shown an increasing willingness and capacity to play a leading role in collectively resolving their regional conflicts. Over the past decade, through its multinational peacekeeping force, ECOMOG, members of ECOWAS have jointly intervened to assist in the settlement of internal conflicts in Liberia and Sierra Leone. More recently, ECOWAS has also acted as a mediator in the ongoing civil conflict in Côte d'Ivoire. The modest success of these initiatives points to the pivotal role that must be played by Nigeria in any move toward greater regional cooperation. With about half of West Africa's population and economic output, a revitalized Nigeria has already demonstrated its potential for regional leadership. But Nigeria's own progress, as well as that of the region as a whole, is dependent on its making further progress toward overcoming its own internal political and economic weaknesses.

Benin (Republic of Benin)

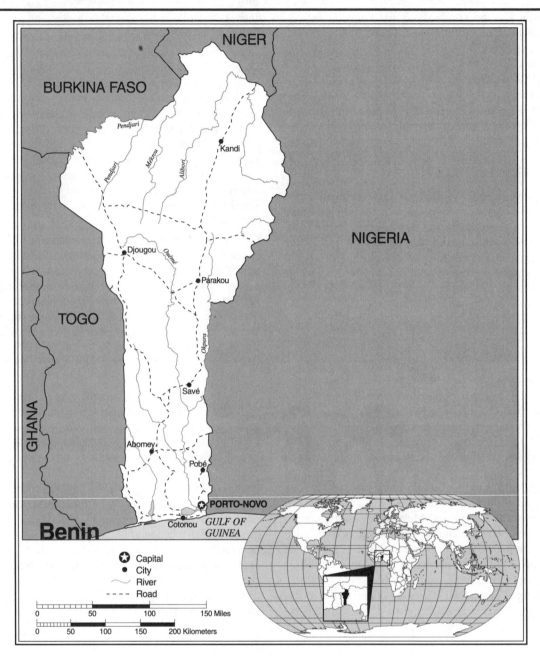

Benin Statistics

GEOGRAPHY

Area in Square Miles (Kilometers):
43,483 (112,620) (about the size of Pennsylvania)

Capital (Population): official: Porto-Novo (225,000); de facto: Cotonou (750,000)

Environmental Concerns: drought; insufficient potable water; poaching; deforestation; desertification

Geographical Features: mostly flat to undulating plain; some hills and low mountains

Climate: tropical to semiarid

PEOPLE

Population

Total: 6,788,000

Annual Growth Rate: 2.91%

Rural/Urban Population Ratio: 58/42

Major Languages: French; Fon; Yoruba; others

Ethnic Makeup: 99% African (most important groupings Fon, Adja, Yoruba, and Bariba); 1% European

Religions: 70% indigenous beliefs; 15% Muslim; 15% Christian

Health

Life Expectancy at Birth: 49 years (male);
  51 years (female)
Infant Mortality: 88.5/1,000 live births
Physicians Available: 1/14,216 people
HIV/AIDS Rate in Adults: 4.1%

Education

Adult Literacy Rate: 37%
Compulsory (Ages): 6–12; free

COMMUNICATION

Telephones: 51,000 main lines
Daily Newspaper Circulation: 2/1,000
  people
Televisions: 4/1,000 people
Internet Users: 50,000 (2002)

TRANSPORTATION

Highways in Miles (Kilometers): 4,208
  (6,787)
Railroads in Miles (Kilometers): 360
  (578)
Usable Airports: 5

Motor Vehicles in Use: 55,000

GOVERNMENT

Type: republic
Independence Date: August 1, 1960 (from
  France)
Head of State/Government: President
  Mathieu Kérékou is both head of state
  and head of government
Political Parties: Alliance for Democracy
  and Progress; Front for Renewal and
  Development; African Movement for
  Democracy and Progress; many others
Suffrage: universal at 18

MILITARY

Military Expenditures (% of GDP): 1.2%
Current Disputes: territorial disputes with
  Niger and Nigeria

ECONOMY

Currency ($ U.S. Equivalent): 752 CFA
  francs = $1

Per Capita Income/GDP: $1,040/$6.8
  billion
GDP Growth Rate: 5.4%
Inflation Rate: 3%
Population Below Poverty Line: 37%
Natural Resources: small offshore oil
  deposits; limestone; marble; timber
Agriculture: palm products; cotton; corn;
  rice; yams; cassava; beans; sorghum;
  livestock
Industry: textiles; construction materials;
  food production; chemical production
Exports: $35.3 million (primary partners
  Brazil, France, Indonesia)
Imports: $437 million (primary partners
  France, China, United States)

SUGGESTED WEB SITES

http://www.benindaily.com
http://www.sas.upenn.edu/
  African_Studies/
  Country_Specific/Benin.htm

Benin Country Report

Over the past decade, Benin has emerged as one of one of Africa's most stable and democratic states. This has coincided with improved economic growth, though the country remains among the world's poorest in terms of both per capita income and human development. Since gaining its independence from France in 1960, Benin has experienced a series of shifts in political and economic policy that have so far failed to lift most Beninois out of chronic poverty. In this respect, the country's ongoing struggle for development can be seen as a microcosm of the challenges facing much of the African continent.

DEVELOPMENT

Palm-oil plantations were established in Benin by Africans in the mid-nineteenth century. They have continued to be African-owned and capitalist-oriented. Today, there are some 30 million trees in Benin, and palm-oil products are a major export used for cooking, lighting, soap, margarine, and lubricants.

Politically, Benin has been in the forefront of those nations on the continent making the transition away from an authoritarian centralized state toward greater democracy

and market reforms. This process has not as yet been accompanied by a decisive shift toward a new generation of leadership. In March 1996, former president Mathieu Kérékou returned to power with 52 percent of the vote, defeating incumbent Nicephore Soglo in Benin's second ballot since the 1990 restoration of multiparty democracy. Five years earlier, Soglo had defeated Kérékou, who had ruled the country as a virtual dictator for 17 years before agreeing to a democratic transition. In the past, Kérékou styled himself as a Marxist Leninist and presided over a one-party state. Today, he presents himself as a "Christian Democrat," affirming that there can be no turning back to the old order. In Parliament, his Popular Revolutionary Party of Benin (PRPB) shares power with other groupings whose existence is primarily a reflection of ethnoregional rather than ideological divisions.

Kérékou's restoration did not result in any significant moves away from his predecessor's economic reforms, which had resulted in a modest rise in gross domestic product, increased investment, reduced inflation, and an easing of the country's debt burden. He is under pressure, however, to raise the living standards of Benin's impoverished masses.

THE OLD ORDER FALLS

Kérékou's first reign began to unravel in late 1989. Unable to pay its bills, his government found itself increasingly vulnerable to mounting internal opposition and, to a lesser extent, to external pressure to institute sweeping political and economic reforms.

FREEDOM

Since 1990, political restrictions have been lifted and prisoners of conscience freed. More recently, however, a number of citizens have been arrested for supposedly inciting people against the government and encouraging them not to pay taxes.

A wave of strikes and mass demonstrations swept through Cotonou, the country's largest city, in December 1989. This upsurge in prodemocracy agitation was partially inspired by the overthrow of Central/Eastern Europe's Marxist-Leninist regimes; ironically, the Stalinist underground Communist Party of Dahomey (PCD) also played a role in organizing much of the unrest. Attempts to quell the demonstrations with force only increased public anger toward the authorities.

(United Nations photo)

Benin is one of the least-developed countries in the world. Beninois must often fend for themselves in innovative ways. The peddler pictured above moves among the lake dwellings of a fishing village, selling cigarettes, spices, rice, and other commodities.

In an attempt to defuse the crisis, the PRPB's state structures were forced to give up their monopoly of power by allowing a representative gathering to convene with the task of drawing up a new constitution. For 10 days in February 1990, the Beninois gathered around their television sets and radios to listen to live broadcasts of the "National Conference of Active Forces of the Nation." The conference quickly turned into a public trial of Kérékou and his PRPB. With the eyes and ears of the nation tuned in, critics of the regime, who had until recently been exiled, were able to pressure Kérékou into handing over effective power to a transitional government. The major task of this new, civilian administration was to prepare Benin for multiparty elections while trying to stabilize the deteriorating economy.

The political success of Benin's "civilian coup d'etat" placed the nation in the forefront of the democratization process then sweeping Africa. But liberating a nation from poverty is a much more difficult process.

A COUNTRY OF MIGRANTS

Benin is one of the least-developed countries in the world. Having for decades experienced only limited economic growth, in recent years the nation's real gross domestic product has actually declined.

HEALTH/WELFARE

One third of the national budget of Benin goes to education, and the percentage of students receiving primary education has risen to 50% of the school-age population. College graduates serve as temporary teachers through the National Service System, but more teachers and higher salaries are needed.

Emigration has become a way of life for many. The migration of Beninois in search of opportunities in neighboring states is not a new phenomenon. Before 1960, educated people from the then-French colony of Dahomey (as Benin was called until 1975) were prominent in junior administrative positions throughout other parts of French West Africa. But as the region's newly independent states began to localize their civil-service staffs, most of the Beninois expatriates lost their jobs. Their return increased bureaucratic competition within Benin, which, in turn, led to heightened political rivalry among ethnic and regional groups. Such local antagonisms contributed to a series of military coups between 1963 and 1972. These culminated in Kérékou's seizure of power.

While Beninois professionals can be found in many parts of West Africa, the destination of most recent emigrants has been Nigeria. The movement from Benin to Nigeria is facilitated by the close links that exist among the large Yoruba-speaking communities on both sides of the border. After Nigeria, the most popular destination has been Côte d'Ivoire. This may change, however, as economic recession in both of those states has led to heightened hostility against the migrants.

THE ECONOMY

Nigeria's urban areas have also been major markets for food exports. This has encouraged Beninois farmers to switch from cash crops (such as cotton, palm oil, cocoa beans, and coffee) to food crops (such as yams and cassava), which are smuggled across the border to Nigeria. The emergence of this parallel export economy has been encouraged by the former regime's practice of paying its farmers among the lowest official produce prices in the region. Given that agriculture, in terms of both employment and income generation, forms the largest sector of the Beninois economy, the rise in smuggling activities has inevitably contributed to a growth of graft and corruption.

ACHIEVEMENTS

Fon appliquéd cloths have been described as "one of the gayest and liveliest of the contemporary African art forms." Formerly these cloths were used by Dahomeyan kings. Now they are sold to tourists, but they still portray the motifs and symbols of past rulers and the society they ruled.

Benin's small industrial sector is primarily geared toward processing primary products, such as palm oil and cotton, for export. It has thus been adversely affected by the shift away from producing these cash crops for the local market. Small-scale manufacturing has centered around the production of basic consumer goods and construction materials. The biggest enterprises are state-owned cement plants. One source of hope is that with privatization and new exploration, the country's small oil industry will undergo expansion.

Transport and trade are other important activities. Many Beninois find legal as well as illegal employment carrying goods. Due to the relative absence of rain forest (an impediment to travel), Benin's territory has historically served as a trade corridor between the coastal and inland savanna regions of West Africa. Today the nation's roads are comparatively well developed, and the railroad carries goods from the port at Cotonou to northern areas of the country. An extension of the railroad will eventually reach Niamey, the capital of Niger. The government has also tried, with little success, to attract tourists in recent years, through such gambits as selling itself as the "home of Voodoo."

POLITICS AND RELIGION

Kérékou's narrow victory margin in 1996 amid charges and countercharges of electoral fraud underscored the continuing north–south division of Beninois politics and society. Although he is now a self-proclaimed Christian, Kérékou's political base remains the mainly Muslim north, while Soglo enjoyed majority support in the more Christianized south. Religious allegiance in Benin is complicated, however, by the prominence of the indigenous belief system known as Voodoo. Having originated in Benin, belief in Voodoo spirits has taken root in the Americas, especially Haiti, as well as elsewhere in West Africa. During his first presidency, Kérékou sought to suppress Voodoo, which he branded as "witchcraft." Soglo, on the other hand, publicly embraced Voodoo, which was credited with helping him recover from a serious illness in 1992. On the eve of the 1996 election, Soglo recognized Voodoo as an official religion, proclaiming January 10 as "Voodoo National Day." (This move may have politically backfired, however, as it was condemned by the influential Catholic archbishop of Cotonou.)

A more decisive factor in Soglo's fall was the failure of his free-market reforms to revive the Beninois economy. Modest initial growth was seriously undermined in 1994 by the massive devaluation of the CFA franc, while privatization led to the retrenchment of 10,000 public workers. It was unlikely, however, that Kérékou would be tempted to return to his own failed policies of "Marxist-Beninism."

Timeline: PAST

1625
The kingdom of Dahomey is established

1892
The French conquer Dahomey and declare it a French protectorate

1960
Dahomey becomes independent

1972
Mathieu Kérékou comes to power in the sixth attempted military coup since independence

1975
The name of Dahomey is changed to Benin

1990s
Kérékou announces the abandonment of Marxism-Leninism as Benin's guiding ideology; multiparty elections are held; Kérékou loses power to Nicephore Soglo; Kérékou is reelected 5 years later

PRESENT

2000s
Benin marks its 40th year of independence

Poverty remains an overwhelming problem

Burkina Faso

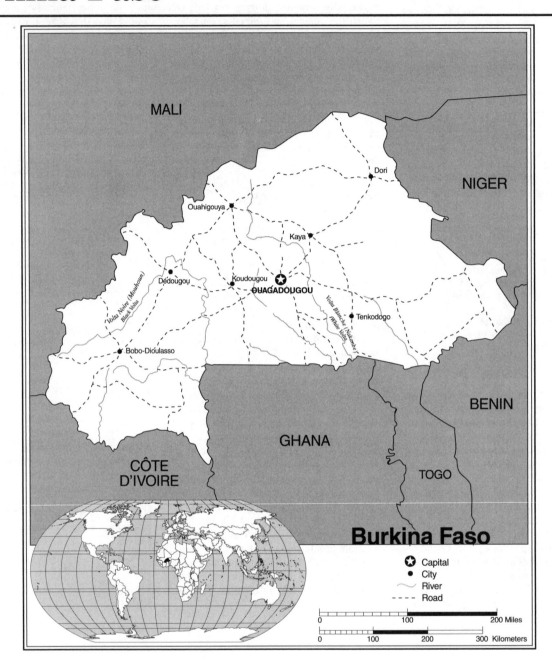

Burkina Faso Statistics

GEOGRAPHY

Area in Square Miles (Kilometers):
106,000 (274,500) (about the size of Colorado)

Capital (Population): Ouagadougou (862,000)

Environmental Concerns: drought; desertification; overgrazing; soil erosion; deforestation

Geographical Features: mostly flat to dissected, undulating plains; hills in west and southeast; landlocked

Climate: tropical; semiarid

PEOPLE

Population

Total: 12,604,000

Annual Growth Rate: 2.64%

Rural/Urban Population Ratio: 82/18

Major Languages: French; Mossi; Senufo; Fula; Bobo; Mande; Gurunsi; Lobi

Ethnic Makeup: about 40% Mossi; Gurunsi; Senufo; Lobi; Bobo; Mande; Fulani

Religions: 50% Muslim; 40% indigenous beliefs; 10% Christian

Health

Life Expectancy at Birth: 45 years (male);
47 years (female)
Infant Mortality: 105.3/1,000 live births
Physicians Available: 1/27,158 people
HIV/AIDS Rate in Adults: 6.44%

Education

Adult Literacy Rate: 36%
Compulsory (Ages): 7–14; free

COMMUNICATION

Telephones: 53,200 main lines
Televisions: 4.4/1,000 people
Internet Users: 10,000 (2001)

TRANSPORTATION

Highways in Miles (Kilometers): 7,504
(12,506)
Railroads in Miles (Kilometers): 385
(622)
Usable Airfields: 33
Motor Vehicles in Use: 55,000

GOVERNMENT

Type: parliamentary
Independence Date: August 5, 1960 (from France)
Head of State/Government: President Blaise Compaoré; Prime Minister Ernest Paramanga Yonli
Political Parties: Congress for Democracy and Progress; African Democratic Rally—Alliance for Democracy and Federation; others
Suffrage: universal

MILITARY

Military Expenditures (% of GDP): 1.4%
Current Disputes: two villages are in a dispute with Benin

ECONOMY

Currency ($ U.S. Equivalent): 752 CFA francs = $1
Per Capita Income/GDP: $1,040/$12.8 billion

GDP Growth Rate: 4.7%
Inflation Rate: 3.5%
Labor Force by Occupation: 90% agriculture
Population Below Poverty Line: 45%
Natural Resources: manganese; limestone; marble; gold; antimony; copper; bauxite; nickel; lead; phosphates; zinc; silver
Agriculture: peanuts; shea nuts; cotton; sesame; millet; sorghum; corn; rice; livestock
Industry: cotton lint; beverages; agricultural processing; soap; cigarettes; textiles; gold
Exports: $265 million (primary partners Venezuela, Benelux, Italy)
Imports: $580 million (primary partners Côte d'Ivoire, Venezuela, France)

SUGGESTED WEB SITES

http://burkinaembassy-usa.org
http://www.sas.upenn.edu/
African_Studies/
Country_Specific/Burkina.html

Burkina Faso Country Report

Notwithstanding some notable achievements, especially in the utilization of the Volta River and in the promotion of indigenous culture, Burkina Faso (formerly called Upper Volta) remains an impoverished country searching for a governing consensus. Recently its government has faced both domestic and external criticism over the state of the economy, human rights, and allegations that it has been involved in the smuggling of arms for diamonds ("blood diamonds") to the now-defeated rebel movements in Sierra Leone and Angola. In response to the latter allegation, in 2001 a UN–supervised body was set up to monitor the country's trade in weapons. Since falling gold prices forced the closure of its biggest gold mine, Burkina Faso has had little in the way of legitimate exports, leaving the landlocked, semiarid country with few economic prospects, and causing many of its citizens to seek opportunities elsewhere.

Much of Burkina Faso's four decades of independence has been characterized by chronic political instability, with civilian rule being interrupted by the military on seven different occasions. The restoration of multiparty democracy in 1991 under the firm guidance of former military leader Blaise Compaoré seemed to usher in an era of greater political stability. Compaoré's reelection in November 1998 was accepted

as legitimate by international observers. But the assassination a month later of independent journalist Robert Zongo touched off a wave of violent strikes and protests. Since then, there have been sustained calls for more fundamental political and social reform from an emerging generation of activists within civil society, including the traditionally powerful trade unions. An umbrella body known as the Collective of Democratic Organizations for the Masses and Political Parties has been formed to challenge the status quo.

DEVELOPMENT

Despite political turbulence, Burkina Faso's economy has recorded positive, albeit modest, annual growth rates for more than a decade. Most of the growth has been in agriculture. New hydroelectric projects have significantly reduced the country's dependence on imported energy.

But power has remained in the hands of Compaoré's hands. Along with his party, the Popular Democratic Organization–Worker's Movement (ODP–MT), he won elections against fragmented opposition in 1991 and 1995, as well as 1998.

Before adopting the mantle of democracy, Compaoré rose to power through a

series of coups, the last of which resulted in the overthrow and assassination of the charismatic and controversial Thomas Sankara. A man of immense populist appeal for many Burkinabé, Sankara remains as a martyr to their unfulfilled hopes. By the time of its overthrow, his radical regime had become the focus of a great deal of external as well as internal opposition.

Of the three men directly responsible for Sankara's toppling, two—Boukari Lingani and Henri Zongo—were executed following a power struggle with the third—Compaoré. It is in this context of sanguinary political competition that the assassination of a prominent media critic has once more called into question the government's commitment to political pluralism.

DEBILITATING DROUGHTS

At the time of its independence from France, in 1960, the landlocked country then named the Republic of Upper Volta inherited little in the way of colonial infrastructure. Since independence, progress has been hampered by prolonged periods of severe drought. Much of the country has been forced at times to depend on international food aid. To counteract some of the negative effects of this circumstance, efforts have been made to integrate relief donations into local development schemes.

(United Nations photo by John Isaac)

Since Burkina Faso gained its independence from France, its progress has been hampered by prolonged periods of drought. Local cooperatives have been responsible for small-scale improvements, such as the construction of the water barrage or barricade pictured above.

Of particular note have been projects instituted by the traditional rural cooperatives known as *naam,* which have been responsible for such small-scale but often invaluable local improvements as new wells and pumps, better grinding mills, and distribution of tools and medical supplies.

FREEDOM

There has been a surprisingly strong tradition of pluralism in Burkina Faso despite the circumscribed nature of human rights under successive military regimes. Freedoms of speech and association are still curtailed, and political detentions are common. The Burkinabé Movement for Human Rights has challenged the government.

Despite such community action, the effects of drought have been devastating. Particularly hard-hit has been pastoral production, long a mainstay of the local economy, especially in the north. It is estimated that a recent drought destroyed about 90 percent of the livestock in Burkina Faso.

To counteract the effects of drought while promoting greater development, the Burkinabé government has developed two major hydroelectric and agricultural projects over the past decade. The Bagre and Kompienga Dams, located east of Ouagadougou, have significantly reduced the country's dependence on imported energy, while also supplying water for large-scale irrigation projects. This has already greatly reduced the need for imported food.

Most Burkinabé continue to survive as agriculturalists and herders, but many people are dependent on wage labor. In the urban centers, there exists a significant working-class population that supports the nation's politically powerful trade-union movement. The division between this urban community and rural population is not absolute, for it is common for individuals to combine wage labor with farming activities. Another population category—whose numbers exceed those of the local wage-labor force—are individuals who seek employment outside of the country. At least 1 million Burkinabé work as migrant laborers in other parts of West Africa. This is part of a pattern that dates back to the early 1900s. Approximately 700,000 of these Burkinabé regularly migrate to Côte d'Ivoire. Returning workers have infused the rural areas with consumer goods and a working-class consciousness.

HEALTH/WELFARE

The inadequacy of the country's public health measures is reflected in the low Burkinabé life expectancy. Mass immunization campaigns have been successfully carried out, but in an era of structural economic adjustment, the prospects for a dramatic improvement in health appear bleak.

UNIONS FORCE CHANGE

As is the case in much of Africa, it is the salaried urban population (at least, next to the army) who have exercised the greatest influence over successive Burkinabé regimes. Trade-union leaders representing these workers have been instrumental in forcing changes in government. They have spoken out vigorously against government efforts to ban strikes and restrain unions. They have also demanded that they be shielded from downturns in the local economy. Although many unionists have championed various shades of Marxist-Leninist ideology, they, along with their natural allies in the civil service, arguably constitute a conservative element within the local society. During the mid-1980s, they became increasingly concerned about the dynamic Sankara's efforts to promote a nationwide network of grassroots Committees for the Defense of the Revolution (CDRs) as vehicles for empowering the nation's largely rural masses.

ACHIEVEMENTS

In 1997, a record total of 19 feature films competed for the Etalon du Yennenga award, the highest distinction of the biannual Pan-African Film Festival, hosted in Ouagadougou. Over the past 3 decades, this festival has contributed significantly to the development of the film industry in Africa. Burkina Faso has nationalized its movie houses, and the government has encouraged the showing of films by African filmmakers.

To many unionists, the mobilization and arming of the CDRs was perceived as a direct challenge to their own status. This threat seemed all the more apparent when Sankara began to cut urban salaries, in the name of a more equitable flow of revenue to the rural areas. When several union leaders challenged this move, they were arrested on charges of sedition. Sankara's

subsequent overthrow thus had strong backing from within organized labor and the civil service. These groups, along with the military, remain the principal supporters of Compaoré's ODP–MT and its policy of "national rectification." Yet despite this support base, the government has moved to restructure the until recently all-encompassing public sector of the economy by reducing its wage bill. This effort has impressed international creditors.

Beyond its core of support, the ODP–MT government has generally been met with sentiments ranging from hostility to indifference. While Compaoré claimed—with some justification—that Sankara's rule had become too arbitrary and that he had resisted forming a party with a set of rules, many people mourned the fallen leader's death. In the aftermath of the coup, the widespread use of a new cloth pattern, known locally as "homage to Sankara," became an informal barometer of popular dissatisfaction. Compaoré has also been challenged by the high regard that has been accorded Sankara outside Burkina Faso, as a symbol of a new generation of African radicalism.

Compaoré, like Sankara, has sometimes resorted to sharp anti-imperialist rhetoric. However, his government has generally sought to cultivate good relations with France (the former colonial power) and other members of the Organization for Economic Cooperation and Development, as well as the major international financial institutions. But he has alienated himself from some of his West African neighbors, as well as the Euro–North American diplomatic consensus, through his close ties to Libya and past military support for Charles Taylor's National Patriotic Front in Liberia. Along with Taylor, Compaoré has more recently been accused of, but denies,

providing support for the Revolutionary United Front rebels in Sierra Leone. To many outsiders, as well as the Burkinabé people themselves, the course of Compaoré's government remains ambiguous.

Timeline: PAST

1313
The first Mossi kingdom is founded

1896
The French overcome Mossi resistance and claim Upper Volta

1932
Upper Volta is divided among adjoining French colonies

1947
Upper Volta is reconstituted as a colony

1960
Independence under President Maurice Yameogo

1980s
Captain Thomas Sankara seizes power and changes the country's name to *Burkina* (Mossi for "land of honest men") *Faso* (Dioula for "democratic republic"); Sankara is assassinated in a coup; Blaise Compaoré succeeds as head of state

1990s
Compaoré introduces multipartyism, but his critics are skeptical

PRESENT

2000s
The country marks 4 decades of independence

Burkina Faso is believed to be involved in the "blood diamonds" trade

Cape Verde (Republic of Cape Verde)

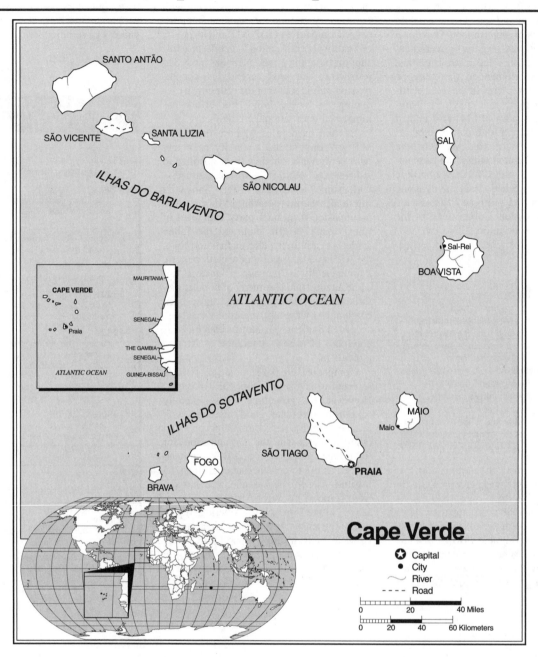

Cape Verde Statistics

GEOGRAPHY

Area in Square Miles (Kilometers): 1,557 (4,033) (about the size of Rhode Island)

Capital (Population): Praia (82,000)

Environmental Concerns: soil erosion; overgrazing; deforestation; desertification; threats to wildlife populations; overfishing

Geographical Features: steep; rugged; rocky; volcanic

Climate: temperate; precipitation meager and very erratic

PEOPLE

Population

Total: 409,000
Annual Growth Rate: 0.85%

Rural/Urban Population Ratio: 39/61
Major Languages: Portuguese; Kriolu
Ethnic Makeup: 71% Creole (mixed); 28% African; 1% European
Religions: Roman Catholicism fused with indigenous beliefs

Health

Life Expectancy at Birth: 66 years (male); 73 years (female)

Infant Mortality: 51.8/1,000 live births
Physicians Available: 1/4,208 people
HIV/AIDS Rate in Adults: 0.04%

Education

Adult Literacy Rate: 72%
Compulsory (Ages): 7–11

COMMUNICATION

Telephones: 61,000 main lines
Televisions: 2.6/1,000 people
Internet Users: 8,000 (2001)

TRANSPORTATION

Highways in Miles (Kilometers): 686 (1,100)
Railroads in Miles (Kilometers): none
Usable Airfields: 6
Motor Vehicles in Use: 18,000

GOVERNMENT

Type: republic
Independence Date: July 5, 1975 (from Portugal)
Head of State/Government: President Pedro Pires; Prime Minister José Maria Neves
Political Parties: Movement for Democracy; African Party for Independence of Cape Verde; Party for Democratic Convergence; Party of Work and Solidarity; others
Suffrage: universal at 18

MILITARY

Military Expenditures (% of GDP): 1.6%
Current Disputes: none

ECONOMY

Currency ($ U.S. Equivalent): 104 escudos = $1

Per Capita Income/GDP: $1,500/$600 million
GDP Growth Rate: 3%
Inflation Rate: 3%
Unemployment Rate: 21%
Population Below Poverty Line: 30%
Natural Resources: salt; basalt rock; pozzuolana; limestone; kaolin; fish
Agriculture: bananas; corn; beans; sweet potatoes; sugarcane; coffee; fish
Industry: fish processing; salt mining; shoes and garments; ship repair; food and beverages
Exports: $27.3 million (primary partners Portugal, United Kingdom, Germany)
Imports: $218 million (primary partners Portugal, Germany, France)

SUGGESTED WEB SITES

http://www.sas.upenn.edu/
African_Studies/
Country_Specific/C_Verde.html
http://virtualcapeverde.net

Cape Verde Country Report

On July 5, 2000, Cape Verdeans proudly celebrated a quarter-century of economic, political, and social progress since independence from Portugal. Despite a late start and unfavorable environmental conditions, the country has emerged as one of postcolonial Africa's tangible success stories.

Over the past decade this has been accompanied by a change in political direction. In 1992, Cape Verde adopted a new flag and Constitution, reflecting the country's transition to political pluralism. After 15 years of single-party rule by the African Party for the Independence of Cape Verde (PAICV), rising agitation led to the legalization of opposition groups in 1990. In January 1991, a quickly assembled antigovernment coalition, the Movement for Democracy (MPD), stunned the political establishment by gaining 68 percent of the votes and 56 out of 79 National Assembly seats. A month later, the MPD candidate, Antonio Mascarenhas Monteiro, defeated the long-serving incumbent, Aristides Pereira, in the presidential elections. It is a credit to both the outgoing administration and its opponents that this dramatic political transformation occurred without significant violence or rancor.

Parliamentary elections in December 1995 resulted in the MPD retaining power, albeit with a reduced majority. In 2001, the political pendulum swung back to PAICV in both the legislative and presidential elec-

tions, with Pedro Pires becoming the country's new leader.

The Republic of Cape Verde is an archipelago located about 400 miles west of the Senegalese Cape Verde, or "Green Cape," after which it is named. Unfortunately, green is a color that is often absent in the lives of the islands' citizens. Throughout its history, Cape Verde has suffered from periods of prolonged drought, which before the twentieth century were often accompanied by extremely high mortality rates (up to 50 percent). The last severe drought lasted from 1968 to 1984. Even in normal years, though, rainfall is often inadequate.

DEVELOPMENT

In a move designed to attract greater investment from overseas, especially from Cape Verdean Americans, the country has joined the International Finance Corporation. Efforts are under way to promote the islands as an offshore banking center for the West African (ECOWAS) region.

When the country gained independence, in 1975, there was little in the way of nonagricultural production. As a result, the new nation had to rely for its survival on foreign aid and the remittances of Cape Verdeans working abroad, but the post-independence period has been marked by a genuine im-

provement in the lives of most Cape Verdeans.

Cape Verde was ruled by Portugal for nearly 500 years. Most of the islanders are the descendants of Portuguese colonists, many of whom arrived as convicts, and African slaves who began to settle on the islands shortly after their discovery by Portuguese mariners in 1456. The merging of these two groups gave rise to the distinct Cape Verdean Kriolu language (which is also spoken in Guinea-Bissau). Under Portuguese rule, Cape Verdeans were generally treated as second-class citizens, although a few rose to positions of prominence in other parts of the Portuguese colonial empire. Economic stagnation, exacerbated by cycles of severe drought, caused many islanders to emigrate elsewhere in Africa, Western Europe, and the Americas.

FREEDOM

The new Constitution should entrench the country's recent political liberalization. Opposition publications have emerged to complement the state- and Catholic Church–sponsored media

In 1956, the African Party for the Independence of Guinea-Bissau and Cape Verde (PAIGC) was formed under the dynamic leadership of Amilcar Cabral, a Cape Verdean revolutionary who hoped to see the two Portuguese colonies form a united nation. Between 1963 and 1974, PAIGC waged a successful war of liberation in Guinea-Bissau that led to the independence of both territories. Although Cabral was assassinated by the Portuguese in 1973, his vision was preserved during the late 1970s by his successors, who, while ruling the two countries separately, maintained the unity of the PAIGC. This arrangement, however, began to break down in the aftermath of a 1980 coup in Guinea-Bissau and resulted in the party's division along national lines. In 1981, the Cape Verdean PAIGC formally renounced its Guinean links, becoming the PAICV.

HEALTH/WELFARE

Greater access to health facilities has resulted in a sharp drop in infant mortality and a rise in life expectancy. Clinics have begun to encourage family planning. Since independence, great progress has taken place in social services. Nutrition levels have been raised, and basic health care is now provided to the entire population.

After independence, the PAIGC/CV government was challenged by the colonial legacy of economic underdevelopment, exacerbated by drought. Massive famine was warded off through a reliance on imported foodstuffs, mostly received as aid. The government attempted to strengthen local food production and assist the 70 percent of the local population engaged in subsistence agriculture. Its efforts took the forms of drilling for underground water, terracing, irrigating, and building a water-desalinization plant with U.S. assistance. Major efforts were also devoted to tree-planting schemes as a way to cut back on soil erosion and eventually make the country self-sufficient in wood fuel.

ACHIEVEMENTS

Cape Verdean Kriolu culture has a rich literary and musical tradition. With emigrant support, Cape Verde bands have acquired modest followings in Western Europe, Lusophone Africa, Brazil, and the United States. Local drama, poetry, and music are showcased on the national television service.

With no more than 15 percent of the islands' territory potentially suitable for cultivation, the prospect of Cape Verde developing self-sufficiency in food appears remote. The few factories that exist on Cape Verde are small-scale operations catering to local needs. Only textiles have enjoyed modest success as an export. Another promising area is fishing.

Timeline: PAST

1462
Cape Verdean settlement begins

1869
Slavery is abolished

1940s
Thousands of Cape Verdeans die of starvation during World War II

1956
The PAIGC is founded

1973
Warfare begins in Guinea-Bissau; Amilcar Cabral is assassinated

1974
A coup in Lisbon initiates the Portuguese decolonization process

1975
Independence

1990s
The PAICV is defeated by the MPD in the country's first multiparty elections; Cape Verde adopts a new Constitution

PRESENT

2000s
Cape Verdeans celebrate 25 years of independence

Power shifts back to PAICV, with Pedro Pires becoming president

Côte d'Ivoire (Republic of Côte d'Ivoire)

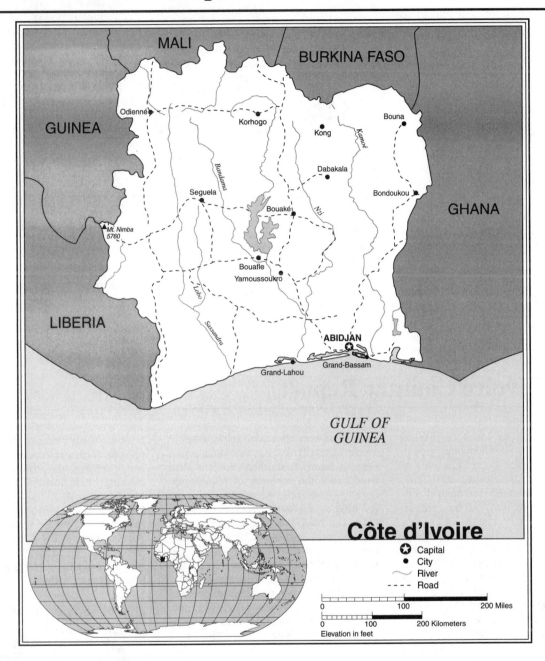

Côte d'Ivoire Statistics

GEOGRAPHY

Area in Square Miles (Kilometers):
124,503 (323,750) (about the size of New Mexico)

Capital (Population): Abidjan (administrative) (3,956,000); Yamoussoukro (political) (120,000)

Environmental Concerns: water pollution; deforestation

Geographical Features: mostly flat to undulating plains; mountains in the northwest

Climate: tropical to semiarid

PEOPLE

Population

Total: 16,805,000
Annual Growth Rate: 2.45%

Rural/Urban Population Ratio: 54/46

Major Languages: French; Dioula; many indigenous dialects

Ethnic Makeup: 42% Akan; 18% Voltaics or Gur; 11% Krous; 16% Northern Mandes; 10% Southern Mandes; 3% others

Religions: 60% Muslim; 22% Christian; 18% indigenous

Health

Life Expectancy at Birth: 43 years (male);
46 years (female)
Infant Mortality: 92.2/1,000 live births
Physicians Available: 1/11,745 people
HIV/AIDS Rate in Adults: 10.76%

Education

Adult Literacy Rate: 48.5%
Compulsory (Ages): 7–13; free

COMMUNICATION

Telephones: 264,000 main lines
Televisions: 57/1,000 people
Internet Users: 10,000 (2001)

TRANSPORTATION

Highways in Miles (Kilometers): 30,240
(50,400)
Railroads in Miles (Kilometers): 408
(660)
Usable Airfields: 36
Motor Vehicles in Use: 255,000

GOVERNMENT

Type: republic
Independence Date: August 7, 1960 (from
France)
Head of State/Government: President
Laurent Gbagbo; Prime Minister Affi
N'Guessan
Political Parties: Democratic Party of
Côte d'Ivoire; Ivoirian Popular Front;
Rally of the Republicans; Ivoirian
Workers' Party; others
Suffrage: universal at 18

MILITARY

Military Expenditures (% of GDP): 1.3%
Current Disputes: civil war

ECONOMY

Currency ($ U.S. Equivalent): 752 CFA
francs = $1
Per Capita Income/GDP: $1,550/$25.5
billion
GDP Growth Rate: -1%
Inflation Rate: 2.5%

Unemployment Rate: 13%
Natural Resources: petroleum; diamonds;
manganese; iron ore; cobalt; bauxite;
copper; hydropower
Agriculture: coffee; cocoa beans; bananas;
palm kernels; corn; rice; manioc; sweet
potatoes; sugar; cotton; rubber; timber
Industry: foodstuffs; beverages; oil
refining; wood products; textiles;
automobile assembly; fertilizer;
construction materials; electricity
Exports: $3.6 billion (primary partners
France, the Netherlands, United States)
Imports: $2.4 billion (primary partners
France, Nigeria, China)

SUGGESTED WEB SITES

http://www.sas.upenn.edu/
    African_Studies/
    Country_Specific/Cote.html

http://geography.miningco.com/
    library/maps/blcote.htm

http://www.africanews.org/west/
    ivorycoast/

Côte d'Ivoire Country Report

Once considered an island of political stability and a model of economic growth in West Africa, since the death in 1993 of its first president Félix Houphouët-Boigny, Côte d'Ivoire (previously known by its English name, Ivory Coast) has been shaken by a series of military crises as well as sustained economic decline. In Sptember 2002, a military mutiny sparked fighting that has left the country divided along regional and sectarian lines. The predominantly Muslim northern half of the country has come under the control of rebel soldiers, while the mainly Christian southern half has remained under the rule of the embattled government of Laurent Ghagbo. Although a partial truce between the two sides, upheld by French troops, was negotiated in October, intensive mediation efforts by neighboring states failed to reconcile the two sides. By the end of 2002, the opportunity for a quick end to the crisis appeared to be fading.

POLITICAL POLARIZATION

Religious and ethnic divisions among Ivoirians in recent years have been aggravated by growing xenophobia against immigrants, who make up at least one third of the country's total population. Under Houphouët-Boigny, people from other Af-

rican states were allowed to settle and even vote in Côte d'Ivoire. For more than a half-century, Ivoirians and non-Ivoirians alike lived under the certainty of Houphouët-Boigny's leadership. Known by friend and foe alike in his latter years as *Le Vieux* ("The Old Man"), he was a dominant figure not only in Côte d'Ivoire but also throughout Francophone Africa. A pioneering Pan-Africanist, he had served for three years as a French cabinet minister before leading his country to independence in 1960. During his subsequent 33-year rule, the Côte d'Ivoire was seemingly conspicuous for its social harmony as well as economic growth. But his paternalistic autocracy, exercised through the Democratic Party of Côte d'Ivoire (PDCI), had begun to break down before his death.

Political life entered a new phase in 1990. Months of mounting prodemocracy protests and labor unrest had led to the legalization of opposition parties, previously banned under the country's single-party government; and to the emergence within the PDCI itself, of a reformist wing seemingly committed to the liberalization process. Although the first multiparty presidential and legislative elections in October–November 1990 were widely regarded as having been less than free and fair by outside observers as well as the op-

position, many believed that the path was open for further reform. But as Le Vieux's health declined, the reform process was increasingly held hostage by tensions within the PDCI as well as between it and opposition movements, of which the most prominent was Laurent Gbagbo's Ivoirian Popular Front (FPI). Gbagbo and others were briefly jailed in 1992 on charges of inciting violence after mass demonstrations turned to rioting.

DEVELOPMENT

It has been said that Côte d'Ivoire is "power hungry." The Soubre Dam, being developed on the Sassandra River, is the sixth and largest hydroelectric project in Côte d'Ivoire. It will serve the eastern area of the country. Another dam is planned for the Cavalla River, between Côte d'Ivoire and Liberia.

Houphouët-Boigny was succeeded by Konan Bédié, a Christian southerner who came out ahead in a power struggle with Allassane Ouatarra, a northern technocrat who had occupied the post of prime minister. Once in power, Bédié stirred up ethnic discord and xenophobia against Muslim northerners. In 1995, he retained the presidency in elections boycotted by supporters of Ouatarra, who was banned from running for office due to a new law mandating that both parents of any presidential candidate must have been born in Côte d'Ivoire. The boycott enjoyed widespread support in the north. For the first time, immigrants were also banned from voting. Violent protests prior to the poll were met with repression.

FREEDOM

Former president Konan Bédié showed little tolerance for dissent, within either the PDCI or society as a whole. Journalists by the score were jailed for such "offenses" as writing "insulting" articles. Six Ivoirian gendarmes were charged in connection with a mass grave discovered near Abidjan in 2000.

Bédié's increasingly unpopular rule came to an abrupt end on December 24, 1999, when General Robert Guei assumed power following the country's first coup d'état. Initial international condemnation of the end of 39 years of uninterrupted civilian rule was muted by the obvious jubilation with which many greeted Bédié's overthrow. There was hope that the divisions of the Bédié era might be laid to rest. Guei reached out to Ouatarra and his supporters as well as other opposition and PDCI members. A new Constitution was drafted and accepted in a referendum. But political goodwill evaporated when, in a move designed to assure his own election, Guei excluded Ouatarra from running for president by reintroducing the provision that both parents of candidates must be Ivoirian. An attempted second coup by northern officers was then crushed, further increasing tensions.

With Bédié and others barred, Gbagbo was the only serious contender allowed to run against Guei in the October 2000 elections. Having lost the ballot, Guei's further attempts to rig the election results were frustrated by a popular uprising that led to Gbagbo's assumption of power. Many of Ouattara's supporters were killed following the rejection of their leader's call for new elections. Opposition boycotts of the December 2000 legislative elections resulted in Gbagbo's Ivoirian Popular Front emerging as the biggest single party in Parliament, with a turnout of only 33 percent. This was followed by another failed coup attempt in January 2001 and subsequent security clapdowns. Nevertheless, there were further calls for fresh presidential and legislative elections in March after Ouattara's party gained a majority at local polls.

In a move toward reconciliation, Gbagvo set up a "National Reconciliation Forum" in October 2001. This resulted in Outtara's return from a year-long exile and a subsequent, January 2002, meeting between the country's "big four"—President Gbagbo, Ouatarra, Guei, and Bédié. But the goodwill created by the talks has since collapsed in the wake of fighting, which in its first days resulted in Guei's death, in disputed circumstances.

ECONOMIC DOWNTURN

The on-going cycle of reform, repression, and increasingly violent political conflict has been taking place against the backdrop of a prolonged deterioration in Côte d'Ivoire's once-vibrant economy. The primary explanation for this downturn is the decline in revenue from cocoa and coffee, which have long been the country's principal export earners. This has led to mounting state debt, which in turn has pressured the government to adopt unpopular austerity measures.

The economy's current problems and prospects are best understood in the context of its past performance. During its first two decades of independence, Côte d'Ivoire enjoyed one of the highest economic growth rates in the world. This growth was all the more notable in that, in contrast to many other developing-world "success stories" during the same period, it had been fueled by the expansion of commercial agriculture. The nation had become the world's leading producer of cocoa and third-largest coffee producer.

HEALTH/WELFARE

Côte d'Ivoire has one of the lowest soldier-to teacher ratios in Africa. Education absorbs about 40% of the national budget. The National Commission to Combat AIDS has reported significant success in its campaign to promote condom use, by targeting especially vulnerable groups.

Although prosperity gave way to recession during the 1980s, the average per capita income of the country remained one of Africa's highest. Statistics also indicated that, on average, Ivoirians lived longer and better than people in many neighboring states. But the creation of a productive, market-oriented economy did not eliminate the reality of widespread poverty, leading some to question whether the majority of Ivoirians have derived reasonable benefit from their nation's wealth.

To the further dismay of many young Ivoirians struggling to enter the country's tight job market, much of the political and economic life of Côte d'Ivoire is controlled by its large and growing expatriate population, largely comprised of French and Lebanese. The size of these communities has multiplied since independence. Many foreigners are now quasi-permanent residents who have thrived while managing plantations, factories, and commercial enterprises. Others can be found in senior civil-service positions.

Another group who have until recently prospered are the commercial farmers, who include millions of medium- and small-scale producers. About two thirds of the workforce are employed in agriculture, with coffee alone being the principal source of income for some 2.5 million people. In addition to coffee, Ivoirian planters grow cocoa, bananas, pineapples, sugar, cotton, palm oil, and other cash crops for export. While some of these farmers are quite wealthy, most have only modest incomes.

ACHIEVEMENTS

Ivoirian textiles are varied and prized. Block printing and dyeing produce brilliant designs; woven cloths made strip by strip and sewn together include the white Korhogo tapestries, covered with Ivoirian figures, birds, and symbols drawn in black. The Ivoirian singer Alpha Blondy has become an international superstar as the leading exponent of West African reggae.

In recent years, the circumstance of Ivoirian coffee and cocoa planters has become much more precarious, due to fluctuations in commodities prices. In this respect, the growers, along with their colleagues elsewhere, are to some extent victims of their own success. Their productivity, in response to international demand, has been a factor in depressing prices through increased supply. In 1988, Houphouët-Boigny held cocoa in storage in an attempt to force a price rise, but the effort failed, aggravating the nation's economic downturn. As a result, the government has taken a new approach, scrapping plans for future expansion in cocoa production in favor of promoting food crops such as yams, corn, and plantains, for which there is a regional as well as a domestic market.

Until recently, Ivoirian planters continued to hire low-paid laborers from other West African countries. At any given time, there have been about 2 million migrant laborers in Côte d'Ivoire, employed throughout the economy. Their presence is not a new phenomenon but goes back to colonial times. Many laborers come from Burkina Faso, which was once a part of Côte d'Ivoire. A good road system and the Ivoirian railroad (which extends to the Burkinabé capital of Ouagadougou) facilitated the travel of migrant workers to rural as well as urban areas.

DEBT AND DISCONTENT

Other factors may determine how much an Ivoirian benefits from the country's development. Residents of Abidjan, the capital, and its environs near the coast receive more services than do citizens of interior areas. Professionals in the cities make better salaries than do laborers on farms or in small industries. Yet persistent inflation and recession have made daily life difficult for the middle class as well as poorer peasants and workers.

The nonagricultural sectors of the national economy have also been experiencing difficulties. Many state industries are unable to make a profit due to their heavy indebtedness. Serious brush fires, mismanagement, and the clearing of forests for cash-crop plantations have put the nation's once-sizable timber industry in jeopardy. Out of a former total of 12 million hectares of forest, 10 1/2 million have been lost. Plans for expansion of offshore oil production have not been implemented due to an inability to raise investment capital.

Difficulty in raising capital for oil development is a reflection of the debt crisis that has plagued the country since the collapse of its cocoa and coffee earnings. Finding itself in the desperate situation of being forced to borrow to pay interest on its previous loans, the government suspended most debt repayments in 1987. Subsequent rescheduling of negotiations with international creditors resulted in a Structural Adjustment Plan (SAP). This plan has resulted in a reduction in the prices paid to farmers and a drastic curtailment in public spending, leading to severe salary cuts for public and parastatal workers. Recent pressure on the part of the international lending agencies for the Ivoirian government to cut back further on its commitment to cash crops is particularly ironic, given the praise that they bestowed on the same policies in the not-too-distant past. Many also see the imposition of such conditions as hypocritical given the heavy rate of agricultural subsidies within the European Union and United States. Those subsidies have further disadvantaged Ivoirian, along with other African, commercial farmers. With the country now on the brink of full-scale civil war, for most Ivoirians the harsh economic conditions are likely to continue to get worse before they get better.

THE SEARCH FOR STABILITY

The ability of various Côte d'Ivoire governments to gain acceptance for austerity measures has been compromised by corruption and extravagance at the top. A notorious example of the latter was the basilica that was constructed at Yamoussoukro, the home village of Houphouët-Boigny (which also at great cost was made the nation's new capital city before his death). The air-conditioned basilica, patterned after the papal seat of St. Peter's in Rome, is the largest Christian church building in the world. Supposedly a personal gift of Houphouët-Boigny to the Vatican—a most reluctant recipient—its three-year construction is believed to have cost hundreds of millions of U.S. dollars.

The course of domestic conflict in Côte d'Ivoire is being watched elsewhere. For decades Houphouët-Boigny was the doyen of the more conservative, pro-Western leaders in Africa. Hostile to Libya and receptive to both Israel and to "dialogue" with South Africa, Côte d'Ivoire has remained especially close to France, which continues to maintains a military presence in the country. During the recent fighting, U.S. marines also intervened to evacuate foreign nationals.

Timeline: PAST

1700s
Agni and Baoulé peoples migrate to the Ivory Coast from the East

1893
The Ivory Coast officially becomes a French colony

1898
Samori Touré, a Malinke Muslim leader and an empire builder, is defeated by the French

1915
The final French pacification of the country takes place

1960
Côte d'Ivoire becomes independent under Félix Houphouët-Boigny's leadership

1980s
The PDCI approves a plan to move the capital from Abidjan to Houphouët-Boigny's home village of Yamoussoukro

1990s
Prodemocracy demonstrations lead to multiparty elections; Houphouët-Boigny dies

PRESENT

2000s
Côte d'Ivoire adjusts after the startling coup in late 1999

Laurent Gbagbo becomes president

A mass grave of 57 bullet-ridden bodies is discovered near Abidjan

The Gambia (Republic of The Gambia)

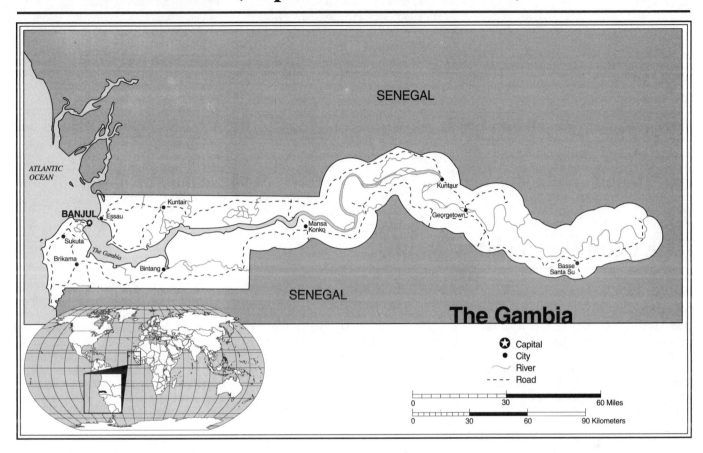

The Gambia Statistics

GEOGRAPHY

Area in Square Miles (Kilometers):
(4,361) (11,295) (about twice the size of
Delaware)
Capital (Population): Banjul (418,000)
Environmental Concerns: deforestation;
desertification; water-borne diseases
Geographical Features: floodplain of The
Gambia River flanked by some low hills
Climate: tropical; hot rainy season, cooler
dry season

PEOPLE

Population

Total: 1,456,000
Annual Growth Rate: 3.09%
Rural/Urban Population Ratio: 68/32
Major Languages: English; Mandinka;
Wolof; Fula; Sarakola; Diula; others

Ethnic Makeup: 42% Mandinka; 18%
Fula; 16% Wolof; 24% others (99%
African; 1% non-Gambian)
Religions: 90% Muslim; 9% Christian; 1%
indigenous beliefs

Health

Life Expectancy at Birth: 52 years (male);
56 years (female)
Infant Mortality: 76.3/1,000 live births
Physicians Available: 1/14,536 people
HIV/AIDS Rate in Adults: 1.95%

Education

Adult Literacy Rate: 47.5%
Compulsory (Ages): 7–13; free

COMMUNICATION

Telephones: 32,000 main lines
Internet Users: 5,000 (2001)

TRANSPORTATION

Highways in Miles (Kilometers): 1,584
(2,640)
Railroads in Miles (Kilometers): none
Usable Airfield: 1
Motor Vehicles in Use: 9,000

GOVERNMENT

Type: republic
Independence Date: February 18, 1965
(from the United Kingdom)
Head of State/Government: President
Yahya Jammeh is both head of state and
head of government
Political Parties: Alliance for Patriotic
Reorientation and Construction;
National Reconciliation Party; People's
Democratic Organization for
Independence and Socialism; others

Suffrage: universal at 18

MILITARY

Military Expenditures (% of GDP): 0.3%
Current Disputes: internal conflicts;
boundary dispute with Senegal

ECONOMY

Currency ($ U.S. Equivalent): 25.46
dalasis = $1
Per Capita Income/GDP: $1,770/$2.5
billion

GDP Growth Rate: 5.7%
Inflation Rate: 4%
Labor Force by Occupation: 75%
agriculture; 19% industry and services;
6% government
Natural Resources: fish
Agriculture: peanuts; millet; sorghum;
rice; corn, cassava; livestock; fish and
forest resources
Industry: processing peanuts, fish, and
hides; tourism; beverages; agricultural
machinery assembly; wood- and
metalworking; clothing

Exports: $139 million (primary partners
Benelux, Japan, United Kingdom)
Imports: $200 million (primary partners
China, Hong Kong, United Kingdom, the
Netherlands)

SUGGESTED WEB SITES

http://www.gambianews.com
http://www.gambianet.com
http://www.gambia.com/gambia.html
http://www.sas.upenn.edu/
    African_Studies/
    Country_Specific/Gambia.html

The Gambia Country Report

Since his seizure of power in a 1994 coup, Yahya Jammeh has dominated politics in The Gambia. In 2001, he was reelected president in what international election monitors generally viewed as a free and fair poll. The government called for the elections, however, only shortly after Jammeh lifted a ban on politicians whom he had ousted from power.

DEVELOPMENT

Since independence, The Gambia has developed a tourist industry. Whereas in 1966 only 300 individuals were recorded as having visited the country, the figure for 1988–1989 was over 112,000. Tourism is now the second-biggest sector of the economy. Still, tourism has declined since 2000.

Subsequent parliamentary elections, in January 2002, were boycotted by most of the opposition, allowing Jammeh's Alliance for Patriotic Reorientation and Construction to win by a landslide. The Gambian opposition has otherwise been weakened by its internal divisions, all attempts to bring about a united front having failed. Opposition to Jammeh's rule has become more open.

In April 2000, Gambians were shocked when student protests in the capital city, Banjul, resulted in the killing of 14 people and the wounding of many more by government security forces. Many interpreted the violence as an ominous official response to the re-emergence of independent voices within the media and civil society, which have been pushing for greater openness and accountability in government.

Jammeh came to power in July 1994, after The Gambia's armed forces overthrew the government of Sir Dawda Jawara, bringing to an abrupt end what had been postcolonial West Africa's only example of uninterrupted multiparty democracy. Under international pressure, elections were held in September 1996 and January 1997, resulting in victories for Jammeh and the Alliance for Patriotic Reorientation and Construction. But the process was marred by the regime's continuing intolerance of genuine opposition. Since the failure of an alleged coup attempt in January 1995, critical voices have been largely silenced by an increasingly powerful National Intelligence Agency. Meanwhile, The Gambia's already weak economy has suffered from reduced revenues from tourism and foreign donors.

FREEDOM

Despite the imposition of martial law in the aftermath of the 1981 coup attempt, The Gambia has had a strong record of respect for individual liberty and human rights. Under its current regime, The Gambia has forfeited its model record of respect for freedoms of speech and association.

The Gambia is Africa's smallest noninsular nation. Except for a small seacoast, it is entirely surrounded by its much larger neighbor, Senegal. The two nations' separate existence is rooted in the activities of British slave traders who, in 1618, established a fort at the mouth of The Gambia River, from which they gradually spread their commercial and, later, political dominance upstream. Gambians have much in common with Senegalese. The Gambia's three major ethnolinguistic groups—the Mandinka, Wolof, and Fula (or Peul)—are found on both sides of the border. The Wolof language serves as a lingua franca in both the Gambian capital of Banjul and the urban areas of Senegal. Islam is the major religion of both countries, while each also has a substantial Christian minority. The economies of the two countries are also similar, with each being heavily reliant on the cultivation of groundnuts as a cash crop.

HEALTH/WELFARE

Forty percent of Gambian children remain outside the primary-school setup. Economic Recovery Program austerity has made it harder for the government to achieve its goal of education for all.

In 1981, the Senegalese and Gambian governments were drawn closer together by an attempted coup in Banjul. While Jawara was in London, dissident elements within his Paramilitary Field Force joined in a coup attempt with members of two small, self-styled revolutionary parties. Based on a 1965 mutual-defense agreement, Jawara received assistance from Senegal in putting down the rebels. Constitutional rule was restored, but the killing of 400 to 500 people during the uprising and the subsequent mass arrest of suspected accomplices left Gambians bitter and divided.

In the immediate aftermath of the coup, The Gambia agreed to join Senegal in a loose confederation, which some hoped would lead to a full political union. But from the beginning, the Senegambia Confederation was marred by the circumstances of its formation. The continued presence of Senegalese soldiers in their country led Gambians to speak of a "shotgun wedding." Beyond fears of losing their local identity, many believed that proposals for closer economic integration, through a proposed monetary and customs union, would be to The Gambia's disadvantage. Underlying this concern was the role played by Gambian traders in providing imports to Senegal's market. Other squabbles, such as a long-standing dispute over the financing of a bridge across The Gambia River, finally led to the Confederation's formal demise in 1989. But the two countries still recognize a need to develop alternative forms of cooperation.

The Gambia was modestly successful in rebuilding its politics in the aftermath of the 1981 coup attempt. Whereas the 1982 elections were arguably compromised by the detention of the main opposition leader,

Sherif Mustapha Dibba, on (later dismissed) charges of complicity in the revolt, the 1987 and 1992 polls restored most people's confidence in Gambian democracy. In both elections, opposition parties significantly increased their share of the vote, while Jawara's People's Progressive Party retained majority support.

Instances of official corruption had compromised the Jawara government's ability to use its electoral mandate to implement an Economic Recovery Program (ERP), which included austerity measures. The Gambia has always been a poor country. During the 1980s, conditions worsened as a result of bad harvests and falling prices for groundnuts, which usually account for half of the nation's export earnings. The tourist industry was also disrupted by the 1981 coup attempt. Faced with mounting debt, the government submitted to International Monetary Fund pressure by cutting back its civil service and drastically devaluing the local currency. The latter step initially led to high inflation, but prices have become more stable in recent years, and the economy as a whole has begun to enjoy a gross domestic product growth rate of up to 5 percent per year. As elsewhere, the negative impact of Structural Adjustment has proved especially burdensome to urban dwellers.

Timeline: PAST

1618
The British build Fort James at the current site of Banjul, on the Gambia River

1807
The Gambia is ruled by the United Kingdom through Sierra Leone

1965
Independence

1970
Dawda Jawara comes to power

1980s
An attempted coup against President Dawda Jawara; the rise and fall of the Senegambia Confederation

1990s
Jawara is overthrown by a military coup; Yahya Jammeh becomes head of state

PRESENT

2000s
Government security forces kill 14 people during student protests

Jammeh is reelected president, but the opposition gains in parliamentary elections

Ghana (Republic of Ghana)

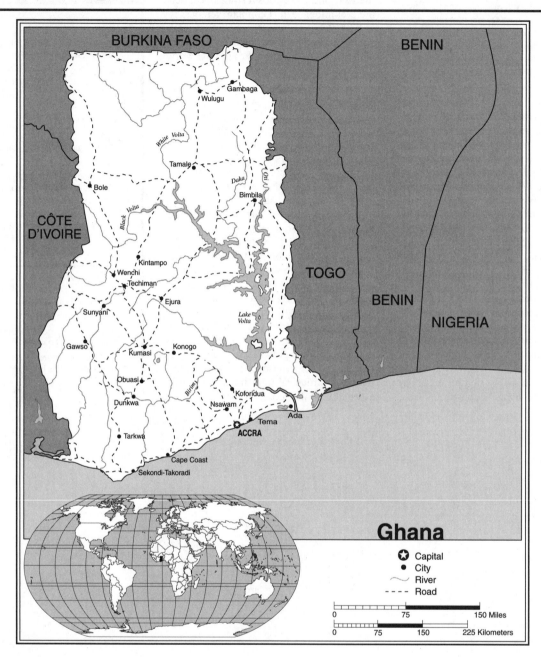

Ghana Statistics

GEOGRAPHY

Area in Square Miles (Kilometers):
92,100 (238,536) (about the size of Oregon)

Capital (Population): Accra (1,925,000)

Environmental Concerns: drought; deforestation; overgrazing; soil erosion; threatened wildlife populations; water pollution; insufficient potable water

Geographical Features: low plains with dissected plateau

Climate: tropical

PEOPLE

Population

Total: 20,245,000

Annual Growth Rate: 1.7%

Rural/Urban Population Ratio: 62/38

Major Languages: English; Akan; Ewe; Ga

Ethnic Makeup: nearly 100% African

Religions: 63% Christian; 21% indigenous beliefs; 16% Muslim

Health

Life Expectancy at Birth: 56 years (male); 59 years (female)

Infant Mortality: 55.6/1,000 live births

Physicians Available: 1/22,452 people

HIV/AIDS Rate in Adults: 3.6%

Education

Adult Literacy Rate: 64.5%

Compulsory (Ages): 6–16

COMMUNICATION

Telephones: 240,000 main lines
Daily Newspaper Circulation: 64/1,000 people
Televisions: 15/1,000 people
Internet Users: 200,000 (2002)

TRANSPORTATION

Highways in Miles (Kilometers): 24,428 (39,400)
Railroads in Miles (Kilometers): 592 (953)
Usable Airfields: 12
Motor Vehicles in Use: 135,000

GOVERNMENT

Type: constitutional democracy
Independence Date: March 6, 1957 (from the United Kingdom)

Head of State/Government: President John Agyekum Kufuor is both head of state and head of government
Political Parties: National Democratic Congress; New Patriotic Party; People's Convention Party; Every Ghanaian Living Everywhere; others
Suffrage: universal at 18

MILITARY

Military Expenditures (% of GDP): 0.7%
Current Disputes: internal conflicts

ECONOMY

Currency ($ U.S. Equivalent): 8,863 cedis = $1
Per Capita Income/GDP: $1,980/$39.4 billion
GDP Growth Rate: .3%
Inflation Rate: 25%
Unemployment Rate: 20%
Labor Force by Occupation: 39% services; 36% agriculture; 25% industry

Population Below Poverty Line: 31.4%
Natural Resources: gold; timber; industrial diamonds; bauxite; manganese; fish; rubber; hydropower
Agriculture: cocoa beans; rice; coffee; cassava; peanuts; corn; shea nuts; bananas; timber; fish
Industry: mining; lumbering; light manufacturing; aluminum smelting; food processing
Exports: $1.94 billion (primary partners Togo, United Kingdom, Italy)
Imports: $2.83 billion (primary partners United Kingdom, Nigeria, United States)

SUGGESTED WEB SITES

http://www.ghana.gov.gh
http://www.ghana.com
http://www.state.gov/www/ background_notes/ ghana_0298_bgn.html
http://www.ghana-embassy.org/
http://www.ghanaweb.com/ GhanaHomePage/ghana.html

Ghana Country Report

In December 2000, John Kufuor defeated then–vice-president John Atta Mills in a ballot that marked the first real transfer of power through elections in Ghana in 4 1/2 decades as an independent republic. The presidential elections also brought to an end two decades of rule by the incumbent Jerry Rawlings, who agreed to step down in accordance with the Constitution. Known as the "Gentle Giant," President Kufuor has sought to promote reconciliation without recrimination in a nation with a violent history of political division.

DEVELOPMENT

In the 1960s, Ghana invested heavily in schooling, resulting in perhaps the best-educated population in Africa. Today, hundreds of thousands of professionals who began their schooling under Nkrumah work overseas, annually remitting an estimated $1 billion to the Ghanaian economy.

In the 1990s, Ghana made gradual but steady progress in rebuilding its economy as well as its political culture, after decades of decline. For many Ghanaians, these gains have been of modest benefit. The country has achieved economic growth, while implementing a socially painful World Bank/International Monetary Fund–sponsored Economic Recovery Program

(ERP) that in 2001 was rewarded with a debt-relief package. Yet when adjusted for inflation, per capita income remains below the level that existed in 1957, when Ghana became the first colony in sub-Saharan Africa to obtain independence.

The country is also overcoming the legacy of political instability brought about by revolving-door military coups. In March 1992, Rawlings marked the country's 35th anniversary of independence by announcing an accelerated return to multiparty rule. Eight months later, he was elected by a large majority as the president of what has been hailed as Ghana's "Fourth Republic." Although the election received the qualified endorsement of international monitors, its result was rejected by the main opposition, the New Patriotic Party (NPP). The NPP subsequently boycotted parliamentary elections, allowing an easy victory for Rawlings's National Democratic Congress (NDC), which captured 189 out of 200 seats, with a voter turnout of just under 30 percent. After months of bitter standoff, the political climate has eased since December 1993, when the NPP agreed to enter into a dialogue with the government about its basic demand that the interests of the ruling party be more clearly separated from those of the state.

Ghana's political transformation was a triumph for Rawlings, who ruled since

1981, when he and other junior military officers seized power as the Provisional National Defense Council (PNDC). In the name of ending corruption, they overthrew Ghana's previous freely elected government after it had been in office for less than two years. Rawlings was dismissive of elections at that time: "What does it mean to stuff bits of paper into boxes?" But political success seems to have altered his opinion.

FREEDOM

The move to multipartyism has promoted freedom of speech and assembly. Dozens of independent periodicals have emerged. In April 2002, a state of emergency was declared in the north, after a prominent local leader and more than 30 others were killed in clan violence.

At its independence, Ghana assumed a leadership role in the struggle against colonial rule elsewhere on the continent. Both its citizens and many outside observers were optimistic about the country's future. As compared to many other former colonies, the country seemed to have a sound infrastructure for future progress. Unfortunately, economic development and political democracy have proven to be elusive goals.

HEALTH/WELFARE

In 1991, the African Commission of Health and Human Rights Promoters established a branch in Accra to help rehabilitate victims of human-rights violations from throughout Anglophone Africa. The staff deals with both the psychological and physiological after-effects of abused ex-detainees.

The "First Republic," led by the charismatic Kwame Nkrumah, degenerated into a bankrupt and an increasingly authoritarian one-party state. Nkrumah had pinned his hopes on an ambitious policy of industrial development. When substantial overseas investment failed to materialize, he turned to socialism. His efforts led to a modest rise in local manufacturing, but the sector's productivity was compromised by inefficient planning, limited resources, expensive inputs, and mounting government corruption. The new state enterprises ended up being financed largely from the export earnings of cocoa, which had emerged as Ghana's principal cash crop during the colonial period. Following colonial precedent, Nkrumah resorted to paying local cocoa farmers well below the world market price for their output in an attempt to expand state revenues.

Nkrumah was overthrown by the military in 1966. Despite his regime's shortcomings, he is still revered by many as the leading pan-African nationalist of his generation. His warnings about the dangers of neo-imperialism have proved prophetic.

ACHIEVEMENTS

In 1993, Ghana celebrated the 30th anniversary of the School of the Performing Arts at the University of Legon. Integrating the world of dance, drama, and music, the school has trained a generation of artists committed to perpetuating Ghanaian, African, and international traditions in the arts.

Since Nkrumah's fall, the army has been Ghana's dominant political institution, although there were brief returns to civilian control in 1969–1972 and again in 1979–1981. Both the military and the civilian governments abandoned much of Nkrumah's socialist commitment, but for years they continued his policy of squeezing the cocoa farmers, with the long-term result of encouraging planters both to cut back on their production and to attempt to circumvent the official prices through smuggling. This situation, coupled with falling cocoa prices on the world market and rising import costs, helped to push Ghana into a state of severe economic depression during the 1970s. During that period, real wages fell by some 80 percent. Ghana's crisis was then aggravated by an unwillingness on the part of successive governments to devalue the country's currency, the cedi, which encouraged black-market trading.

RAWLINGS'S RENEWAL

By 1981, many Ghanaians welcomed the PNDC, seeing in Rawlings's populist rhetoric the promise of change after years of corruption and stagnation. The PNDC initially tried to rule through People's Defense Committees, which were formed throughout the country to act as both official watchdogs and instruments of mass mobilization. Motivated by a combination of idealism and frustration with the status quo, the vigilantism of these institutions threatened the country with anarchy until, in 1983, they were reined in. Also in 1983, the country faced a new crisis, when the Nigerian government suddenly expelled nearly 1 million Ghanaian expatriates, who had to be resettled quickly.

Faced with an increasingly desperate situation, the PNDC, in a move that surprised many, given its leftist leanings, began to implement the Economic Recovery Program. Some 100,000 public and parastatal employees were retrenched, the cedi was progressively devalued, and wages and prices began to reflect more nearly their market value. These steps have led to some economic growth, while annually attracting $500 million in foreign aid and soft loans and perhaps double that amount in cash remittances from the more than 1 million Ghanaians living abroad.

The human costs of ERP have been a source of criticism. Many ordinary Ghanaians, especially urban salary-earners, have suffered from falling wages coupled with rising inflation. Unemployment has also increased in many areas (today it is estimated at about 20 percent). Yet a recent survey found surprisingly strong support among "urban lower income groups" for ERP and the government in general. In the countryside, farmers have benefited from higher crop prices and investments in rural infrastructure, while there has been a countrywide boom in legitimate retailing.

ERP continues to have its critics, but it gained substantial support from politicians aligned with Ghana's three principal political tendencies: the Nkrumahists, loyal to the first president's pan-African socialist vision; the Danquah-Busia grouping, named after two past statesmen who struggled against Nkrumah for more liberal economic and political policies; and those loyal to the PNDC. In the November 1992 presidential elections, the NPP emerged as the main voice of the Danquah-Busia camp, while Rawlings's NDC attracted substantial support from Nkrumahists as well as those sympathetic to his own legacy. There was also a body of opinion that was critical of all three historic tendencies, characterizing the NPP and NDC as fronts for power-hungry men fighting yesterday's battles. During the April 1992 referendum to approve the new Constitution, more than half the registered voters (many Ghanaians complained that they were denied registration) failed to participate, despite the government and opposition's joint call for a large "yes" vote. Many also boycotted the November presidential elections. In December 1996, in a poll widely judged to have been fair, Rawlings was reelected. He narrowly defeated his former vice-president, John Kufuor, who enjoyed the backing of both the New Patriotic Party and the Convention People's Party—the same man who eventually would come to succeed him as president.

Timeline: PAST

1482
A Portuguese fort is built at Elmina

1690s
The establishment of the Asante Confederation under Osei Tutu

1844
The "Bonds" of 1844 signed by British officials and Fante chiefs as equals

1901
The British conquer the Asante, a final step in British control of the region

1957
Ghana is the first of the colonial territories in sub-Saharan Africa to become independent

1966
Kwame Nkrumah is overthrown by a military coup

1979
The first coup of Flight Lieutenant Jerry Rawlings

1980s
The second coup by Rawlings; the PNDC is formed

1990s
World Bank and IMF austerity measures; prodemocracy agitation leads to a transition to multiparty rule

PRESENT

2000s
Ghanaians elect John Kufuor of the opposition NPP as president, ending Rawlings's 2-decade rule

Guinea (Republic of Guinea)

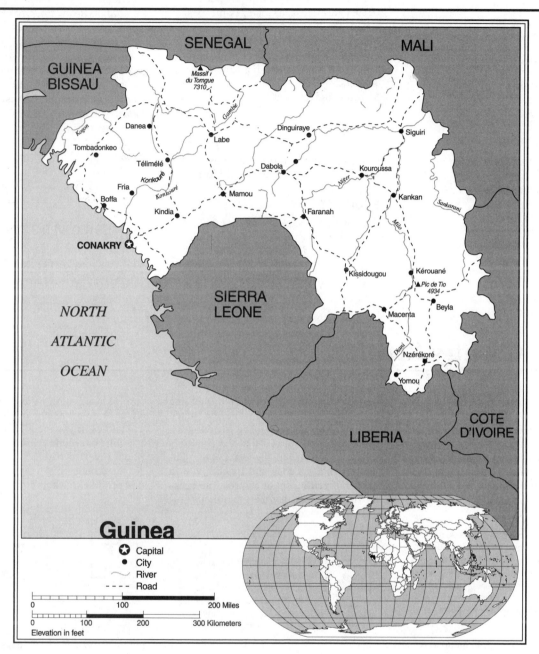

GUINEA BISSAU

SENEGAL

MALI

Massif du Tomgue 7310

Danea

Tombadonkeo

Labe

Dinguiraye

Siguiri

Télimélé

Konkouré

Konkouré

Dabola

Kouroussa

Niger

Fria

Boffa

Mamou

Faranah

Kankan

Sankarani

Kindia

CONAKRY

Milo

NORTH

ATLANTIC

OCEAN

SIERRA LEONE

Kissidougou

Kérouané

Pic de Tio 4934

Beyla

Macenta

Diani

Nzérékoré

LIBERIA

COTE D'IVOIRE

Yomou

Guinea

⭐ Capital
● City
〜 River
--- Road

0   100   200 Miles
0   100   200   300 Kilometers
Elevation in feet

Guinea

GEOGRAPHY

Area in Square Miles (Kilometers):
95,000 (246,048) (about the size of
Oregon)

Capital (Population): Conakry
(1,272,000)

Environmental Concerns: deforestation;
insufficient potable water;
desertification; soil erosion and
contamination; overfishing;
overpopulation

Geographical Features: mostly flat
coastal plain; hilly to mountainous
interior

Climate: tropical

PEOPLE

Population

Total: 7,776,000
Annual Growth Rate: 2.23%
Rural/Urban Population Ratio: 68/32

Major Languages: French; many tribal
languages

Ethnic Makeup: 40% Peuhl; 30% Malinke;
20% Soussou; 10% other African groups

Religions: 85% Muslim; 8% Christian; 7%
indigenous beliefs

Health

Life Expectancy at Birth: 43 years (male);
48 years (female)

Infant Mortality: 127/1,000 live births

187

Physicians Available: 1/9,732 people
HIV/AIDS Rate in Adults: 1.54%

Education

Adult Literacy Rate: 36%
Compulsory (Ages): 7–13; free

COMMUNICATION

Telephones: 37,000 main lines
Televisions: 10/1,000 people
Internet Users: 8,000 (2000)

TRANSPORTATION

Highways in Miles (Kilometers): 18,060
  (30,100)
Railroads in Miles (Kilometers): 651
  (1,086)
Usable Airfields: 15
Motor Vehicles in Use: 33,000

GOVERNMENT

Type: republic

Independence Date: October 2, 1958
  (from France)
Head of State/Government: President
  (General) Lansana Conté; Prime
  Minister Lamine Sidime
Political Parties: Party for Unity and
  Progress; Union for the New Republic;
  Rally for the Guinean People; many
  others
Suffrage: universal at 18

MILITARY

Military Expenditures (% of GDP): 3.3%
Current Disputes: refugee crisis as a result
  of unrest in Sierra Leone and Liberia

ECONOMY

Currency ($ U.S. Equivalent): 2,029
  Guinean francs = $1
Per Capita Income/GDP: $1,970/$15
  billion
GDP Growth Rate: 3.3%

Inflation Rate: 6%
Labor Force by Occupation: 80%
  agriculture; 20% industry and services
Population Below Poverty Line: 40%
Natural Resources: bauxite; iron ore;
  diamonds; gold; uranium; hydropower;
  fish
Agriculture: rice; cassava; millet; sweet
  potatoes; coffee; bananas; palm
  products; pineapples; livestock
Industry: bauxite; gold; diamonds;
  alumina refining; light manufacturing
  and agricultural processing
Exports: $695 million (primary partners
  Belgium, United States, Ireland)
Imports: $555 million (primary partners
  France, United States, Belgium)

SUGGESTED WEB SITES

http://www.sas.upenn.edu/
  African_Studies/
  Country_Specific/Guinea.html

Guinea Country Report

In recent years, Guinea has managed to maintain internal peace in the face of armed conflict along its borders. But renewed fighting in neighboring Sierra Leone, Liberia, and Côte d'Ivoire has revived fears that the country is being dragged into a wider regional conflict.

DEVELOPMENT

A measure of economic growth in Guinea was reflected in the rising traffic in Conakry harbor, whose volume rose 415% over a 4-year period. Plans are being made to improve the port's infrastructure, but regional conflicts threaten further development.

Since the end of 2000, incursions by rebels along Guinea's border regions with Liberia and Sierra Leone have claimed more than 1,000 lives and caused massive population displacement. The government has accused the governments of Liberia and Burkina Faso, the (now-disarmed) Revolutionary United Front (RUF) of Sierra Leone, and former Guinean Army mutineers of working together to destabilize Guinea. With tensions rising, Alpha Conde, leader of the opposition Guinean People's Rally (RPG), was sentenced in September 2000 to five years in prison, charged with endangering state security and recruiting foreign mercenaries. By February 2001, the government began de-

ploying attack helicopters to the front line to fight with rebels. Meanwhile, the United Nations high commissioner for refugees, Ruud Lubbers, warned that the country's refugee crisis, mostly the result of the conflicts in Sierra Leone and Liberia, was in danger of getting out of control. The country shelters more than half a million (estimates vary widely) cross-border refugees.

FREEDOM

Human rights continue to be restricted in Guinea, with the government's security forces being linked to disappearances, abuse of prisoners and detainees, torture by military personnel, and inhumane prison conditions.

At home, the harassment of journalists and opposition leaders has underscored the government's continued insecurity despite the 1992 transition to multiparty politics. In a constitutional referendum that took place in November 2001, voters endorsed President Lansana Conté's proposal to extend the presidential term from five to seven years. But the opposition boycotted the poll, accusing Conté of trying to stay in office for life.

Since coming to power in 1984, Conté has proven adept at surviving challenges to his authority. In April 1992, he announced that a new Constitution guaranteeing free-

dom of association would take immediate effect. Within a month, more than 30 political parties had formed. This initiative was a political second chance for a nation whose potential had been mismanaged for decades, under the dictatorial rule of its first president, Sekou Touré.

HEALTH/WELFARE

The life expectancy of Guineans is among the lowest in the world, reflecting the stagnation of the nation's health service during the Sekou Touré years.

From his early years as a radical trade-union activist in the late 1940s until his death in office in 1984, Sekou Touré was Guinea's dominant personality. A descendent of the nineteenth-century Malinke hero Samori Touré, who fiercely resisted the imposition of French rule, Sekou Touré was a charismatic but repressive leader. In 1958, he inspired Guineans to vote for immediate independence from France. At the time, Guinea was the only territory to opt out of Charles de Gaulle's newly established French Community. The French reacted spitefully, withdrawing all aid, personnel, and equipment from the new nation, an event that heavily influenced Guinea's postindependence path. The ability of Touré's Democratic Party of Guinea

(PDG) to step into the administrative vacuum was the basis for Guinea's quick transformation into the African continent's first one-party socialist state, a process that was encouraged by the Soviet bloc.

ACHIEVEMENTS

More than 80% of the programming broadcast by Guinea's television service is locally produced. This output has included more than 3,000 movies. A network of rural radio stations is currently being installed.

Touré's rule was characterized by economic mismanagement and the widespread abuse of human rights. It is estimated that 2 million people—at the time about one out of every four Guineans—fled the country during his rule. At least 2,900 individuals disappeared under detention by the government.

By the late 1970s, Touré, pressured by rising discontent and his own apparent realization of his country's poor economic performance, began to modify both his domestic and foreign policies. This shift led to better relations with Western countries but little improvement in the lives of his people. In 1982, Amnesty International publicized the Touré regime's dismal record of political killings, detentions, and torture, but the world remained largely indifferent.

On April 3, 1984, a week after Touré's death, the army stepped in, claiming that it wished to end all vestiges of the late president's dictatorial regime. The bloodless coup was well received by Guineans. Hundreds of political prisoners were released; and the once-powerful Democratic Party of

Guinea, which during the Touré years had been reduced from a mass party into a hollow shell, was disbanded. A new government was formed, under the leadership of then-colonel Conté; and a 10-point program for national recovery was set forth, including the restoration of human rights and the renovation of the economy.

Faced with an empty treasury, the new government committed itself to a severe Structural Adjustment Program (SAP). This has led to a dismantling of many of the socialist structures that had been established by the previous government. While international financiers have generally praised it, the government has had to weather periodic unrest and coup attempts. In spite of these challenges, however, it has remained committed to SAP.

Guinea is blessed with mineral resources, which could lead to a more prosperous future. The country is rich in bauxite and has substantial reserves of iron and diamonds. New mining agreements, leading to a flow of foreign investment, have already led to a modest boom in bauxite and diamond exports. Small-scale gold mining is also being developed. These initiatives, however, are being threatened by the conflicts in neighboring states.

Guinea's greatest economic failing has been the poor performance of its agricultural sector. Unlike many of its neighbors, the country enjoys a favorable climate and soils. But, although some 80 percent of Guineans are engaged in subsistence farming, only 3 percent of the land is cultivated, and foodstuffs remain a major import. Blame for this situation largely falls on the Touré regime's legacy of an inefficient, state-controlled system of marketing and distribution. In 1987, the government initi-

ated an ambitious plan of road rehabilitation, which, along with better produce prices, has encouraged farmers to produce more for the domestic market.

Timeline: PAST

1700s
A major Islamic kingdom is established in the Futa Djalon

1898
Samori Touré is defeated by the French

1958
Led by Sekou Touré, Guineans reject continued membership in the French Community; an independent republic is formed

1978
French president Giscard d'Estaing visits Guinea: the beginning of a reconciliation between France and Guinea

1980s
Sekou Touré's death is followed by a military coup; the introduction of SAP leads to urban unrest

1990s
President Lansana Conté begins to establish a multiparty democracy; multiparty elections are held for the presidency; Conté claims victory

PRESENT

2000s
Guinea stays the course of Structural Adjustment despite severe hardships

Fears intensify regarding a regional conflict

Guinea-Bissau (Republic of Guinea-Bissau)

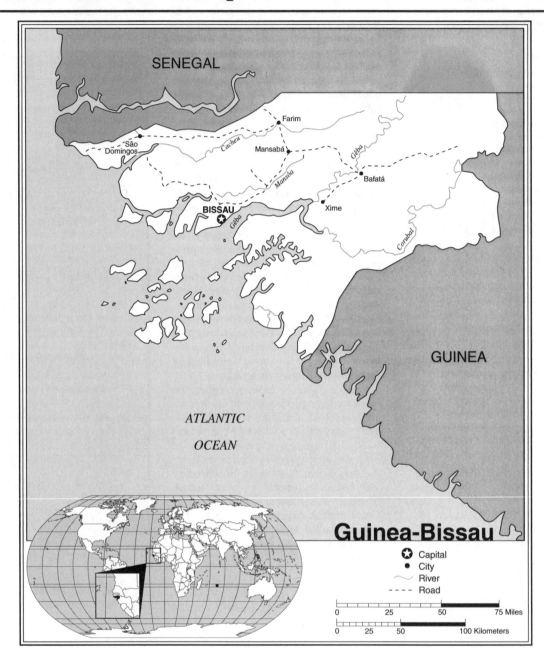

Guinea-Bissau Statistics

GEOGRAPHY

Area in Square Miles (Kilometers):
13,948 (36,125) (about 3 times the size of Connecticut)

Capital (Population): Bissau (292,000)

Environmental Concerns: soil erosion; deforestation; overgrazing; overfishing

Geographical Features: mostly low coastal plain, rising to savanna in the east

Climate: tropical

PEOPLE

Population

Total: 1,346,000

Annual Growth Rate: 2.23%

Rural/Urban Population Ratio: 77/23

Major Languages: Portuguese; Kriolo; various African languages

Ethnic Makeup: 30% Balanta; 20% Fula; 14% Manjaca; 13% Mandinka; 23% others (99% African; 1% others)

Religions: 50% indigenous beliefs; 45% Muslim; 5% Christian

Health

Life Expectancy at Birth: 47 years (male); 52 years (female)

Infant Mortality: 118/1,000 live births

Physicians Available: 1/9,477 people

HIV/AIDS Rate in Adults: 2.5%

Education

Adult Literacy Rate: 54%
Compulsory (Ages): 7–13

COMMUNICATION

Telephones: 10,000 main lines
Internet Users: 1,500 (1999)

TRANSPORTATION

Highways in Miles (Kilometers): 2,610 (4,350)
Railroads in Miles (Kilometers): none
Usable Airfields: 28
Motor Vehicles in Use: 6,000

GOVERNMENT

Type: republic
Independence Date: September 10, 1974 (from Portugal)

Head of State/Government: President Kumba Yala; Prime Minister Alamara Intchia Nhasse
Political Parties: African Party for the Independence of Guinea-Bissau and Cape Verde; Front for the Liberation and Independence of Guinea; United Social Democratic Party; Social Renovation Party; Democratic Convergence; others
Suffrage: universal at 18

MILITARY

Military Expenditures (% of GDP): 2.8%
Current Disputes: trouble along the border with Senegal

ECONOMY

Currency ($ U.S. Equivalent): n/a
Per Capita Income/GDP: $900/$1.2 billion

GDP Growth Rate: 7.2%
Inflation Rate: 5%
Labor Force by Occupation: 82% agriculture
Natural Resources: fish; timber; phosphates; bauxite; petroleum
Agriculture: corn; beans; cassava; cashew nuts; cotton; fish and forest products; peanuts; rice; palm kernels
Industry: agricultural-products processing; beverages
Exports: $80 million (primary partners India, Italy, South Korea)
Imports: $55.2 million (primary partners Portugal, Senegal, Thailand)

SUGGESTED WEB SITES

```
http://www.guineabissau.com
http://www.sas.upenn.edu/
   African_Studies/
   Country_Specific/G_Bissau.html
```

Guinea-Bissau Country Report

In February 2000, Kumba Yala of the Social Renovation Party (PRS) took 72 percent of the vote in the second round of presidential elections, defeating the candidate of the former ruling African Party for the Independence of Guinea-Bissau and Cape Verde (PAIGC). Yala's election was the culmination of an 18-month process that has brought a measure of political peace to the country, which in 1998 had appeared to be heading toward civil war. Two months of fighting, which had resulted in the displacement of up to one third of the country's population, was brought to an end in August 1998 with the signing of a cease-fire accord between the government and rebel soldiers. This was followed up in November of that year by the establishment of a "Government of National Unity," which presided over a transition to genuine multiparty politics.

The Yala government has faced the unenviable challenge of promoting economic development. Since independence, the country has consistently been listed as one of the world's 10 poorest countries. Unfor-

tunately, the period since Yala's installation has been marred by continued political instability. There have been three attempted coups, hundreds of lives lost in war and political violence, and the pulling out of the ruling coalition by one of the major partners due to lack of consultation. Former president General Mane was killed in 2000 after allegedly trying to stage a coup. Secretary-General Kofi Annan of the United Nations intervened to urge political dialogue in order to defuse domestic tensions. An International Monetary Fund team praised improvements in financial controls, but this came after the country had lost tens of millions of dollars in revenue from corrupt practices of government officials. Two prime ministers and the foreign minister were dismissed by the president for various failings. Meanwhile, the head of the Supreme Court and three judges were dismissed by Yala for allegedly overturning the presidents' decision to expel leaders of a Muslim sect from the country.

To many outsiders, the nation has been better known for its prolonged liberation war, from 1962 to 1974, against Portuguese colonial rule. Mobilized by the PAIGC, the freedom struggle in Guinea-Bissau played a major role in the overthrow of the Fascist dictatorship in Portugal itself and the liberation of its other African colonies.

The origins of Portuguese rule in Guinea-Bissau go back to the late 1400s. The area was raided for centuries as a source of

slaves, who were shipped to Portugal and its colonies of Cape Verde and Brazil. With the nineteenth-century abolition of slave trading, the Portuguese began to impose forced labor within Guinea-Bissau itself.

In 1956, six *assimilados*—educated Africans who were officially judged to have assimilated Portuguese culture—led by Amilcar Cabral, founded the PAIGC as a vehicle for the liberation of Cape Verde as well as Guinea-Bissau. From the beginning, many Cape Verdeans, such as Cabral, played a prominent role within the PAIGC. But the group's largest following and main center of activity were in Guinea-Bissau. In 1963, the PAIGC turned to armed resistance and began organizing itself as an alternative government. By the end of the decade, the movement was in control of two thirds of the country. In those areas, the PAIGC was notably successful in establishing its own marketing, judicial, and educational as well as political institutions. Widespread participation throughout Guinea-Bissau in the 1973 election of a National Assembly encouraged a number of countries to formally recognize the PAIGC declaration of state sovereignty.

HEALTH/WELFARE

Guinea-Bissau's health statistics remain appalling: an overall 48-year life expectancy, 12% infant mortality, and more than 90% of the population infected with malaria.

INDEPENDENCE

Since 1974, the leaders of Guinea-Bissau have tried to confront the problems of independence while maintaining the idealism of their liberation struggle. The nation's weak economy has limited their success. Guinea-Bissau has little in the way of mining or manufacturing, although explorations have revealed potentially exploitable reserves of oil, bauxite, and phosphates. More than 80 percent of the people are engaged in agriculture, but urban populations depend on imported foodstuffs. This situation has been generally attributed to the poor infrastructure and a lack of incentives for farmers to grow surpluses. Efforts to improve the rural economy during the early years of independence were hindered by severe drought. Only 8 percent of the small country's land is cultivated.

Under financial pressure, the government adopted a Structural Adjustment Program (SAP) in 1987. The peso was devalued, civil servants were dismissed, and various subsidies were reduced. The painful effects of these SAP reforms on urban workers were cushioned somewhat by external aid.

ACHIEVEMENTS

With Portuguese assistance, a new fiber-optic digital telephone system is being established in Guinea-Bissau.

In 1988, in a desperate move, the government signed an agreement with Intercontract Company, allowing the firm to use its territory for five years as a major dump site for toxic waste from Britain, Switzerland, and the United States. In return, the government was to earn up to $800 million. But the deal was revoked after it was exposed by members of the country's exiled opposition; a major environmental catastrophe would have resulted had it gone through.

POLITICAL DEVELOPMENT

Following the assassination of Amilcar Cabral, in 1973, his brother, Luis Cabral, succeeded him as the leader of the PAICG, thereafter becoming Guinea-Bissau's first president. Before 1980, both Guinea-Bissau and Cape Verde were separately governed by a united PAIGC, which had as its ultimate goal the forging of a political union between the two territories. But in 1980, Luis Cabral was overthrown by the military, which accused him of governing through a "Cape Verdean clique." João Vieira, a popular commander during the liberation war who had also served as prime minister, was appointed as the new head of state. As a result, relations between Cape Verde and Guinea-Bissau deteriorated, leading to a breakup in the political links between the two nations.

The PAIGC under Vieira continued to rule Guinea-Bissau as a one-party state for 10 years. The system's grassroots democracy, which had been fostered in its liberated zones during the war, gave way to a centralization of power around Vieira and other members of his military-dominated Council of State. Several coup attempts resulted in increased authoritarianism.

But the government reversed course, and in 1991, the country formally adopted multipartyism. Progress has been slow. An alleged coup attempt in 1993 led to the detention and subsequent trial of a leading opposition figure, João da Costa, on charges of plotting the government's overthrow. Elections finally occurred in July 1994. The vote resulted in a narrow second-round victory for Vieira against a very divided opposition.

Timeline: PAST

1446
Portuguese ships arrive; claimed as Portuguese Guinea; slave trading develops

1915
Portugal gains effective control over most of the region

1956
The African Party for the Independence of Guinea-Bissau and Cape Verde is formed

1963–1973
Liberation struggle in Guinea-Bissau under the PAIGC and Amilcar Cabral

1973
Amilcar Cabral is assassinated; the PAIGC declares Guinea-Bissau independent

1974
Revolution in Portugal leads to recognition of Guinea-Bissau's independence

1980
João Vieira comes to power through a military coup, ousting Luis Cabral

1990s
The country moves toward multipartyism; "Government of National Unity"

PRESENT

2000s
Kumba Yala of the Social Renovation (or Renewal) Party is elected president

Liberia (Republic of Liberia)

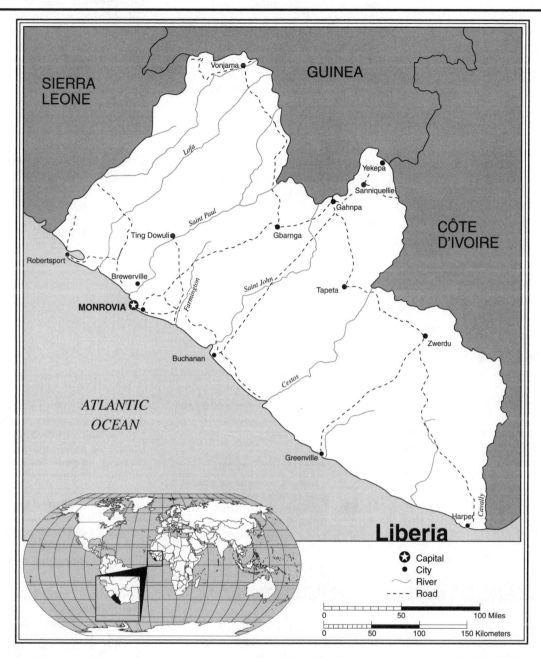

Liberia Statistics

GEOGRAPHY

Area in Square Miles (Kilometers):
43,000 (111,370) (about the size of Tennessee)

Capital (Population): Monrovia (491,000)

Environmental Concerns: soil erosion; deforestation; loss of biodiversity; water pollution

Geographical Features: mostly flat to rolling coastal plains, rising to rolling plateau and low mountains in the northeast

Climate: tropical

PEOPLE

Population

Total: 3,289,000

Annual Growth Rate: 1.91%

Rural/Urban Population Ratio: 56/44

Major Languages: English; Kpelle; Grio; Kru; Krahn; others

Ethnic Makeup: 95% indigenous groups; 5% Americo-Liberian

Religions: 40% indigenous beliefs; 40% Christian; 20% Muslim

Health

Life Expectancy at Birth: 50 years (male); 53 years (female)

Infant Mortality: 130.2/1,000 live births

Physicians Available: 1/8,333 people
HIV/AIDS Rate in Adults: 9%

Education

Adult Literacy Rate: 38.3%
Compulsory (Ages): 7–16; free

COMMUNICATION

Telephones: 6,700 main lines
Daily Newspaper Circulation: 15/1,000 people
Televisions: 20/1,000 people
Internet Users: 500 (2000)

TRANSPORTATION

Highways in Miles (Kilometers): 6,180 (10,300)
Railroads in Miles (Kilometers): 306 (490)
Usable Airfields: 47
Motor Vehicles in Use: 28,000

GOVERNMENT

Type: republic
Independence Date: July 26, 1847
Head of State/Government: President Charles Ghankay Taylor is both head of state and head of government
Political Parties: National Patriotic Party; National Democratic Party of Liberia; Liberian Action Party; Liberian People's Party; United People's Party; others
Suffrage: universal at 18

MILITARY

Military Expenditures (% of GDP): 1.3%
Current Disputes: civil war; border instabilities with Sierra Leone

ECONOMY

Currency ($ U.S. Equivalent): 0.99 Liberian dollar = $1
Per Capita Income/GDP: $1,100/$3.6 billion
GDP Growth Rate: .5%
Inflation Rate: 8%
Unemployment Rate: 70%
Labor Force by Occupation: 70% agriculture; 22% services; 8% industry
Population Below Poverty Line: 80%
Natural Resources: iron ore; timber; diamonds; gold; hydropower
Agriculture: rubber; rice; palm oil; cassava; coffee; cocoa beans; sugarcane; bananas; sheep; goats; timber
Industry: rubber processing; food processing; diamonds
Exports: $55 million (primary partners Belgium, Germany, Italy)
Imports: $170 million (primary partners France, South Korea, Japan)

SUGGESTED WEB SITES

http://www.fol.org
http://www.blackworld.com/country/liberia.htm

Liberia Country Report

In August 2000, Liberian president Charles Taylor declared a state of emergency as fighting intensified in the north of the country. Since then, rebels calling themselves Liberians United for Reconciliation and Democracy (LURD) have made major advances against government forces. During 2002, LURD and Taylor's armies traded control of the strategic town of Tubmanburg, as fighting crept toward the outskirts of the capital city, Monrovia. From their headquarters at Voinjama, on the northwest border with Guinea, the rebels now control about 30 percent of the country. The excesses of President Taylor's troops, who are accused of pillaging the rural countryside, may have strengthened the rebels' hand. There has also been unrest in the urban areas, with Amnesty International acusing government security forces of brutally suppressing civilian protesters. In May 2001, the United Nations Security Council re-imposed an arms embargo on Liberia to punish Taylor for trading weapons for diamonds ("blood diamonds") from rebels in Sierra Leone.

The LURD forces, who are said to be composed of veterans of past civil strife in Sierra Leone as well as Liberia, have also been accused of human-rights abuses. Their provisioning from among the general population certainly contributed to a rise in hunger in rebel-controlled areas, in northeastern Liberia. So far their political program has simply been Taylor's overthrow. In November 2001, they repulsed a major government offensive.

DEVELOPMENT

Liberia's economic and social infrastructure was devastated by the war. Today people are surviving primarily through informal-sector bartering and trading.

Renewed civil war in Liberia has dashed the hopes of those who had believed that the country had achieved peace following the extreme horrors of the previous conflict. Between 1989 and 1996, some 200,000 people were killed. Among the survivors, approximately 750,000 fled the country as refugees, while another 1.2 million were internally displaced, out of a total population of only 2.6 million people. The war ended in July 1997 with Taylor and his National Patriotic Front of Liberia (NPFL) receiving about three quarters of the vote in internationally supervised elections. These elections were the culmination of a year-long process to restore peace to the country. Taylor's subsequent inauguration brought stability to the country—but it is now under renewed threat.

The restoration of peace in Liberia after July 1997 was overseen by a West African regional military force (ECOMOG) of just over 10,000. The force was deployed throughout the country to provide the security, facilitate the disarmament and demobilization of local combatants, and protect returning refugees. Its relative success in carrying out these functions came about only after years of frustration. When the force finally withdrew in 1998, it left behind 10,000 to 25,000 locally fathered children. ECOMOG's security role was taken over by the reconstituted Armed Forces of Liberia, an uneasy mix of former civil-war rivals.

FREEDOM

The Taylor government has tried to reestablish the rule of law. The various government security forces continue to be linked to abuses. In 1998, Taylor accused rivals of plotting a coup and jeopardizing continued efforts to build postwar reconciliation in the country. In 2000, he declared a state of emergency when civil fighting intensified in the north of the country.

A major challenge facing Taylor's new government was Liberia's war-ravaged economy, with little formal-sector employment, some $2.2 billion in debt, and a collapsed infrastructure. In December 1997, the president announced that a new currency would be introduced in 1998 to replace the two separate currencies that were

(United Nations photo by N. van Praag)
The Liberian Civil War of 1989–1996 left the country destitute. Political anarchy destroyed much of the infrastructure, economy, and culture. Nearly a tenth of the population were killed, many more displaced from their homes.

in use in different parts of the country in addition to the U.S. dollar, which is also legal tender.

AFRICAN-AMERICAN-AFRICANS

Among the African states, Liberia shares with Ethiopia the distinction of having avoided European rule. Between 1847 and 1980, Liberia was governed by an elite made up primarily of descendants of African-Americans who had begun settling along its coastline two decades earlier. These "Americo-Liberians" make up only 5 percent of the population. But they dominated politics for decades through their control of the governing True Whig Party (TWP). Although the republic's Constitution was ostensibly democratic, the TWP rigged the electoral process.

HEALTH/WELFARE

Outside aid and local self-help efforts were mobilized against famine in Liberia in 1990–1991. But the long and brutal Civil War of the 1990s took a dreadful toll on Liberians' health and well-being.

Most Liberians belong to indigenous ethnolinguistic groups, such as the Kpelle, Bassa, Grio, Kru, Krahn, and Vai, who were conquered by the Americo-Liberians during the 1800s and early 1900s. Some individuals from these subjugated communities accepted Americo-Liberian norms. Yet book learning, Christianity, and an ability to speak English helped an indigenous person to advance within the state only if he or she accepted its social hierarchy by becoming a "client" of an Americo-Liberian "patron." In a special category were the important interior "chiefs," who were able to maintain their local authority as long as they remained loyal to the republic.

During the twentieth century, Liberia's economy was transformed by vast Firestone rubber plantations, iron-ore mining, and urbanization. President William Tubman (1944–1971) proclaimed a "Unification Policy," to promote national integration, and an "Open-Door Policy," to encourage outside investment in Liberia. However, most of the profits that resulted from the modest external investment that did occur left the country, while the wealth that remained was concentrated in the hands of the TWP elite.

During the administration of Tubman's successor, William Tolbert (1971–1980), Liberians became more conscious of the inability of the TWP to address the inequities of the status quo. Educated youths from all ethnic backgrounds began to join dissident associations rather than the regime's patronage system.

As economic conditions worsened, the top 4 percent of the population came to control 60 percent of the wealth. Rural stagnation drove many to the capital city of Monrovia (named after U.S. president James Monroe), where they suffered from high unemployment and inflation. The inevitable explosion occurred in 1979, when the government announced a 50 percent price increase for rice, the national food staple. Police fired on demonstrators, killing and wounding hundreds. Rioting, which resulted in great property damage, led the government to appeal to neighboring Guinea for troops. It was clear that the TWP was losing its grip. Thus Sergeant Samuel Doe enjoyed widespread support when, in 1980, he led a successful coup.

ACHIEVEMENTS

Through a shrewd policy of diplomacy, Liberia managed to maintain its independence when Great Britain and France conquered neighboring areas during the late nineteenth century. It espoused African causes during the colonial period; for instance, Liberia brought the case of Namibia to the World Court in the 1950s.

DOE DOESN'T DO

Doe came to power as Liberia's first indigenous president, a symbolically important event that many believed would herald substantive changes. Some institutions of the old order, such as the TWP and the Masonic Temple (looked upon as Liberia's secret government) were disbanded. The House of Representatives and Senate were suspended. Offices changed hands, but the old administrative system persisted. Many of those who came to power were members of Doe's own ethnic group, the Krahn, who had long been prominent in the lower ranks of the army.

Doe declared a narrow victory for himself in the October 1985 elections, but there was widespread evidence of ballot tampering. A month later, exiled general Thomas Quiwonkpa led an abortive coup attempt. During and after the uprising, thousands of people were killed, mostly civilians belonging to Quiwonkpa's Grio group who were slaughtered by loyalist (largely Krahn) troops. Doe was inaugurated, but opposition-party members refused to take their seats in the National Assembly. Some, fearing for their lives, went into exile.

During the late 1980s, Doe became increasingly dictatorial. Many called on the U.S. government, in particular, to withhold aid until detainees were freed and new elections held. The U.S. Congress criticized the regime but authorized more than $500 million in financial and military support. Meanwhile, Liberia suffered from a shrinking economy and a growing foreign debt, which by 1987 had reached $1.6 billion.

Doe's government was not entirely to blame for Liberia's dire financial condition. When Doe came to power, the Liberian treasury was already empty, in large part due to the vast expenditures incurred by the previous administration in hosting the 1979 Organization of African Unity Conference. The rising cost of oil and decline in world prices for natural rubber, iron ore, and sugar had further crippled the economy. But government corruption and instability under Doe made the bad situation worse.

DOE'S DOWNFALL

Liberia's descent into violent anarchy began on December 24, 1989, when a small group of insurgents, led by Charles Taylor, who had earlier fled the country amid corruption charges, began a campaign to overthrow Doe. As Taylor's NPFL rebels gained ground, the war developed into an increasingly vicious interethnic struggle among groups who had been either victimized by or associated with the regime. Thousands of civilians were thus massacred by ill-disciplined gunmen on both sides; hundreds of thousands began to flee for their lives. By June 1990, with the rump of Doe's forces besieged in Monrovia, a small but efficient breakaway armed faction of the NPFL, under the ruthless leadership of a former soldier named Prince Johnson, had emerged as a deadly third force.

By August, with the United States unwilling to do more than evacuate foreign nationals from Monrovia (the troops of Doe, Johnson, and Taylor had begun kidnapping

expatriates and violating diplomatic immunity), members of the Economic Community of West African States decided to establish a framework for peace by installing an interim government, with the support of the regional peacekeeping force known as ECOMOG: the ECOWAS Monitoring Group. The predominantly Nigerian force, which also included contingents from Ghana, Guinea, Sierra Leone, The Gambia, and, later, Senegal, landed in Monrovia in late August. This coincided with the nomination, by a broad-based but NPFL–boycotted National Conference, of Amos Sawyer, a respected academic, as the head of the proposed interim administration.

Initial hopes that ECOMOG's presence would end the fighting proved to be naïve. On September 9, 1990, Johnson captured Doe by shooting his way into ECOMOG headquarters. The following day, Doe's gruesome torture and execution were videotaped by his captors. This "outrage for an outrage" did not end the suffering. Protected by a reinforced ECOMOG, Sawyer was able to establish his interim authority over most of Monrovia, but the rest of the country remained in the hands of the NPFL or of local thugs.

Repeated attempts to get Johnson and Taylor to cooperate with Sawyer in establishing an environment conducive to holding elections proved fruitless. While most neighboring states have supported ECOMOG's mediation efforts, some have provided support (and, in the case of Burkina Faso, troops) to the NPFL, which has encouraged Taylor in his on-again, off-again approach toward national reconciliation.

In September 1991, a new, fiercely anti-NPFL force, the United Liberation Movement of Liberia (ULIMO), emerged from bases in Sierra Leone. The group is identified with former Doe supporters. Subsequent clashes between ULIMO and NPFL on both sides of the Liberian–Sierra Leonean border contributed to the April 1992 overthrow of the Sierra Leonean government as well as the failure of an October 1991 peace accord brokered by the Côte d'Ivoire's then-president, Felix Houphouët-Boigny.

In October 1992, ECOWAS agreed to impose sanctions on the NPFL for block-

ing Monrovia. ECOMOG then joined ULIMO and remnants of the Armed Forces of Liberia (AFL) in a counteroffensive. In 1993, yet another armed faction, the Liberia Peace Council (LPC), emerged to challenge Taylor for control of southeastern Liberia. In March 1994, an all-party interim State Council, agreed to in principle eight months earlier, was finally sworn in. But it quickly collapsed, while a violent split in ULIMO contributed to further anarchy. In July 2002, UN secretary-general Kofi Annan warned that the Liberian conflict threatens the United Nations' peacekeeping work in neighboring countries.

Timeline: PAST

1500s
The Vai move onto the Liberian coast from the interior

1822
The first African-American settlers arrive from the United States

1871
The first coup exchanges one Americo-Liberian government for another

1931
The League of Nations investigates forced-labor charges

1944
President William Tubman comes to office

1971
William Tolbert becomes president

1980s
Tolbert is assassinated; a military coup brings Samuel Doe to power

1990s
Civil war leads to the execution of Doe, anarchy, and foreign intervention; Charles Taylor and the NPFL win power in internationally supervised elections; the Civil War ends

PRESENT

2000s
President Taylor declares a state of emergency

Civil war resumes

Mali (Republic of Mali)

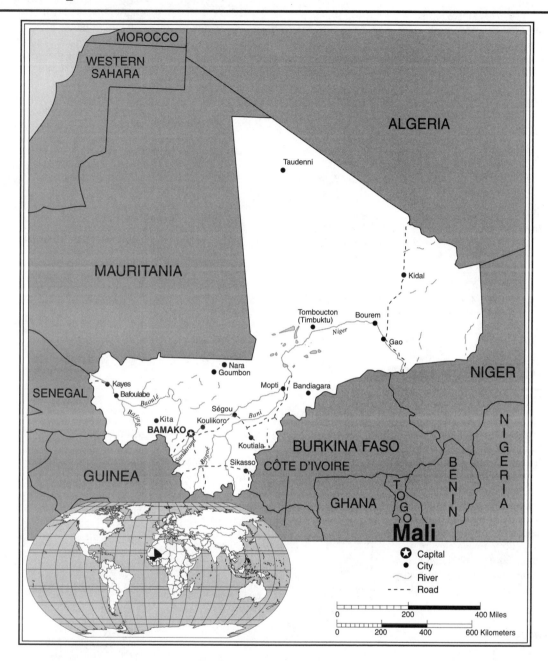

Mali Statistics

GEOGRAPHY

Area in Square Miles (Kilometers):
478,819 (1,240,142) (about twice the size of Texas)

Capital (Population): Bamako (1,161,000)

Environmental Concerns: soil erosion; deforestation; desertification; insufficient potable water; poaching

Geographical Features: mostly flat to rolling northern plains covered by sand; savanna in the south; rugged hills in the northeast

Climate: subtropical to arid

PEOPLE

Population

Total: 11,341,000

Annual Growth Rate: 2.98%

Rural/Urban Population Ratio: 71/29

Major Languages: French; Bambara; numerous African languages

Ethnic Makeup: 50% Mande; 17% Peul; 12% Voltaic; 6% Songhai; 10% Tuareg and Maur (Moor); 5% others

Religions: 90% Muslim; 9% indigenous beliefs; 1% Christian

Health

Life Expectancy at Birth: 46 years (male);
48 years (female)
Infant Mortality: 119.6/1,000 live births
Physicians Available: 1/18,376 people
HIV/AIDS Rate in Adults: 1.7%

Education

Adult Literacy Rate: 38%
Compulsory (Ages): 7–16; free

COMMUNICATION

Telephones: 45,000 main lines
Televisions: 12/1,000 people
Internet Users: 10,000 (2000)

TRANSPORTATION

Highways in Miles (Kilometers): 9,362
(15,100)
Railroads in Miles (Kilometers): 452
(729)
Usable Airfields: 27
Motor Vehicles in Use: 41,000

GOVERNMENT

Type: republic
Independence Date: September 22, 1960
(from France)
Head of State/Government: President
Amadou Toumani Touré; Prime Minister
Modibo Keita
Political Parties: Alliance for Democracy;
National Congress for Democratic
Initiative; Sudanese Union/African
Democratic Rally; others
Suffrage: universal at 18

MILITARY

Military Expenditures (% of GDP): 2%
Current Disputes: none

ECONOMY

Currency ($ U.S. Equivalent): 752 CFA
francs = $1
Per Capita Income/GDP: $840/$9.2
billion
GDP Growth Rate: –1.2%

Inflation Rate: 4.5%
Unemployment Rate: 14.6% in urban areas
Labor Force by Occupation: 80%
agriculture and fishing
Population Below Poverty Line: 64%
Natural Resources: hydropower; bauxite;
iron ore; manganese; tin; phosphates;
kaolin; salt; limestone; gold; uranium;
copper
Agriculture: millet; sorghum; corn; rice;
sugar; cotton; peanuts; livestock
Industry: food processing; construction;
phosphate and gold mining; consumer-
goods production
Exports: $575 million (primary partners
Brazil, South Korea, Italy)
Imports: $600 million (primary partners
Côte d'Ivoire, France, Senegal)

SUGGESTED WEB SITES

http://www.maliembassy-usa.org
http://www.cia.gov/cia/
publications/factbook/geos/ml/
html

Mali Country Report

Amadou Toumani Touré, the army general credited with rescuing Mali from military dictatorship and handing it back to its people, won the presidential elections of May 2002. He entered office on a program of anticorruption, peace, and development aimed at alleviation of poverty. This program resonated strongly with the population because corruption was viewed as rampant. In July 2000, an anticorruption commission published a report highlighting embezzlement and mismanagement within a number of state-owned companies and other public bodies.

DEVELOPMENT

In 1989, the government received international funding to overhaul its energy infrastructure. The opening of new gold mines has provided the economy with a boost.

Touré will be building on the legacy of his immediate predecessor, Dr. Alpha Konare, who stepped down from power after two terms in office that moved Mali away from its authoritarian past. The current democratic order was inaugurated a year after a coup led by Touré ended the dictatorial regime of Moussa Traoré. Like his predecessor, Modibo Keita (the first

president of Mali), Traoré had governed Mali as a single-party state. True to their word, the coup leaders who seized power in 1991 following bloody antigovernment riots presided over a quick transition to civilian rule.

FREEDOM

The human-rights situation in Mali has improved in recent years, though international attention was drawn to the suppression of opposition demonstrations in the run-up to the 1997 elections.

Konare ruled as an activist scholar who, like many Malians, found political inspiration in his country's rich heritage. But his efforts to rebuild Mali were hampered by a weak economy, aggravated by the 1994 collapse in value of the CFA franc. In 1994 and 1995, violence occurred between security forces and university students protesting against economic hardship. The plight of Malian economic refugees in France gained international attention in 1996, when a number sought sanctuary in a Parisian church and went on a hunger strike in protest against attempts to deport them. The government has enjoyed greater success in reaching a (still fragile) settlement

with Tuareg rebels in the country's far north.

AN IMPERIAL PAST

The published epic of Sundiata Keita, the thirteenth-century A.D. founder of the great Mali Empire, is recognized throughout the world as a masterpiece of classical African literature. In Mali itself, Sundiata remains a source of national pride and unity.

HEALTH/WELFARE

About a third of Mali's budget is devoted to education. A special literacy program in Mali teaches rural people how to read and write, by using booklets that concern fertilizers, measles, and measuring fields.

Sundiata's state was one of three great West African empires whose centers lay in modern Mali. Between the fourth and thirteenth centuries, the area was the site of the prosperous, ancient Ghana. The Mali Empire was superseded by that of Songhai, which was conquered by the Moroccans at the end of the sixteenth century. All these empires were in fact confederations. Although they encompassed vast areas united under a single supreme ruler, local communities generally enjoyed a great deal of autonomy.

From the 1890s until 1960, another form of imperial unity was imposed over Mali (then called French Sudan) and the adjacent territories of French West Africa. The legacy of broader colonial and precolonial unity as well as its landlocked position have inspired Mali's postcolonial leaders to seek closer ties with neighboring countries.

Mali formed a brief confederation with Senegal during the transition period to independence. This initial union broke down after only a few months, but since then the two countries have cooperated in the Organization for the Development of the Senegal River and other regional groupings. The Senegalese port of Dakar, which is linked by rail to Mali's capital city, Bamako, remains the major outlet for Malian exports. Mali has also sought to strengthen its ties with nearby Guinea. In 1983, the two countries signed an agreement to harmonize policies and structures.

ENVIRONMENTAL CHALLENGES

Mali is one of the poorest countries in the world. About 80 percent of the people are employed in (mostly subsistence) agriculture and fishing, but the government usually has to rely on international aid to make up for local food deficits. Most of the country lies within either the expanding Sahara Desert or the semiarid region known as the Sahel, which has become drier as a result of recurrent drought. Much of the best land lies along the Senegal and Niger Rivers, which support most of the nation's agropastoral production. In earlier centuries, the Niger was able to sustain great trading cities such as Timbuctu and Djenne, but today, most of its banks do not even support crops. Efforts to increase cultivation have so far been met with limited overall success.

Mali's frequent inability to feed itself has been largely blamed on locust infestation, drought, and desertification. The inefficient state-run marketing and distribution systems, however, have also had a negative impact. Low official produce prices have encouraged farmers either to engage in subsistence agriculture or to sell their crops on the black market. Thus, while some regions of the country remain dependent on international food donations, crops continue to be smuggled across Mali's borders. Recent policy commitments to liberalize agricultural trading, as part of an International Monetary Funding–approved Structural Adjustment Program (SAP), have yet to take hold.

In contrast to agriculture, Mali's mining sector has experienced promising growth. The nation exports modest amounts of gold, phosphates, marble, and uranium. Potentially exploitable deposits of bauxite, manganese, iron, tin, and diamonds exist. The Manantali Dam in southwestern Mali opened in December 2001. It is expected to provide electricity and jobs for thousands of Malians.

For decades, Mali was officially committed to state socialism. Its first president, Keita, a descendant of Sundiata, established a command economy and one-party state during the 1960s. His attempt to go it alone outside the CFA Franc Zone proved to be a major failure. Under Traoré, socialist structures were modified but not abandoned. Agreements with the IMF ended some government monopolies, and the country adopted the CFA franc as its currency. But the lack of a significant class of private entrepreneurs and the role of otherwise unprofitable public enterprises in providing employment discouraged radical privatization.

Timeline: PAST

1250–1400s
The Mali Empire extends over much of the upper regions of West Africa

late 1400s–late 1500s
The Songhai Empire controls the region

1890
The French establish control over Mali

1960
The Mali Confederation

1968
Moussa Traoré and the Military Committee for National Liberation grab power

1979
Traoré's Democratic Union of the Malian People is the single ruling party

1979–1980
School strikes and demonstrations; teachers and students are detained

1990s
The country's first multiparty elections are held; economic problems stir civic unrest

PRESENT

2000s
The Touré government explores ways to strengthen the economy

Mauritania (Islamic Republic of Mauritania)

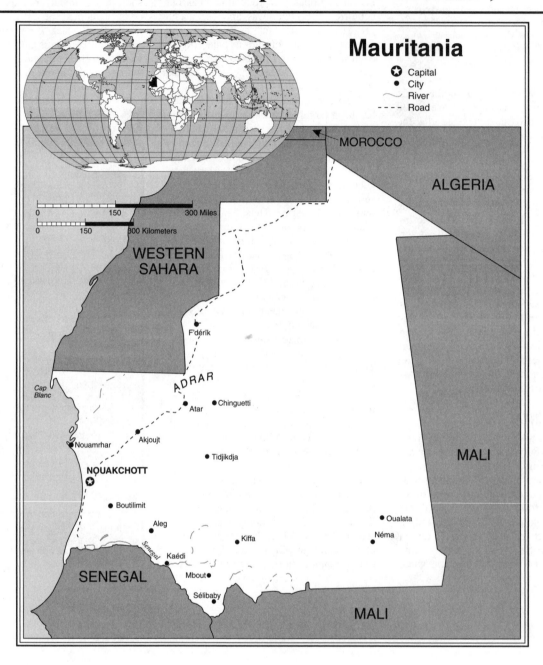

Mauritania Statistics

GEOGRAPHY

Area in Square Miles (Kilometers): 398,000 (1,030,700) (about 3 times the size of New Mexico)

Capital (Population): Nouakchott (626,000)

Environmental Concerns: overgrazing; deforestation; soil erosion; desertification; very limited natural freshwater resources; overfishing

Geographical Features: mostly barren, flat plains of the Sahara; some central hills

Climate: desert

PEOPLE

Population

Total: 2,829,000
Annual Growth Rate: 2.92%

Rural/Urban Population Ratio: 44/56

Major Languages: Hasanixa; Soninke; Arabic; Pular; Wolof

Ethnic Makeup: 40% mixed Maur/ black; perhaps 30% Maur; 30% black

Religion: 100% Muslim

Health

Life Expectancy at Birth: 49 years (male); 54 years (female)

Infant Mortality: 75.2/1,000 live births

Physicians Available: 1/11,085 people

HIV/AIDS Rate in Adults: 1.8%

Education

Adult Literacy Rate: 37.7%

Compulsory (Ages): 6–12

COMMUNICATION

Telephones: 26,500 main lines

Internet Users: 7,500

TRANSPORTATION

Highways in Miles (Kilometers): 4,560 (7,600)

Railroads in Miles (Kilometers): 422 (704)

Usable Airfields: 26

Motor Vehicles in Use: 26,500

GOVERNMENT

Type: republic

Independence Date: November 28, 1960 (from France)

Head of State/Government: President Maaouya Ould Sid Ahmed Taya; Prime Minister Cheikh El Avia Ould Mohamed Khouna

Political Parties: Democratic and Social Republican Party; Union for Democracy and Progress; Popular Social and Democratic Union; others

Suffrage: universal at 18

MILITARY

Military Expenditures (% of GDP): 3.7%

Current Disputes: ethnic tensions

ECONOMY

Currency ($ U.S. equivalent): 276 ouguiyas = $1

Per Capita Income/GDP: $1,800/$5 billion

GDP Growth Rate: 4%

Inflation Rate: 4.4%

Unemployment Rate: 21%

Labor Force by Occupation: 50% agriculture; 40% services; 10% industry

Population Below Poverty Line: 50%

Natural Resources: iron ore; gypsum; fish; copper; phosphates

Agriculture: millet; sorghum; dates; root crops; cattle and sheep; fish products

Industry: iron-ore and gypsum mining; fish processing

Exports: $359 million (primary partners France, Japan, Italy)

Imports: $335 million (primary partners France, United States, Spain)

SUGGESTED WEB SITES

http://www.cia.gov/cia/
publications/ factbook/geos/
mr.html

Mauritania Country Report

Since the adoption of its current Constitution in 1991, Mauritania has legally been a multiparty democracy. But in practice, power remains in the hand of President Ould Taya's Republican Social Democratic Party (PRDS). The ruling party won large majorities in the 1992 and 1997 elections, which were boycotted by the leading opposition groupings. Participation in the 2001 elections resulted in some opposition gains but yet another controversial victory for the PRDS. Action for Change, a new party claiming to represent the Haratine (or Harratin), Mauritania's Arab-oriented black underclass of ex-slaves, won four seats. But in the beginning of 2002, the party was banned. Multiparty politics has thus so far failed to assure either social harmony or a respect for human rights. Neither has it resolved the country's severe social and economic problems.

For decades, Mauritania has grown progressively drier. Today, about 75 percent of the country is covered by sand. Less than 1 percent of the land is suitable for cultivation, 10 percent for grazing. To make matters worse, the surviving arable and pastoral areas have been plagued by grasshoppers and locusts.

In the face of natural disaster, people have moved. Since the mid-1960s, the per-

DEVELOPMENT

Mauritania's coastal waters are among the richest in the world. During the 1980s, the local fishing industry grew at an average annual rate of more than 10%. Many now believe that the annual catch has reached the upper levels of its sustainable potential.

centage of urban dwellers has swelled, from less than 10 percent to 53 percent, while the nomadic population during the same period has dropped, from more than 80 percent to perhaps 20 percent. In Nouakchott, the capital city, vast shanty-towns now house nearly a quarter of the population. As the capital has grown, from a few thousand to 626,000 in a single generation, its poverty—and that of the nation as a whole—has become more obvious. People seek new ways to make a living away from the land, but there are few jobs. The best hope for lifting up the economy may lie in offshore oil exploration. A prospecting report in 2002 has attracted the interest of major international oil companies.

Mauritania's heretofore faltering economy has coincided with an increase in racial and ethnic tensions. Since independence,

the government has been dominated by the Maurs (or Moors), who speak Hasaniya Arabic. This community has historically been divided between the aristocrats and commoners, of Arab and Berber origin, and the Haratine, who were black African slaves who had assimilated Maurish culture but remain socially segregated. Including the Haratine, the Maurs account for perhaps 60 percent of the citizenry (the government has refused to release comprehensive data from the last two censuses).

FREEDOM

The Mauritanian government is especially sensitive to continuing allegations of the existence of chattel slavery in the country. While slavery is outlawed, there is credible evidence of its continued existence. In 1998, five members of a local advocacy group SOS–Esclaves (Slaves) were sentenced to 13 months' imprisonment for "activities within a non-authorized organization."

The other half of Mauritania's population is composed of the "blacks," who mostly speak Pulaar, Soninke, or Wolof. Like the Maurs, all these groups are Muslim. Thus Mauritania's rulers have stressed Islam as a source of national unity. The country proclaimed itself an Islamic republic at independence, and since 1980 the Shari'a—the Islamic penal code—has been the law of the land.

Muslim brotherhood has not been able to overcome the divisions between the northern Maurs and southern blacks. One major source of friction has been official Arabization efforts, which are opposed by most southerners. In recent years, the country's desertification has created new sources of tension. As their pastures turned to sand, many of the Maurish nomads who did not find refuge in the urban areas moved southward. There, with state support, they began in the 1980s to deprive southerners of their land.

HEALTH/WELFARE

There have been some modest improvements in the areas of health and education since the country's independence, but conditions remain poor. Mauritania has received low marks regarding its commitment to human development.

Oppression of blacks has been met with resistance from the underground Front for the Liberation of Africans in Mauritania (FLAM). Black grievances were also linked to an unsuccessful coup attempt in 1987. In 1989, interethnic hostility exploded when a border dispute with Senegal led to race riots that left several hundred "Senegalese" dead in Nouakchott. In response, the "Moorish" trading community

in Senegal became the target of reprisals. Mauritania claimed that 10,000 Maurs were killed, but other sources put the number at about 70. Following this bloodshed, more than 100,000 refugees were repatriated across both sides of the border. Mass deportations of "Mauritanians of Senegalese origin" have fueled charges that the Nouakchott regime is trying to eliminate its non-Maurish population.

ACHIEVEMENTS

There is a current project to restore ancient Mauritanian cities, such as Chinguette, which are located on traditional routes from North Africa to Sudan. These centers of trade and Islamic learning were points of origin for the pilgrimage to Mecca and were well known in the Middle East.

Tensions between Mauritania and Senegal were eased in June 2000 by the newly elected Senegalese president Abdoulaye Wade. This helped to reduce cross-border raids by deported Mauritanians. Genuine peace, however, will require greater reform within Mauritania itself and provision for the return of refugees.

In recent years, the regime in Nouakchott has sent out conflicting signals. Although the government has legalized some opposition parties, it has also continued to pursue its Arabization program and has clamped down on genuine dissent. Maur militias have been armed, and the army has been expanded with assistance from Arab countries.

Timeline: PAST

1035–1055
The Almoravids spread Islam in the Western Sahara areas through conquest

1920
The Mauritanian area becomes a French colony

1960
Mauritania becomes independent under President Moktar Ould Daddah

1978
A military coup brings Khouma Ould Haidalla and the Military Committee for National Recovery to power

1979
The Algiers Agreement: Mauritania makes peace with Polisario and abandons claims to Western Sahara

1980
Slavery is formally abolished

1990s
Multiparty elections are boycotted by the opposition; tensions continue between Mauritania and Senegal

PRESENT

2000s
Desertification takes its toll on the environment and the economy

It is estimated that 90,000 Mauritanians still live in servitude, despite the legal abolishment of slavery

Senegal and Mauritania seek better relations

Niger (Republic of Niger)

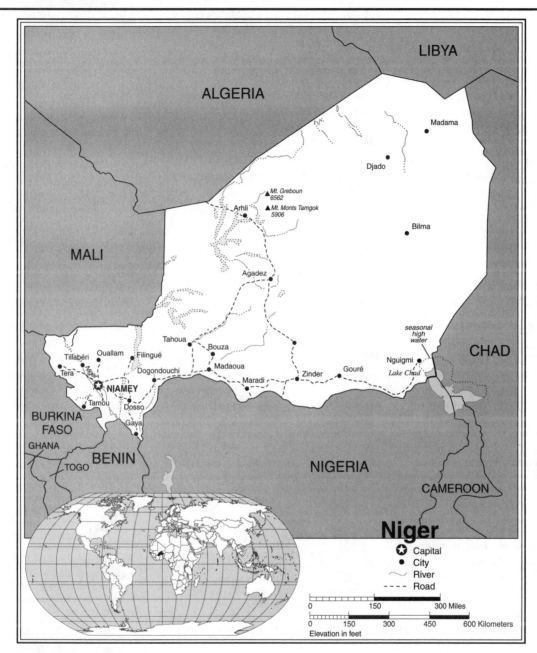

Niger Statistics

GEOGRAPHY

Area in Square Miles (Kilometers): 489,191 (1,267,000) (about twice the size of Texas)

Capital (Population): Niamey (821,000)

Environmental Concerns: overgrazing; deforestation; soil erosion; desertification; poaching and habitat destruction

Geographical Features: mainly desert plains and sand dunes; flat to rolling plains in the south; hills in the north; landlocked

Climate: desert; tropical in the extreme south

PEOPLE

Population

Total: 10,640,000

Annual Growth Rate: 2.5%
Rural/Urban Population Ratio: 80/20
Major Languages: French; Hausa; Djerma
Ethnic Makeup: 56% Hausa; 22% Djerma; 8% Fula; 8% Tuareg; 6% others
Religions: 80% Muslim; 20% indigenous beliefs and Christian

Health

Life Expectancy at Birth: 42 years (male); 42 years (female)

Infant Mortality: 122/1,000 live births
Physicians Available: 1/35,141 people
HIV/AIDS Rate in Adults: 4%

Education

Adult Literacy Rate: 15.3%
Compulsory (Ages): 7–15, free

COMMUNICATION

Telephones: 20,000 main lines
Televisions: 2.8 per 1,000 people
Internet Users: 3,000 (2000)

TRANSPORTATION

Highways in Miles (Kilometers): 6,262
   (10,100)
Railroads in Miles (Kilometers): none
Usable Airfields: 27
Motor Vehicles in Use: 51,500

GOVERNMENT

Type: republic
Independence Date: August 3, 1960 (from
   France)

Head of State/Government: President
   Mamadou Tandja is both head of state
   and head of government
Political Parties: National Movement for
   a Developing Society–Nassara;
   Democratic and Social Convention–
   Rahama; Nigerien Party for Democracy
   and Socialism–Tarayya; Nigerien
   Alliance for Democracy and Social
   Progress– Zaman-lahia; others
Suffrage: universal at 18

MILITARY

Military Expenditures (% of GDP): 1.3%
Current Disputes: territorial dispute with
   Libya; boundary disputes over Lake
   Chad

ECONOMY

Currency ($ U.S. Equivalent): 752 CFA
   francs = $1
Per Capita Income/GDP: $820/$8.4
   billion
GDP Growth Rate: 3.1%

Inflation Rate: 4.2%
Labor Force by Occupation: 90%
   agriculture; 6% industry and commerce;
   4% government
Population Below Poverty Line: 63%
Natural Resources: uranium; coal; iron
   ore; tin; phosphates; gold; petroleum
Agriculture: millet; sorghum; peanuts;
   cotton; cowpeas; cassava; livestock
Industry: cement; brick; textiles;
   chemicals; agricultural products; food
   processing; uranium mining
Exports: $246 million (primary partners
   France, Nigeria, Spain)
Imports: $331 million (primary partners
   France, Côte d'Ivoire, United States)

SUGGESTED WEB SITES

http://www.friendsofniger.org
http://www.cia.gov/cia/
   publications/factbook/geos/
   ng.html

Niger Country Report

Niger is ranked by the United Nations as the world's second-poorest country, after war-ravaged Sierra Leone. This circumstance can in part be blamed on poor governance. For most of the past four decades, since it gained independence from France in 1960, Niger has been governed by a succession of military regimes that have left it bankrupt. This has led to chronic instability, as the government has regularly failed to pay its salaries, resulting in strikes by civil servants and mutinies by soldiers. In November 1999, the current president, Mamadou Tandja, was elected under a new Constitution. But ultimate power remains in the hands of the military, which, in January 1996, overthrew Niger's last elected government. In July 1996, the leader of the coup, Colonel Ibrahim Bare Mainassara, claimed victory in elections that were widely condemned as fraudulent. Mainassara's subsequent assassination in April 1999 opened the door to a return to civilian rule, but at this writing, a military committee under strongman Major Daouda Mallam-Wanke continues to wield influence over the government.

DROUGHT AND DESERTIFICATION

Most Nigeriens subsist through small-scale crop production and herding. Yet farming is especially difficult in Niger. Less than 10 percent of the nation's vast territory is suit-

able for cultivation even during the best of times. Most of the cultivable land lies along the banks of the Niger River. Unfortunately, much of the past four decades has been the worst of times. During this period, Nigeriens have been constantly challenged by recurrent drought and an ongoing process of desertification.

DEVELOPMENT

Nigerien village cooperatives, especially marketing cooperatives, predate independence and have grown in size and importance in recent years. They have successfully competed with well-to-do private traders for control of the grain market.

Drought had an especially catastrophic effect during the 1970s. Most Nigeriens were reduced to dependency on foreign food aid, while about 60 percent of their livestock perished. Some people believe that the ecological disaster that afflicted Africa's Sahel region (which includes southern Niger) during that period was of such severity as to disrupt the delicate long-term balance between desert and savanna. Others, however, have concluded that the intensified desertification of recent years is primarily rooted in human, rather than natural, causes, which can be reversed. In particular, many attribute envi-

ronmental degradation to the introduction of inappropriate forms of cultivation, overgrazing, deforestation, and new patterns of human settlement.

FREEDOM

Nigeriens have been effectively disenfranchised by the 1996 coup and subsequent fraudulent presidential election. Security forces are known to beat and intimidate opposition political figures. The private media are a target of repression, with a number of journalists having been detained. Opposition meetings and demonstrations are often banned.

Ironically, much of the debate on people's negative impact on the Sahel environment has been focused on some of the agricultural-development schemes that once were perceived as the region's salvation. In their attempts to boost local food production, international aid agencies often promoted so-called Green Revolution programs. These were designed to increase per-acre yields, typically through the intensive planting of new seeds and reliance on imported fertilizers and pesticides. Such projects often led to higher initial local outputs that proved unsustainable, largely due to expensive overhead. In addition, many experts promoting the new agricultural techniques failed to appreciate the value of traditional technologies and forms of social organization in limiting desertification while allowing people to cope with drought. It is now appreciated that patterns of cultivation long championed by Nigerien farmers allowed for soil conservation and reduced the risks associated with pests and poor climate.

HEALTH/WELFARE

A national conference on educational reform stimulated a program to use Nigerien languages in primary education and integrated the adult literacy program into the rural development efforts. The National Training Center for Literacy Agents is crucial to literacy efforts.

The government's recent emphasis has been on helping Niger's farmers to help themselves through the extension of credit, better guaranteed minimum prices, and improved communications. Vigorous efforts have been made in certain regions to halt the spread of desert sands by supporting village tree-planting campaigns. Given the local inevitability of drought, the government has also increased its commitment to the stockpiling of food in granaries. But for social and political as much as economic reasons, government policy has continued to discourage the flexible, nomadic pattern of life that has long characterized many Nigerien communities.

The Nigerien government's emphasis on agriculture has, in part, been motivated by the realization that the nation could not rely on its immense uranium deposits for future development. The opening of uranium mines in the 1970s resulted in the country becoming the world's fifth-largest producer. By the end of that decade, uranium exports accounted for some 90 percent of Niger's foreign-exchange earnings. Depressed international demand throughout the 1980s, however, resulted in substantially reduced prices and output. Although uranium still accounts for 75 percent of foreign-exchange earnings, its revenue contribution in recent years is only about a third of what it was prior to the slump.

MILITARY RULE

For nearly half of its existence after its independence, Niger was governed by a ci-vilian administration, under President Hamani Diori. In 1974, during the height of drought, Lieutenant Colonel Seyni Kountché took power in a bloodless coup. Kountché ruled as the leader of a "Supreme Military Council," which met behind closed doors. Ministerial portfolios, appointed by the president, were filled by civilians as well as military personnel. In 1987, Kountché died of natural causes and was succeeded by Colonel Ali Saibou.

ACHIEVEMENTS

Niger has consistently demonstrated a strong commitment to the preservation and development of its national cultures through its media and educational institutions, the National Museum, and events such as the annual youth festival at Agades.

The National Movement for the Development of Society (MNSD) was established in 1989 as the country's sole political party, after a constitutional referendum in which less than 4 percent of the electorate participated. But, as was the case in many other countries in Africa, the year 1990 saw a groundswell of local support for a return to multipartyism. In Niger, this prodemocracy agitation was spearheaded by the nation's labor confederation, which organized a widely observed 48-hour general strike. Having earlier rejected the strikers "as a handful of demagogues," in 1991, President Saibou agreed to the formation of a National Conference to prepare a new constitution.

The conference ended its deliberations with the appointment of an interim government, headed by Amadou Cheffou, which led the country to multiparty elections in February–March 1993. After two rounds of voting, the presidential contest was won by Mahamane Ousmane. Ousmane's Alliance of Forces for Change (AFC) opposition captured 50 seats in the new 83-seat National Assembly, while the MNSD became the major opposition party, with 29 seats.

Ousmane's government made a promising start by reaching peace agreements with two rebel movements, the Tuareg Front for the Liberation of Air and Azaouad and the Organization of Army Resistance. But the nation's economic crisis deepened with the 1994 devaluation of the CFA franc. Naturally, Ousmane's political status was weakened. In February 1995, the opposition coalition, led by Hama Amadou, gained control of the National Assembly, resulting in an uneasy government of "cohabitation." Serious student unrest was followed by the military coup in January 1996 that resulted in the installation of Colonel Ibrahim Bare Mainassara as president. Under international pressure, Mainassara agreed to early elections, which were seen to have been fraudulent.

The political turn is likely to further poison interethnic relations in Niger. Since independence, members of the Zarma group have been especially prominent in the government, MNSD, and military. The deposed Ousmane has been Niger's first Hausa leader (the Hausa constitute the country's largest ethnolinguistic group).

Timeline: PAST

1200s–1400s
The Mali Empire includes territories and peoples of current Niger areas

1400s
Hausa states develop in the south of present-day Niger

1800s
The area is influenced by the Fulani Empire, centered at Sokoto, now in Nigeria

1906
France consolidates rule over Niger

1960
Niger becomes independent

1974
A military coup brings Colonel Seyni Kountché and a Supreme Military Council to power

1987
President Kountché dies and is replaced by Ali Saibou

1990s
The Nigerien National Conference adopts multipartyism; President Ibrahim Bare Mainassara is assassinated

PRESENT

2000s
President Mamadou Tandja holds power under the new Constitution

The military retains significant influence

Nigeria (Federal Republic of Nigeria)

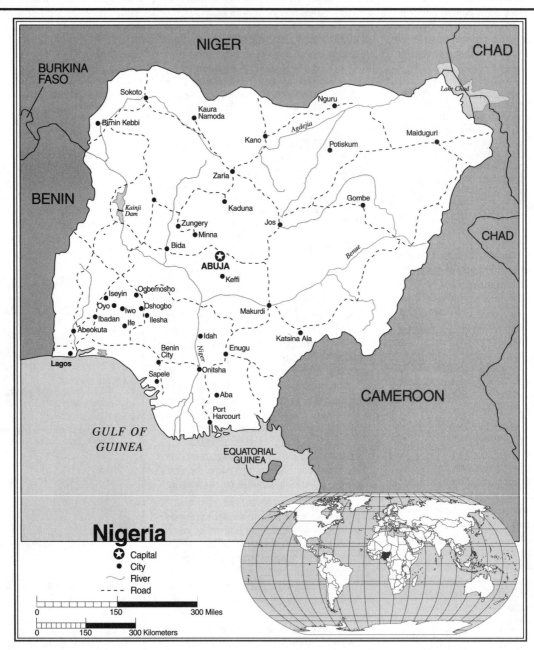

Nigeria Statistics

GEOGRAPHY

Area in Square Miles (Kilometers):
356,669 (923,768) (twice the size of California)

Capital (Population): Abuja (420,000)

Environmental Concerns: soil degradation; deforestation; desertification; drought

Geographical Features: southern lowlands merge into central hills and plateaus; mountains in southeast; plains in north

Climate: varies from equatorial to arid

PEOPLE

Population

Total: 129,935,000

Annual Growth Rate: 2.54%

Rural/Urban Population Ratio: 57/43

Major Languages: English; Hausa; Yoruba; Ibo; Fulani

Ethnic Makeup: about 21% Hausa; 21% Yoruba; 18% Ibo; 9% Fulani; 31% others

Religions: 50% Muslim; 40% Christian; 10% indigenous beliefs

Health

Life Expectancy at Birth: 51 years (male); 51 years (female)

Infant Mortality: 72.5/1,000 live births
Physicians Available: 1/4,496 people
HIV/AIDS Rate in Adults: 5.06%

Education

Adult Literacy Rate: 57%
Compulsory (Ages): 6–15; free

COMMUNICATION

Telephones: 500,000 main lines
Daily Newspaper Circulation: 18 per 1,000 people
Televisions: 38 per 1,000 people
Internet Users: 100,000 (1999)

TRANSPORTATION

Highways in Miles (Kilometers): 120,524 (194,394); but much of the road system is barely usable
Railroads in Miles (Kilometers): 2,226 (3,567)
Usable Airfields: 70
Motor Vehicles in Use: 954,000

GOVERNMENT

Type: republic in transition from military rule

Independence Date: October 1, 1960 (from the United Kingdom)
Head of State/Government: President Olusegun Obasanjo is both head of state and head of government
Political Parties: People's Democratic Party; Alliance for Democracy; All People's Party
Suffrage: universal at 18

MILITARY

Military Expenditures (% of GDP): 1%
Current Disputes: civil strife; various border disputes

ECONOMY

Currency ($ U.S. Equivalent): 128.3 nairas = $1
Per Capita Income/GDP: $840/$106 billion
GDP Growth Rate: 3.5%
Inflation Rate: 15%
Unemployment Rate: 28% (1992 est.)
Labor Force by Occupation: 70% agriculture; 20% services; 10% industry
Population Below Poverty Line: 45%

Natural Resources: petroleum; tin; columbite; iron ore; coal; limestone; lead; zinc; natural gas; hydropower
Agriculture: cocoa; peanuts; rubber; yams; cassava; sorghum; palm oil; millet; corn; rice; livestock; timber; fish
Industry: mining; petroleum; food processing; textiles; cement; building materials; chemicals; agriculture products; printing; steel
Exports: $20.3 billion (primary partners United States, Spain, India)
Imports: $13.7 billion (primary partners United Kingdom, United States, France)

SUGGESTED WEB SITES

http://www.nigeria.com
http://nigeria.indymedia.org
http://www.nigeriatoday.com
http://www.nigeriadaily.com
http://nigeriaworld.com
http://www.africanews.org/west/nigeria/
http://www.sas.upenn.edu/African_Studies/Country_Specific/Nigeria.html

Nigeria Country Report

Nigeria, the "sleeping giant of Africa," is showing signs of waking up. Its president, Olusegun Obasanjo, has been a driving force behind the New Partnership for African Development (NEPAD), which commits African states to good governance, respect for human rights, and efforts to end regional conflicts. In return, the continent is seeking more aid and foreign investment, and a lifting of trade barriers that impede African exports. The Nigerian government has also been playing a leading role in attempts to bring peace to West Africa, where President Obasanjo is generally referred to as an honest broker. Nigeria's military has been active in the settlement of regional disputes in Liberia, Côte d'Ivoire, and Sierra Leone. But the country's regional diplomatic standing was compromised in October 2002, when it reneged on a previous agreement to abide by the judgment of the International Court of Justice in a long-running border dispute with Cameroon.

With a vast population of nearly 130 million, Nigeria's human resources are yet to be fully tapped in the interest of the country. Poverty and inequality between the rich and the poor remain extreme. Nigeria's industrious people hope that the restoration of democracy will allow them to make renewed progress in the face of these challenges. In February 1999, Obasanjo was elected Nigeria's first civilian president in 15 years. His government has since struggled to push forward with the immense task of governing the diverse communities that make up Africa's most populous country. But ethnic/religious tension and corruption have continued to plague Nigeria. Transparency International has ranked the country as the second most corrupt in the world. In January 2002, a blast at a munitions dump in the principal city, Lagos, left more than 1,000 people dead, sparking renewed calls for public accountability.

DEVELOPMENT

Nigeria hopes to mobilize its substantial human and natural resources to encourage labor-intensive production and self-sufficient agriculture. Recent bans on food imports will increase local production, and restrictions on imported raw materials should encourage research and local input for industry.

Since early 2000, attempts to introduce Shari'a (Islamic law) in northern areas of the country have touched off severe violence between Muslim and Christian communities, as well as international condemnation for stoning sentences against single mothers convicted of adultery. The most serious violence occurred in 2001, when some 40,000 people were reportedly displaced following ethnic fighting between the Tiv people and several smaller ethnic groups in the central Nasarawa state. In February 2002, some 100 people were killed and thousands more left temporarily homeless in Lagos, as a result of further ethnic strife, which the city's governor suggested was orchestrated by retired army officers seeking to restore military rule.

Despite such incidents, there are signs that Nigeria's civil society is being rebuilt after almost two decades of military rule. Nigeria's transition to civilian rule followed the unexpected deaths of its last dictator, General Sani Abacha, and his most famous political prisoner, Chief Mashood Abiola. (The latter was considered by many to have won annulled 1993 presidential elections.) In the wake of the deaths, a caretaker administration under General Abdulsalem Abubakar, in cooperation with previously repressed sections of civil society, promised to restore democracy.

The Nigerian government's credibility at home and abroad was enhanced by the freeing of political prisoners and Abubakar's personal paying of respects to the late Abiola. Among those freed was Obasanjo, who had briefly served as Nigeria's military ruler in the 1970s, before handing over power to civilians. Besides the jailing of such figures as Abiola and Obasanjo, the Abacha regime had reduced Nigeria to the status of a pariah state through its execution of dozens of political opponents, including the internationally prominent writer, ecologist, and human-rights activist Ken Saro Wiwa.

FREEDOM

Under Abacha, Nigeria had one of the world's worst human-rights records. In 1998, the Nigerian Advocacy Group for Human Rights joined other international groups in issuing a statement insisting that nothing essentially changed after Abubakar succeeded Abacha. With the transition to civilian rule under former political detainee Obasanjo, the situation should improve.

Since Nigeria's independence, in 1960, its citizens have been through an emotional, political, and material rollercoaster ride. It has been a period marred by inter-ethnic violence, economic downturns, and mostly military rule. But there have also been impressive levels of economic growth, cultural achievement, and human development. To some people, this land of great extremes typifies both the hopes and frustrations of its continent.

Nigeria's hard-working population is responsible for one of Africa's largest economies. But per capita income is still only $840 per year, which is about average for the globe's most impoverished continent but down from Nigeria's estimated 1980 per capita income of $1,500.

A decade ago, it was common to equate Nigeria's wealth with its status as Africa's leading oil producer, but oil earnings have since plummeted. Although hydrocarbons still account for about 90 percent of the country's export earnings and 75 percent of its government revenue, the sector's current contribution to total gross domestic product is a more modest 20 percent.

NIGERIA'S ROOTS

For centuries, the River Niger, which cuts across much of Nigeria, has facilitated long-distance communication among vari-

ous communities of West Africa's forest and savanna regions. This fact helps to account for the rich variety of cultures that have emerged within the territory of Nigeria over the past millennium. Archaeologists and historians have illuminated the rise and fall of many states whose cultural legacies continue to define the nation.

Precolonial Nigeria produced a wide range of craft goods, including leather, glass, and metalware. The cultivation of cotton and indigo supported the growth of a local textiles industry. During the mid-nineteenth century, southern Nigeria prospered through palm-oil exports, which lubricated the wheels of Europe's Industrial Revolution. Earlier, much of the country was disrupted through its participation in the slave trade. Most African-Americans have Nigerian roots.

Today, more than 250 languages are spoken in Nigeria. Pidgin, which combines an English-based vocabulary with local grammar, is widely used as a lingua franca in the cities and towns. Roughly two thirds of Nigerians speak either Hausa, Yoruba, or Igbo as a home language. During and after the colonial era, the leaders of these three major ethnolinguistic groups clashed politically from their separate regional bases.

The British, who conquered Nigeria in the late nineteenth and early twentieth centuries, administered the country through a policy of divide-and-rule. In the predominantly Muslim, Hausa-speaking north, they co-opted the old ruling class while virtually excluding Christian missionaries. But in the south, the missionaries, along with their schools, were encouraged, and Christianity and formal education spread rapidly. Many Yoruba farmers of the southwest profited through their cultivation of cocoa. Although most remained as farmers, many of the Igbo of the southeast became prominent in nonagricultural pursuits, such as state employees, artisans, wage workers, and traders. As a result, the Igbo tended to migrate in relatively large numbers to other parts of the colony.

HEALTH/WELFARE

Nigeria's infant mortality rate is now believed to have dropped to about 72 per 1,000 live births. (Some estimate it to be as high as 150 per 1,000.) While social services grew rapidly during the 1970s, Nigeria's strained economy since then has led to cutbacks in health and education.

REGIONAL CONFLICTS

At independence, the Federal Republic of Nigeria was composed of three states: the

Northern Region, dominated by Hausa speakers; the Western Region, of the Yoruba; and the predominantly Igbo Eastern Region. National politics quickly deteriorated into conflict among these three regions. At one time or another, politicians in each of the areas threatened to secede from the federation. In 1966, this strained situation turned into a crisis following the overthrow by the military of the first civilian government.

In the coup's aftermath, the army itself was divided along ethnic lines; its ranks soon became embroiled in an increasingly violent power struggle. The unleashed tensions culminated in the massacre of up to 30,000 Igbo living in the north. In response, the Eastern Region declared its independence, as the Republic of Biafra. The ensuing civil war between Biafran partisans and federal forces lasted for three years, claiming an estimated 2 million lives. During this time, much of the outside world's attention became focused on the conflict through visual images of the mass starvation that was occurring in rebel-controlled areas under federal blockade. Despite the extent of the war's tragedy, the collapse of Biafran resistance was followed by a largely successful process of national reconciliation. The military government of Yakubu Gowon (1966–1975) succeeded in diffusing ethnic politics, through a restructured federal system based on the creation of new states. The oil boom, which began soon after the conflict, helped the nation-building process by concentrating vast resources in the hands of the federal government in Lagos.

ACHIEVEMENTS

When many of their leading writers, artists, and intellectuals were exiled and the once-lively press was suppressed, Nigerians found some solace in the success of their world-class soccer team and other athletes.

CIVILIAN POLITICS

Thirteen years of military rule ended in 1979. A new Constitution was implemented, which abandoned the British parliamentary model and instead adopted a modified version of the U.S balance-of-powers system. In order to encourage a national outlook, Nigerian presidential candidates needed to win a plurality that included at least one-fourth of the vote in two-thirds of the states.

Five political parties competed in the 1979 elections. They all had similar platforms, promising social welfare for the masses, support for Nigerian business, and a foreign policy based on nonalignment and anti-imperialism. Ideological differences tended to exist within the parties as much as among them, although the People's Redemption Party (PRP) of Aminu Kano became the political home for many Socialists. The most successful party was the somewhat right-of-center National Party of Nigeria (NPN), whose candidate, Shehu Shagari, won the presidency.

New national elections took place in August and September 1983, in which Shagari received more than 12 million of 25.5 million votes. However, the reelected government did not survive long. On December 31, 1983, there was a military coup, led by Major General Muhammad Buhari. The 1979 Constitution was suspended, Shagari and others were arrested, and a federal military government was reestablished. Although no referendum was ever taken on the matter, it is clear that many Nigerians welcomed the coup: this initial response was a reflection of widespread disillusionment with the Second (civilian) Republic.

The political picture seemed very bright in the early 1980s. A commitment to national unity was well established. Although marred by incidents of political violence, two elections had successfully taken place. Due process of law, judicial independence, and press freedom—never entirely eliminated under previous military rulers—had been extended and were seemingly entrenched. But the state was increasingly seen as an instrument of the privileged that offered little to the impoverished masses, with an electoral system that, while balancing the interests of the elite in different sections of the country, failed to empower ordinary citizens. A major reason for this failing was pervasive corruption. People lost confidence as certain officials and their cronies became millionaires overnight. Transparent abuses of power had also occurred under the previous military regime. Conspicuous kleptocracy (rule by thieves) had been tolerated during the oil-boom years of the 1970s, but it became the focus of popular anger as Nigeria's economy contracted during the 1980s.

OIL BOOM—AND BUST

Nigeria, as a leading member of the Organization of Petroleum Exporting Countries (OPEC), experienced a period of rapid social and economic change during the 1970s. The recovery of oil production after the Civil War and the subsequent hike in its prices led to a massive increase in government revenue. This allowed for the expansion of certain types of social services. Universal primary education was introduced, and the number of universities increased from five (in 1970) to 21 (in 1983).

A few Nigerians became very wealthy, while a growing middle class was able to afford what previously had been luxuries.

Timeline: PAST

1100–1400
Ancient life flourishes

1851
The first British protectorate is established at Lagos

1960
Nigeria becomes independent as a unified federal state

1966–1970
Military seizure of power; proclamation of Biafra; civil war

1979
Elections restore civilian government

1980s
Muhammed Buhari's military coup ends the Second Republic; later, Buhari is toppled by Ibrahim Babanguida; lean times

1990s
Babanguida resigns; Sani Abacha takes the reins; civil unrest and violence intensify; military strongman Abdulsalam Abubakar takes power; elections bring civilian Olusegun Obasanjo to power

PRESENT

2000s
Ethnic and religious conflict intensifies

Hopes for democratic pluralism in Nigeria revive

Oil revenues had already begun to fall off when the NPN government embarked upon a dream list of new prestige projects, most notably the construction of a new federal capital at Abuja, in the center of the country. While such expenditures provided lucrative opportunities for many businesspeople and politicians, they did little to promote local production.

Agriculture, burdened by inflationary costs and low prices, entered a period of crisis, leaving the rapidly growing cities dependent on foreign food. Nonpetroleum exports, once the mainstay of the economy, either virtually disappeared or declined drastically.

While gross indicators appeared to report impressive industrial growth in Nigeria, most of the new industry depended heavily on foreign inputs and was geared toward direct consumption rather than the production of machines or spare parts. Selective import bans led merely to the growth of smuggling.

The golden years of the 1970s were also banner years for inappropriate expenditures, corruption, and waste. For a while, given the scale of incoming revenues, it looked as if these were manageable problems. But GDP fell drastically in the 1980s with the collapse of oil prices. As the economy worsened, populist resentment grew.

In 1980, an Islamic movement condemning corruption, wealth, and private property defied authorities in the northern metropolis of Kano. The army was called in, killing nearly 4,000. Similar riots subsequently occurred in the cities of Maiduguri, Yola, and Gombe. Attempts by the government to control organized labor by reorganizing the union movement into one centralized federation sparked unofficial strikes (including a general strike in 1981). In an attempt to placate the growing number of unemployed Nigerians, more than 1 million expatriate West Africans, mostly Ghanaians, were suddenly expelled, a domestically popular but essentially futile gesture.

REFORM OR RETRIBUTION?

Buhari justified the military's return to power on the basis of the need to take drastic steps to rescue the economy, whose poor performance he blamed almost exclusively on official corruption. A "War Against Indiscipline" was declared, which initially resulted in the trial of a number of political leaders, some of whose economic crimes were indeed staggering. The discovery of large private caches of naira (the Nigerian currency) and foreign exchange fueled public outrage. Tribunals sentenced former politicians to long jail terms. In its zeal, the government looked for more and more culprits, while jailing journalists and others who questioned aspects of its program. In 1985, Major General Ibrahim Babanguida led a successful military coup, charging Buhari with human-rights abuses, autocracy, and economic mismanagement.

Babanguida released political detainees. In a clever strategy, he also encouraged all Nigerians to participate in national forums on the benefits of an International Monetary Fund loan and Structural Adjustment Program (SAP). The government turned down the loan but used the consultations to legitimize the implementation of "homegrown" austerity measures consistent with IMF and World Bank prescriptions.

The 1986 budget signaled the beginning of SAP. The naira was devalued, budgets were restricted, and the privatization of many state-run industries was planned. Because salaries remained the same while prices rose, the cost of basic goods rose dramatically, with painful consequences for middle- and working-class Nigerians as well as for the poor.

Although the international price of oil improved somewhat in the late 1980s, there was no immediate return to prosperity. Continued budgetary excesses on the part of the government (which heaped perks on its officer corps and created more state governments to soak up public coffers), coupled with instability, undermined SAP sacrifices. In 1988, the government attempted a moderate reduction in local fuel subsidies. But when, as a result, some transport owners raised fares by 50 to 100 percent, students and workers protested, and bank staff and other workers went on strike. Police killed demonstrators in Jos. Domestic fuel prices have since remained among the lowest in the world, encouraging a massive smuggling of petroleum to neighboring states. This has recently led to the ironic situation of a severe local petroleum shortage.

The Babanguida government faced additional internal challenges while seeking to project an image of stability to foreign investors. Coup attempts were foiled in 1985 and 1991, while chronic student unrest led to the repeated closure of university campuses. Religious riots between Christians and Muslims became endemic in many areas, leading to hundreds, if not thousands, of deaths.

In 1986, Babanguida promised a phased return to full civilian control. But his program of guided democratization degenerated into a farce. Local nonparty elections were held in 1987, and a (mostly elected) Constituent Assembly subsequently met and approved modifications to Nigeria's 1979 Constitution. Despite the trappings of electoral involvement, the Transitional

Program was tightly controlled. Many politicians were banned as Babanguida tried to impose a two-party system on what traditionally had been a multiparty political culture. When none of 13 potential parties gained his approval, the general decided to create two new parties of his own: the "a little to the left" Social Democratic Party (SDP) and the "a little to the right" National Republican Convention (NRC).

Doubts about the military in general and Babanguida's grasp on power in particular were raised in 1990, when a group of dissident officers launched yet another coup. In radio broadcasts, the insurrectionists announced the expulsion of five northern states from the federal republic, thus raising the specter of a return to interethnic civil war. The uprising was crushed.

A series of national elections were held in 1992 between the two officially sponsored parties. But public indifference and/or fear of intimidation, institutionalized by the replacement of the (ideally, secret) ballot with a procedure of publicly lining up for one's candidate, compromised the results. Allegations of gross irregularities led to the voiding of first-round presidential primary elections and the banning of all the candidates. After additional delays, accompanied by a serious antigovernment rioting in Lagos and other urban areas, escalating intercommunal violence, and further clampdowns on dissent, a presidential poll was finally held in June 1993 between two government-approved candidates: Mashood Abiola and Bashir Tofa. The result was a convincing 58 percent victory for the SDP's Abiola, though an estimated 70 percent of the electorate refused to participate in the charade.

Babanguida annulled the results before they had been officially counted (the final results were released by local officials in defiance of Babanguida's regime). Instead, in August 1992, he resigned and installed an interim government led by an ineffectual businessman, Ernest Shonekan. Growing unrest—aggravated by an overnight 600 percent increase in domestic fuel prices and a dramatic airline hijacking by a group calling itself the Movement for the Advancement of Democracy (MAD)—led to the interim regime's rapid collapse. In November, the defense minister, General Sani Abacha, reimposed full military rule.

Resistance to military rule steadily increased throughout 1994. Abiola was arrested in June after proclaiming himself president. His detention touched off nationwide strikes, which shut down the oil industry and other key sectors of the economy.

CULTURAL PROMINENCE

Nigeria is renowned for its arts. Contemporary giants include Wole Solyinka, who received the Nobel Prize for Literature for his work—plays such as "The Trials of Brother Jero" and "The Road," novels such as *The Interpreters,* and poems and nonfiction works. Two other literary giants are Chinua Achebe, author of *Things Fall Apart, A Man of the People,* and *Anthills of the Savannah;* and the feminist writer Buchi Emecheta, whose works include *The Joy of Motherhood.* The legendary Fela Anikulado Kuti's "Afro-Beat" sound and critical lyrics have made him a local hero and international music megastar. Also prominent is "King" Sunny Ade, who has brought Nigeria's distinctive Juju music to international audiences.

Senegal (Republic of Senegal)

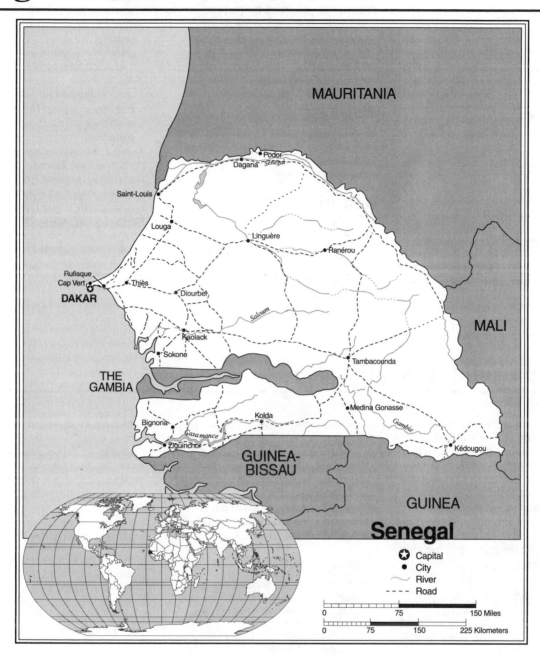

Senegal Statistics

GEOGRAPHY

Area in Square Miles (Kilometers):
76,000 (196,840) (about the size of South Dakota)
Capital (Population): Dakar (2,160,000)
Environmental Concerns: poaching; deforestation; overgrazing; soil erosion; desertification; overfishing
Geographical Features: low, rolling plains, foothills in the southeast; Gambia is almost an enclave of Senegal

Climate: tropical

PEOPLE

Population

Total: 10,590,000
Annual Growth Rate: 2.91%
Rural/Urban Population Ratio: 53/47
Major Languages: French; Wolof; Pulaar; Diola; Mandinka

Ethnic Makeup: 43% Wolof; 24% Pular; 15% Serer; 18% others
Religions: 94% Muslim; 5% Christian; 1% indigenous beliefs

Health

Life Expectancy at Birth: 61 years (male); 65 years (female)
Infant Mortality: 55.4/1,000 live births
Physicians Available: 1/14,825 people
HIV/AIDS Rate in Adults: 1.4%

Education

Adult Literacy Rate: 39%
Compulsory (Ages): 7–13

COMMUNICATION

Telephones: 235,000 main lines
Televisions: 6.9/1,000 people
Internet Users: 40,000 (2001)

TRANSPORTATION

Highways in Miles (Kilometers): 8,746 (14,576)
Railroads in Miles (Kilometers): 565 (905)
Usable Airfields: 20
Motor Vehicles in Use: 160,000

GOVERNMENT

Type: republic
Independence Date: April 4, 1960 (from France)

Head of State/Government: President Abdoulaye Wade; Prime Minister Madior Boye
Political Parties: Socialist Party; Senegalese Democratic Party; Democratic League–Labor Party Movement; Independence and Labor Party; others
Suffrage: universal at 18

MILITARY

Military Expenditures (% of GDP): 1.4%
Current Disputes: civil unrest; issue with The Gambia; tensions with Mauritania and Guinea-Bissau

ECONOMY

Currency ($ U.S. equivalent): 752 CFA francs = $1
Per Capita Income/GDP: $1,580/$16.2 billion
GDP Growth Rate: 5.7%
Inflation Rate: 3.3%

Unemployment Rate: 48%
Labor Force by Occupation: 70% agriculture
Population Below Poverty Line: 54%
Natural Resources: fish; phosphates; iron ore
Agriculture: peanuts; millet; sorghum; corn; rice; cotton; vegetables; livestock; fish
Industry: agricultural and fish processing; phosphate mining; fertilizer production; petroleum refining; construction materials
Exports: $1 billion (primary partners France, Italy, Spain)
Imports: $1.3 billion (primary partners France, Nigeria, Germany)

SUGGESTED WEB SITES

http://www.senegal-online.com/
    anglais/index.html
http://www.sas.upenn.edu/
    African_Studies/
    Country_Specific/Senegal.html

Senegal Country Report

The year 2002 was marked by both triumph and tragedy for the people of Senegal. At the World Cup, the national soccer team beat world champions France in the opening game before going to qualify for the quarterfinals. But the nation was subsequently thrown into grief when hundreds of lives were lost in a ferryboat disaster. Both incidents took attention away from the country's broader struggle to pull itself out of chronic poverty.

DEVELOPMENT

The recently built Diama and Manantali Dams will allow for the irrigation of many thousands of acres for domestic rice production. At the moment, large amounts of rice are imported to Senegal, mostly to feed the urban population.

In March 2000, Senegalese politics entered a new era with the electoral victory of veteran opposition politician Abdoulaye Wade over incumbent Abdou Diouf. Wade became the country's third president. Like his predecessors, Wade faces daunting challenges. Much of Senegal's youthful, relatively well-educated population remains unemployed. Widespread corruption and a long-running separatist rebellion in the southern region of Casamance will

also test the new regime. Taking a step-by-step approach, Wade has been able to bring about a reduction in the separatist rebellion, and corruption is on the decline. But, also like his predecessors, Wade should be able to draw upon the underlying strength of Senegal's culturally rich multiethnic society, which has maintained its cohesion through decades of adversity.

FREEDOM

Senegal's generally favorable human-rights record is marred by persistent violence in its southern region of Casamance, where rebels are continuing to fight for independence. A 2-year cease-fire broke down in 1995 after an army offensive was launched against the rebel Movement of Democratic Forces of Casamance.

THE IMPACT OF ISLAM

The vast majority of Senegalese are Muslim. Islam was introduced into the region by the eleventh century A.D. and was spread through trade, evangelism, and the establishment of a series of theocratic Islamic states from the 1600s to the 1800s.

Today, most Muslims are associated with one or another of the Islamic Brotherhoods. The leaders of these Brotherhoods, known as marabouts, often act as rural

spokespeople as well as the spiritual directors of their followers. The Brotherhoods also play an important economic role. For example, the members of the Mouride Brotherhood, who number about 700,000, cooperate in the growing of the nation's cash crops.

FRENCH INFLUENCE

In the 1600s, French merchants established coastal bases to facilitate their trade in slaves and gum. As a result, the coastal communities have been influenced by French culture for generations. More territory in the interior gradually fell under French political control.

Although Wolof is used by many as a lingua franca, French continues to be the common language of the country, and the educational system maintains a French character. Many Senegalese migrate to France, usually to work as low-paid laborers. The French maintain a military force near the capital, Dakar, and are major investors in the Senegalese economy. Senegal's judiciary and bureaucracy are modeled after those of France.

POLITICS

Under Diouf, Senegal strengthened its commitment to multipartyism. After succeeding Leopold Senghor, the nation's scholarly first president, Diouf liberalized the political process by allowing an increased number of opposition parties effectively to compete against his own ruling Socialist Party (PS). He also restructured his administration in ways that were credited with making it less corrupt and more efficient. Some say that these moves did not go far enough, but Diouf, who inclined toward reform, had to struggle against reactionary elements within his own party.

In national elections in 1988, Diouf won 77 percent of the vote, while the Socialists took 103 out of 120 seats. Outside observers believed that the elections had been plagued by fewer irregularities than in the past. However, opposition protests against alleged fraud touched off serious rioting in Dakar. As a result, the city was placed under a three-month state of emergency. Diouf's principal opponent, Abdoulaye Wade of the Democratic Party (PD), was among those arrested and tried for incitement. But subsequent meetings between Diouf and Wade resulted in an easing of tensions. Indeed, in 1991, Wade shocked many by accepting the post of minister of state in Diouf's cabinet. Elections in 1993 were less controversial, with Diouf being reelected with 58 percent of the vote. PS representation dropped to 84 seats.

In March 1995, a new, multiparty "Government of National Unity" was formed, which survived despite the defection of one of its members, the Independent Labor Party, in September 1996. But interparty tension grew in the face of Diouf's failure to appoint an independent elections commission in preparation for elections in November 1996.

THE ECONOMY

Many believe that the *Sopi* (Wolof for "change") riots of 1988 were primarily motivated by popular frustration with Senegal's weak economy, especially among its youth (about half of the Senegalese are under age 21), who face an uncertain future. Senegal's relatively large (47 percent) urban population has suffered from rising rates of unemployment and inflation, which have been aggravated by the country's attempt to implement an International Monetary Fund–approved Structural Adjustment Program (SAP). In recent years, the economy has grown modestly but has so far failed to attract the investment needed to meet ambitious privatization goals. Among rural dwellers, drought and locusts have also made life difficult. Fluctuating world market prices and disease as well as drought have undermined groundnut exports.

Senegal has also been beset by difficulties in its relations with neighboring states. The Senegambia Confederation, which many hoped would lead to greater cooperation with The Gambia, was dissolved in September 1989. Relations with Guinea-Bissau are strained as a result of that nation's failure to recognize the result of international arbitration over disputed, potentially oil-rich waters. Senegalese further suspect that individuals in Guinea-Bissau may be linked to the separatist unrest in Senegal's Casamance region. There some 1,000 people died in an insurgency campaign between the Senegalese Army and the guerrillas of the Movement of Democratic Forces of Casamance. In July 1993, the rebels agreed to a cease-fire, but the cease-fire collapsed in 1995. In August 2000, the rebels agreed to reopen talks with Wade's new administration.

But the major source of cross-border tension has been Mauritania. In 1989, long-standing border disputes between the two countries led to a massacre of Senegalese in Mauritania, setting off widespread revenge attacks against Mauritanians in Senegal. More than 200,000 Senegalese and Mauritanians were repatriated. Relations between the two countries have remained tense, in large part due to the persecution of Mauritania's "black" communities by its Maur-dominated military government. Many Mauritanians belonging to the persecuted groups have been pushed into Senegal, leading to calls for war, but in 1992, the two countries agreed to restore diplomatic, air, and postal links.

Timeline: PAST

1659
The French occupy present-day St. Louis and, later, Gorée Island

1700s
The Jolof kingdom controls much of the region

1848
All Africans in four towns of the coast vote for a representative to the French Parliament

1889
Interior areas are added to the French colonial territory

1960
Senegal becomes independent as part of the Mali Confederation; shortly afterward, it breaks from the Confederation

1980s
President Leopold Senghor retires and is replaced by Abdou Diouf; Senegalese political leaders unite in the face of threats from Mauritania

1990s
Serious rioting breaks out in Dakar protesting the devaluation of the CFA franc

PRESENT

2000s
Tensions remain with Guinea-Bissau and Mauritania

Abdoulaye Wade wins the presidency

Sierra Leone (Republic of Sierra Leone)

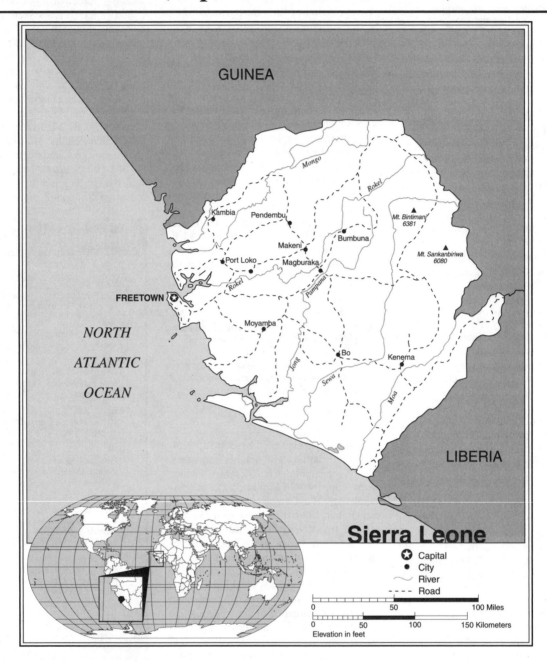

Sierra Leone Statistics

GEOGRAPHY

Area in Square Miles (Kilometers):
27,925 (72,325) (about the size of South Carolina)

Capital (Population): Freetown (837,000)

Environmental Concerns: soil exhaustion; deforestation; overfishing; population pressures

Geographical Features: a coastal belt of mangroves; wooded, hilly country; upland plateau; mountainous east

Climate: tropical; hot, humid

PEOPLE

Population

Total: 5,615,000

Annual Growth Rate: 3.12%

Rural/Urban Population Ratio: 64/36

Major Languages: English, Krio, Temne, Mende

Ethnic Makeup: 30% Temne; 30% Mende; 30% other African; 10% others

Religions: 60% Muslim; 30% indigenous beliefs; 10% Christian

Health

Life Expectancy at Birth: 43 years (male); 49 years (female)

Infant Mortality: 144.3/1,000 live births

Physicians Available: 1/10,832 people

HIV/AIDS Rate in Adults: 2.99%

Education

Adult Literacy Rate: 31.4%

COMMUNICATION

Telephones: 25,000 main lines
Internet Users: 20,000 (2001)

TRANSPORTATION

Highways in Miles (Kilometers): 7,020 (11,700)
Railroads in Miles (Kilometers): 52 (84)
Usable Airfields: 10
Motor Vehicles in Use: 44,000

GOVERNMENT

Type: constitutional democracy
Independence Date: April 27, 1961 (from the United Kingdom)

Head of State/Government: President Ahmad Tejan Kabbah is both head of state and head of government
Political Parties: Sierra Leone People's Party; National Unity Party; others
Suffrage: universal at 18

MILITARY

Military Expenditures (% of GDP): 1.5%
Current Disputes: hopes for a lasting peace after a decade of civil war

ECONOMY

Currency ($ U.S. Equivalent): 1,930 leones = $1
Per Capita Income/GDP: $500/$2.7 billion
GDP Growth Rate: 3%
Inflation Rate: 15%
Population Below Poverty Line: 68%

Natural Resources: diamonds; titanium ore; bauxite; gold; iron ore; chromite
Agriculture: coffee; cocoa; palm kernels; rice; palm oil; peanuts; livestock; fish
Industry: mining; petroleum refining; small-scale manufacturing
Exports: $65 million (primary partners New Zealand, Belgium, United States)
Imports: $145 million (primary partners Czech Republic, United Kingdom, United States)

SUGGESTED WEB SITES

http://www.sierra-leone.org
http://www.sierraleonenews.com
http://www.fosalone.org
http://www.africanews.org/west/sierraleone/
http://www.sas.upenn.edu/African_Studies/Country_Specific/S_Leone.html

Sierra Leone Country Report

In 2002, Sierra Leone emerged from a decade of civil war with the help of Britain (its former colonial power), a large United Nations peacekeeping mission, and other international elements. More than 17,000 UN troops disarmed tens of thousands of rebels and militia fighters. It was the biggest UN peacekeeping success in Africa for many years, following debacles in the 1990s in Angola, Rwanda, and Somalia. Currently the country is rebuilding its infrastructure and civil society. President Ahmed Tejan Kabbah won a landslide victory in May elections, in which his Sierra Leone People's Party also secured a majority in Parliament. In July 2002, a "Truth and Reconciliation Commission" was established to help the people of Sierra Leone overcome the trauma of the war, which was characterized by widespread atrocities.

Sierra Leone's period of political instability began in April 1992, when army Captain Valentine Strasser announced the overthrow of the long-governing All People's Congress (APC). The coup was initially welcomed, as the APC governments of the deposed president Joseph Momoh and is similarly deposed predecessor Siaka Stevens had been renowned for their institutionalized corruption and economic incompetence. But disillusionment grew as the Strasser-led National Provisional Ruling Council postponed holding multiparty elections, while sinking into its own pattern of corruption. The emergence of the

RUF insurgency brought further misery, with both the rebels and army being accused of abuses.

DEVELOPMENT

The recently relaunched Bumbuna hydroelectric project should reduce Sierra Leone's dependence on foreign oil, which has accounted for nearly a third of its imports. In response to threats of boycotting, the country's Lungi International Airport was upgraded. Persistent inflation and unemployment have taken a severe toll on the country's people.

Hopes that the (in many quarters unexpected) successful holding of democratic elections in February–March 1996 would lead to peace and reconciliation were dashed in May 1997, when dissident junior officers overthrew the elected government of President Kabbah. An Armed Forces Revolutionary Council (AFRC), led by Major Johnny Paul Koroma (who had been awaiting trial on charges stemming from an earlier coup attempt) banned political parties and all public demonstrations and meetings and announced that all legislation would be made by military decree. The AFRC soon revealed itself to be a vehicle of the rebel Revolutionary United Front (RUF) as well as of elements within the military unwilling to accept a return to civilian control.

The AFRC/RUF regime attracted overwhelming regional condemnation, with the international community sanctioning efforts by the Economic Community of West African States (ECOWAS) to restore Kabbah to power. This was finally achieved in February 1998, when military units of ECOMOG, the Nigerian-led ECOWAS peace-monitoring force, attacked and routed the junta's forces in the capital city, Freetown, after Koroma abandoned his agreement to step down peacefully.

FREEDOM

The deposed AFRC/RUF regime unleashed a terror campaign, including extra-judicial killings, torture, mutilation, rape, beatings, arbitrary arrest, and the detention of unarmed civilians. Junta forces killed and/or amputated the arms of detainees. Prior to the coup, RUF was infamous for its murderous attacks on civilians during raids in which children were commonly abducted and forced to commit atrocities against their relatives as a form of psychological conditioning.

In January 1999, rebels backing the RUF seized parts of the capital city, Freetown, from ECOMOG. After weeks of bitter fighting they were driven out, but 5,000 people had been killed, and the city was devastated. A cease-fire was declared that May, following a further ECOMOG offensive against the RUF. In July, after six weeks of talks in Lomé, Togo, a new peace agreement was signed under which the rebels were to receive posts in government and assurances that they would not be prosecuted for war crimes. In accordance with the agreement, the RUF leader, Foday Sankoh, was brought into a transitional government pledged to restoring democracy, law, and order, with UN peacekeeper assistance.

HEALTH/WELFARE

Life expectancy for both males and females in Sierra Leone is only in the 40s, while the infant mortality rate, 144.3 per 1,000 live births, remains appalling. In 1990, hundreds, possibly thousands, of Sierra Leone children were reported to have been exported to Lebanon on what amounted to slave contracts. The UNEP Human Development Index rates Sierra Leone last, out of 174 countries.

In November–December 1999, UN forces arrived to police the agreement, but ECOMOG troops continued to be attacked outside Freetown. In April–May 2000, as rebel troops attacked the capital, UN forces came under attack in the eastern part of the country, but far worse was in store when first 50, then several hundred UN troops were abducted. To protect and evacuate British citizens, 800 British paratroopers were sent to Freetown. Working alongside the UN, these troops helped to recapture hostages and secure the airport, while Sankoh was captured. In January 2001, presidential and parliamentary elections were postponed due to continuing strife. But by March the rebel army had begun to surrender, allowing for its forces' disarmament and participation in the elections.

ACHIEVEMENTS

The Sande Society, a women's organization that trains young Mende women for adult responsibilities, has contributed positively to life in Sierra Leone. Beautifully carved wooden helmet masks are worn by women leaders in the society's rituals. Ninety-five percent of Mende women join the Society.

Sierra Leone is the product of a unique colonial history. Freetown was founded by waves of black settlers who were brought there by the British. The first to arrive were the so-called "Black Poor," a group of 400

people sent from England in 1787. Shortly thereafter, former slaves from Jamaica and Nova Scotia arrived; they had gained their freedom by fighting with the British, against their American masters, in the U.S. War of Independence. About 40,000 Africans who were liberated by the British and others from slave ships captured along the West African coast were also settled in Freetown and the surrounding areas in the first half of the nineteenth century.

The descendants of Sierra Leone's various black settlers blended African and British ways into a distinctive *Krio,* or Creole, culture. Besides speaking English, they developed their own Krio language, which has become the nation's lingua franca. Today, the Krio make up only about 5 percent of Sierra Leone's multiethnic population.

As more people were given the vote in the 1950s, the indigenous communities ended Krio domination in local politics. The first party to win broad national support was the Sierra Leone People's Party (SLPP), under Sir Milton Margai, which led the country to independence in 1961. During the 1967 national elections, the SLPP was narrowly defeated by Stevens' APC. From 1968 to 1985, Stevens presided over a steady erosion of Sierra Leone's economy and civil society. The APC's increasingly authoritarian control coincided with the country's economic decline. Although rich in its human as well as natural resources at independence, today Sierra Leone is ranked as one of the world's poorest countries.

Revenues from diamonds (which formed the basis for prosperity during the 1950s) and gold have steadily fallen due to the depletion of old diggings and massive smuggling. The two thirds of Sierra Leone's labor force employed in agriculture have suffered the most from the nation's faltering economy. Poor producer prices, coupled with an international slump in demand for cocoa and robusta coffee, have cut into rural incomes. The promise by Stevens' successor, Momoh, to improve producer prices as part of a "Green Revolution" program went unfulfilled. Like its minerals, much of Sierra Leone's agricultural production has been smuggled out of the country. In 1989, the cost of servicing Sierra Leone's foreign debt was estimated to be 130 percent of the total value of its exports. This grim figure led to the introduction of an International Monetary Fund–supported Structural Adjustment

Program (SAP), whose austerity measures made life even more difficult for urban dwellers.

In some ways the situation in Sierra Leone provides lessons for Africa. Disciplined armies can be forces of stability, just as undisciplined armies can give rise to chaos. The final phase of the conflict also demonstrated the potential of obtaining peace through concerted efforts of the regional states accompanied by external powers. Perhaps another lesson was the usefulness of targeted economic sanctions. In the case of Sierra Leone, the RUF survived for many years through profits gained through diamond smuggling. The relative success in separating such "blood diamonds" from legitimate exports has given rise to greater international control over the marketing of the gems, which may prove to be a model for similar situations in the future.

Timeline: PAST

1400–1750
Early inhabitants arrive from Africa's interior

1787
Settlement by people from the New World and recaptured slave ships

1801
Sierra Leone is a Crown colony

1898
Mende peoples unsuccessfully resist the British in the Hut Tax War

1961
Independence

1978
The new Constitution makes Sierra Leone a one-party state

1985
President Siaka Stevens steps down; Joseph Momoh, the sole candidate, is elected

1990s
Debt-servicing cost mounts; SAP; Liberian rebels destabilize Sierra Leone; Momoh is overthrown

PRESENT

2000s
Civil war ends

Ahmed Kabbah wins reelection

Togo (Togolese Republic)

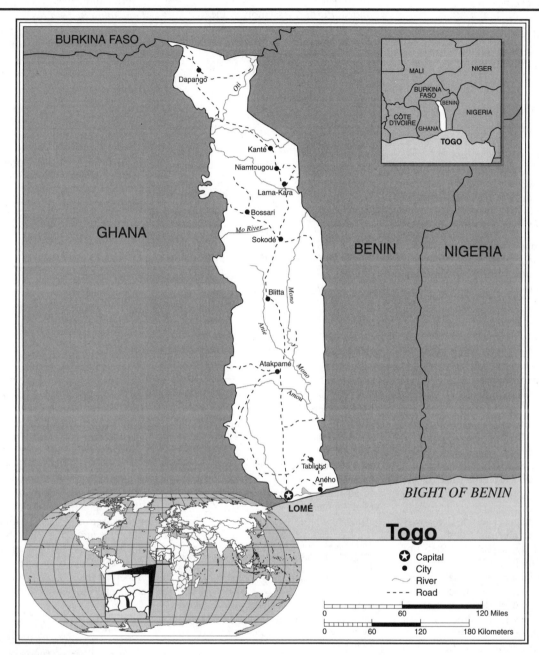

Togo Statistics

GEOGRAPHY

Area in Square Miles (Kilometers):
21,853 (56,600) (about the size of West Virginia)

Capital (Population): Lomé (732,000)

Environmental Concerns: drought; deforestation

Geographical Features: gently rolling savanna in north; central hills; southern plateau; low coastal plain with extensive lagoons and marshes

Climate: tropical to semiarid

PEOPLE

Population

Total: 5,286,000

Annual Growth Rate: 2.48%

Rural/Urban Population Ratio: 67/33

Major Languages: French; Ewe; Mina; Dagomba; Kabye; Dasomsa

Ethnic Makeup: 99% African—Ewe; Mina; Kabye; many others

Religions: 70% indigenous beliefs; 20% Christian; 10% Muslim

Health

Life Expectancy at Birth: 52 years (male); 56 years (female)

Infant Mortality: 69.3/1,000 live births

Physicians Available: 1/11,270 people

HIV/AIDS Rate in Adults: 5.98%

217

Education

Adult Literacy Rate: 51.7%
Compulsory (Ages): 6–12

COMMUNICATION

Telephones: 25,000 main lines
Televisions: 36/1,000 people
Internet Users: 20,000 (2001)

TRANSPORTATION

Highways in Miles (Kilometers): 4,512 (7,520)
Railroads in Miles (Kilometers): 352 (532)
Usable Airfields: 9
Motor Vehicles in Use: 109,000

GOVERNMENT

Type: republic under transition to multiparty democratic rule
Independence Date: April 27, 1960 (from French-administered UN trusteeship)

Head of State/Government: President Gnassingbé Eyadéma; Prime Minister Koffi Sama
Political Parties: Assembly of the Togolese People; Coordination des Forces Nouvelles; Action Committee for Renewal; Patriotic Pan-African Convergence; Union of Forces for Change; others
Suffrage: universal for adults

MILITARY

Military Expenditures (% of GDP): 1.8%
Current Disputes: civil unrest; tensions with Benin

ECONOMY

Currency ($ U.S. Equivalent): 752 CFA francs = $1
Per Capita Income/GDP: $1,500/$7.6 billion
GDP Growth Rate: 2.2%

Inflation Rate: 2.3%
Labor Force by Occupation: 65% agriculture; 30% servies; 5% industry
Population Below Poverty Level: 32%
Natural Resources: phosphates; limestone; marble; arable land
Agriculture: coffee; cocoa; yams; cassava; millet; sorghum; rice; livestock; fish
Industry: phosphates mining; textiles; handicrafts; agricultural processing; cement; beverages
Exports: $306 million (primary partners Benin, Nigeria, Ghana)
Imports: $420 million (primary partners Ghana, France, Côte d'Ivoire)

SUGGESTED WEB SITES

http://www.republicoftogo.com/
    english/index.htm
http://www.republicoftogo.com/
http://www.sas.upenn.edu/
    African_Studies/
    Country_Specific/Togo.html

Togo Country Report

In recent years, Togo has become a prime example of the difficulty of achieving democratic reform in the face of determined resistance by a ruling clique that enjoys military backing and a strong ethnic support base. For the past four decades, the country has been politically dominated by supporters of its long-serving president, Gnassingbé Eyadéma. Their grip on power was evident once more in October 2002, when Eyadéma's Assembly (or Rally) of the Togolese People (RPT) party won another landslide victory in legislative elections, even in the face of continuing allegations of vote rigging and human-rights abuses. The true political contest may have occurred earlier in the year, when an apparent power struggle resulted in Eyadéma's sacking his prime minister and long-time ally, Agbeyome Kodjo. The outcome reconfirmed Eyadéma's status as one of Africa's true political survivors.

Emerging from the ranks of the military, Eyadéma first seized power in 1967. This followed a period of instability in the wake of the assassination of the country's first president, Sylvanius Olympio, by the Togolese military. In 1969, Eyadéma institutionalized his increasingly dictatorial regime as a one-party state. All Togolese have been required to belong to the RPT. But in 1991, faced with mass prodemocracy demonstrations in Lomé, the capital

city, Eyadéma acquiesced to opposition calls for a "National Conference" that would end the RPT's monopoly of power. Since then, Eyadéma has survived Togo's turbulent return to multiparty politics with characteristic ruthlessness, skillfully taking advantage of the weakness of his divided opponents.

DEVELOPMENT

Much hope for the future of Togo is riding on the recently created Free Trade Zone at Lomé. Firms within the zone are promised a 10-year tax holiday if they export at least three quarters of their output. The project is backed by the U.S. Overseas Private Investment Corporation.

DEMOCRACY VS. DICTATORSHIP

Meeting in July–August 1991, the National Conference turned into a public trial of the abuses of the ruling regime. Resisting the president's attempts to dissolve it, the Conference appointed Kokou Koffigoh as the head of an interim government, charged with preparing the country for multiparty elections. The RPT was to be disbanded, and Eyadéma himself was barred from standing for reelection.

In November–December 1991, however, soldiers loyal to Eyadéma launched a bloody attack on Koffigoh's residence.

The French, whose troops had intervened in the past to keep Eyadéma in power, refused Koffigoh's plea for help. Instead, the coup attempt ended with the now-almost-irrelevant Koffigoh and Eyadéma agreeing to maintain their uneasy cohabitation. Elections were henceforth to include the RPT. Despite the "compromise," there was an upsurge in political violence in 1992, which included the May shooting of Gilchrist Olympio (the son of Sylvanus) and other Eyadéma opponents. In September, "rebel" soldiers once more held the government hostage.

FREEDOM

Togo continues to have a poor human-rights record. Its progovernment security forces have been responsible for extrajudicial killings, beatings, arbitrary detentions, and interference with citizens' rights to movement and privacy. Freedoms of speech and of the press are restricted. Interethnic killings have led to major population displacements.

A January 1993 massacre of the prodemocracy demonstrators pushed the country even further to the brink. Some 300,000 southern Togolese, mostly Ewe-speakers, fled the country, fearing "ethnic cleansing" by the largely northern, Kabye-speaking

(United Nations photo by Anthony Fisher)

Potable water is not universally available in Togo. While food production in Togo is officially said to be adequate, outside observers contend that the drought-prone north is uncomfortably reliant on the more agriculturally productive southern areas.

army. In 1993–1994, exiled anti-Eyadéma militants—many of whom coalesced as the Front of the National Committee for the Liberation of the Togolese People (FNCL)—began to fight back. The army chief of staff was among those killed during a daring raid on the main military headquarters in Lomé, in which grenades were also thrown into Eyadéma's bedroom.

In July 1993, Eyadéma and his more moderate opponents signed a peace accord in Burkina Faso, pledging renewed movement toward election. A month later, however, the opposition boycotted a snap presidential poll. Thereafter, Eyadéma gave ground, agreeing to internationally supervised legislative elections in Febru-

HEALTH/WELFARE

The nation's health service has declined as a result of austerity measures. Juvenile mortality is 15%. Self-induced abortion now causes approximately 17% of the deaths among Togolese women of child-bearing age. School attendance has dropped in recent years.

ary 1994. After two rounds of voting, amid escalating violence, Eyadéma's RPT and the main opposition—Action Committee for Renewal (CAR), led by Yaovi Agboyibor—each controlled about 35 seats in the 75-seat Assembly. (The situation was clouded by judicial reviews of the results in

five constituencies.) The balance of power rested with former Organization of African Unity secretary-general Edem Kodjo's Togo Union for Democracy (UTD), which had entered the election allied with CAR. But in May, Kodjo became prime minister, with Eyadéma's backing. The failure of the moderate opposition to capitalize on its apparent victory in undoubtedly flawed elections strengthened the determination of the militants to carry on by other means.

STRUCTURAL ADJUSTMENT

Togo's political crisis has taken place against a backdrop of economic restructuring. In 1979, Togo adopted an economic-recovery strategy that many consider to have been a forerunner of other Structural Adjustment Programs (SAPs) introduced throughout most of the rest of Africa. Faced with mounting debts as a result of falling export revenue, the government began to loosen the state's grip over the local economy. Since 1982 a more rigorous International Monetary Fund/World Bank–supported program of privatization and other market-oriented reforms has been pursued. Given this chronology, Togo's economic prospects have become a focus of attention for those looking for lessons about the possible effects of SAP. Both proponents and opponents of SAP have grounds for debate.

ACHIEVEMENTS

The name of Togo's capital, Lomé, is well known in international circles for its association with the Lomé Convention, a periodically renegotiated accord through which products from various African, Caribbean, and Pacific countries are given favorable access to European markets.

Supporters of Togo's SAP point out that since 1985, the country has enjoyed an average growth in gross domestic product of 3.3 percent per year. While this statistic is an improvement over the 1.7 percent rate recorded between 1973 and 1980, however, it is well below the 7.2 percent growth that prevailed from 1965 to 1972. (The GDP growth rate in 2002 was estimated at 2.2 percent.) During the 1980s, there was a rise in private consumption, 7.6 percent per year, and a drop in inflation, from about 13 percent in 1980 to an estimated 2 percent in 1989. A rate of 2.3 percent was estimated for 2002.

The livelihoods of certain segments of the Togolese population have materially improved during the past decade. Beneficiaries include some of the two thirds of the workforce employed in agriculture. Encouraged by increased official purchase prices, cash-crop farmers have expanded their outputs of cotton and coffee. This is especially true in the case of cotton production, which tripled between 1983 and 1989. Nearly half the nation's small farmers now grow the crop.

Balanced against the growth of cotton has been a decline in cocoa, which emerged as the country's principal cash crop under colonialism. Despite better producer prices during the mid-1980s, output fell as a result of past decisions not to plant new trees. Given the continuing uncertainty of cocoa prices, this earlier shift may prove to have been opportune. The long-term prospects of coffee are also in doubt, due to a growing global preference for the arabica beans of Latin America over the robusta beans that thrive throughout much of West Africa. As a result, the government had to reverse course in 1988, drastically reducing its prices for both coffee and cocoa, a move that it hopes will prove to be only temporary.

Eyadéma's regime has claimed great success in food production, but its critics have long countered official reports of food self-sufficiency by citing the importation of large quantities of rice, a decline in food production in the cotton-growing regions, and widespread childhood malnutrition. The country's food situation is complicated by an imbalance between the drought-prone northern areas and the more productive south. In 1992, famine threatened 250,000 Togolese, mostly northerners.

There have been improvements in transport and telecommunications. The national highway system, largely built by the European Development Fund, has allowed the port of Lomé to develop as a transshipment center for exports from neighboring states as well as Togo's interior. At the same time, there has been modest progress in cutting the budget deficit. But it is in precisely this area that the cost of Togo's SAP is most apparent. Public expenditures in health and education declined by about 50 percent between 1982 and 1985. Whereas school enrollment rose from 40 percent to more than 70 percent during the 1970s, it has slipped back below 60 percent in recent years.

The ultimate justification for Togo's SAP has been to attract overseas capital investment. In addition to sweeping privatization, a Free Trade Zone has been established. But overseas investment in Togo has always been modest. There have also been complaints that many foreign investors have simply bought former state industries on the cheap rather than starting up new enterprises. Furthermore, privatization and austerity measures are blamed for unemployment and wage cuts among urban workers. One third of the state-divested enterprises have been liquidated.

Whatever the long-term merits of Togo's SAP, it is clear that it has so far resulted in neither a pattern of sustainable growth nor an improved standard of living for most Togolese. For the foreseeable future, the health of Togo's economy will continue to be tied to export earnings derived from three commodities—phosphates, coffee, and cocoa—whose price fluctuations have been responsible for the nation's previous cycles of boom and bust.

Timeline: PAST

1884
Germany occupies Togo

1919
Togo is mandated to Britain and France by the League of Nations following Germany's defeat in World War I

1956–1957
UN plebiscites result in the independence of French Togo and incorporation of British Togo into Ghana

1960
Independence is achieved

1963
Murder of President Sylvanus Olympio; a new civilian government is organized after the coup

1967
The coup of Colonel Etienne Eyadéma, now President Gnassingbé Eyadéma

1969
The Coalition of the Togolese People becomes the only legal party in Togo

1990s
Prodemocracy demonstrations lead to interim government and the promise of multiparty elections; Eyadéma survives escalating violence and controversial elections

PRESENT

2000s
Eyadéma retains power

Eyadéma is named chairman of the OAU

Articles from the World Press

Regional Articles

Africa united: Not hopeless, not helpless.

Chris Brazier champions the Pan-African alternative

Boys with guns. A rebel army which cuts off people's limbs. A civil war only kept under control by heroic white soldiers (Sierra Leone). Aggressive ex-combatants seizing farms. White farmers murdered. A wild-eyed leader who will do anything to hold on to power (Zimbabwe). Devastating floods. People stranded and starving. Western helicopters rushing to the rescue (Mozambique).

These images come from the last few months of media coverage of Africa. They are important stories which deserve our attention and concern. And of course the media thrive on bad news. But negative news stories about our own countries will be balanced by culture, sport, political intrigue and humour. They also appear in the context of ordinary life as we experience it—a life in which we take our children to school, do our shopping, and walk in the park; a life generally free from danger and violence.

Right across Africa, the vast majority of people also do these things every day but you would never know it from the reporting: news items full of violence and disaster are the only stories told. Small wonder that the average Westerner might feel inclined to give up on the continent.

> ## The Sierra Leone story no more represents the entire African continent today than the Mandela story did in 1994

Indeed sometimes we are even told to do so. On the cover of *The Economist* in May there appeared a photo of a man with a rocket-launcher over his shoulder. He fills the whole outline of Africa; the issue title is 'The hopeless continent'. With one stark phrase, Africans, from diligent farmers in the Sahel to wise elders in the Rift Valley, from *fellahin* in Egypt to bankers in Botswana, are summarily dismissed from the ranks of our common humanity. Africa is reduced to one single, terrible reality: violence.

To understand the outrage such dismissiveness provokes in Africa itself you should perhaps imagine what it would be like to have a copy of *African Agenda* drop through your letterbox with a picture of a violent British soccer fan superimposed on an outline of Europe, and the title 'The hooligan continent'. Except that, of course, *African Agenda*, good magazine though it may be, could never be thought to have the global establishment behind it and does not carry *The Economist*'s dead weight of power.

When the *Economist* issue emerged I was subscribing to an e-mail conference called 'African Realities' organized by the UN Economic Commission on Africa. This involved contributions from people all over the world interested in debating the state of Africa from economics to education, democracy to peace. But these more sober considerations were forgotten for a fortnight as people expressed their outrage at the grotesque misrepresentation. Eventually a collective protest letter was composed and mailed to the magazine.

'Your articles,' it said, 'reflect the tendency for one sensational story to "epitomize" the continent, as you note Sierra Leone does today. This is precisely what prevents the policy-makers and the public from understanding the diversity of a continent more than three times the size of Europe. The "Sierra Leone" story no more represents the entire African continent today than the "Mandela" story did in 1994.

'It is important to provide readers with a balanced view of both African success stories and crises. Reinforcing stereotypes of "backwardness" and "hopelessness" is not conducive to finding solutions to any of Africa's problems. It is also not an honest portrayal of the complex realities.'

It is not an honest portrayal. But it certainly reflects the Western world's dominant attitude to Africa.

It is true that the situation is bleak, even to those of us who love Africa. Development there has conspicuously failed. Over the two decades that I have worked for the *New Internationalist* if there has been one consistently encouraging global trend it has been the decline in child mortality and rise in life expectancy right across the developing world. We have become accustomed to immunization and generally higher standards of public health and nutrition delivering on the most fundamental level: giving people more years of life.

And that is broadly still the case across the developing world—except in Africa south of the Sahara. Here, average life expectancy increased from 40 years in 1960 to 52 in 1990. But it has now gone into reverse, currently standing at around 48, just as it did in 1980. The average sub-Saharan African now can expect to live 14 years less than someone in the next-poorest region, South Asia—and fully 30 years less than someone in the rich world. In Zimbabwe and Uganda, devastated by AIDS, life expectancy has sunk to below what it was when the 'development decades' began, in 1960.[1]

In economic terms the indicators have also been going into reverse. Gross national product is a thoroughly inadequate measure of human well-being. But it is the chosen yardstick of the World Bank and the International Monetary Fund (IMF) and, by its lights, their own policies and programmes in Africa have demonstrably failed.

While every other region of the world has made steady progress on this front over the last two decades, the income per head for Africa as a whole is, at $665, lower than it was in 1980. Meanwhile average income in the rich world has shown stratospheric growth. The average Westerner in 1980 earned 15 times more than the average African; now they earn 50 times more.[2]

Africans looking at these figures could be forgiven for concluding that there has been a conspiracy at work to boost the wealth of the West at the expense of its former colonial subjects. These, after all, were the two decades in which African nations were forced by debt and desperation to submit to the unyielding economic orthodoxy of the World Bank and the IMF. They have followed the rules, more or less to the letter. By cutting public expenditure and government subsidies, opening up their country to transnational corporations, they have put their citizens through enormous pain, supposedly in the name of future gain. Such suffering has even passed into the language. In francophone West Africa, since the devaluation of the local currency, the popular term used to describe any person in difficulty is 'adjusted'.[3]

There may be no conspiracy as such but the effect is pretty much the same. Each IMF-inspired budget cut, each market liberalization, serves to turn the local economy into a more effective supplier of raw materials, of export crops and minerals—and indeed of debt interest payments—to the industrialized world.

The finesse would be almost admirable for its style were it not so morally repugnant. Direct colonial exploitation being now unfashionable, the West has evolved a way of forcing African countries to service its interests—*apparently of their own free will*. 'Look,' the visiting financiers say: 'the national flag flies, the President is there in

his pomp; we are here by invitation. What's more, we are only here because we are trying to help.'

But finally the mirror has cracked. After four decades of national independence, African governments are being forced by the depth and scale of the continent's crisis to accept that this cannot go on. The dream that independent nation-states could aspire, via heavy doses of education and industrialization, to affluence and influence is in tatters. Nation-states in Africa are barely worthy of the term 'independent'; they now have no room for manoeuvre at all. Sad to say, in all too many cases, the only room for manoeuvre many of its leaders have wished for has involved use of a torture cell and a Swiss bank account.

As a result, in a phenomenon unlooked for even as recently as 1990, another African dream has been revived: that of Pan-African unity. The ideas of Pan-Africanism have been so discounted, so marginalized since the 1960s that few people in the West are aware of them and of the rich heritage that lies behind them.

Born of the yearning of exiled slaves for their ancestral land, Pan-Africanism had by the dawn of the twentieth century become a nascent movement resisting the colonial dismemberment of the continent. By the 1950s, in the hands of Kwame Nkrumah, it had become an alternative path to freedom.

When Nkrumah led Ghana (the former Gold Coast) to independence from Britain in 1957, he inspired Africans resisting colonialism and seeking freedom all over the continent. His firm conviction, however, was that national independence was not enough and he spent much of his energy pursuing the possibility of a united Africa.

As another great leader, Julius Nyerere, recalled, full of regret in the last decade of his life for the missed opportunity of a united Africa: 'Kwame Nkrumah was the great crusader for African unity. He wanted the Accra summit of 1965 to establish a Union Government for the whole of independent Africa. But we failed. The minor reason is that Kwame, like all great believers, underestimated the degree of suspicion and animosity which his crusading passion had created among a substantial number of his fellow heads of state. The major reason was linked to the first: already too many of us had a vested interest in keeping Africa divided…'[4]

Nkrumah was toppled by a coup in 1966 and lived thereafter in exile in Guinea until his death [from] cancer in 1972; his Pan-African ambitions seemed lost with him. During the 1970s and 1980s the influence and coherence of the Pan-African movement waned. The Organization of African Unity (OAU) became a painfully ineffective regional body, too often presided over by dictators who made a mockery of paper concern for human rights and social justice. National leaders ferociously defended the colonial borders they had inherited—there was a fear that if once these were broken open all hell might let loose. The idea of a united Africa seemed to have been consigned to the past.

Yet in the 1990s the notion of Pan-African union was born again. The liberation of South Africa helped it back to life. The failure to overcome the apartheid state had been one of the OAU's worst failures. But when freedom finally came, it unlocked new possibilities—for the first time the unity of the continent from Cape Town to Cairo could at least be conceived.

The 'sanctity' of the colonial borders has now been breached too, by Eritrea's long-sought independence from Ethiopia in 1993—though the OAU continues to resist the plausible case for another such breach in Somaliland, which has formed its own coherent but utterly unrecognized government in a breakaway from the still-chaotic Somalia.

The sheer absurdity of the existing borders has become ever more evident. In many parts of the continent these borders are plainly and quite rightly ignored, as people trade the way they have for centuries, within and between ethnic groupings, irrespective of the lines on the colonizers' map that have divided them. From one point of view these traders are 'smugglers'; from another they are 'Pan-African entrepreneurs'.[5] African leaders, too, have become more conscious than ever before that their own nation-states are unsustainable—at least within the current world economic and political arrangements. Yes, they still have their ten-gun salutes and their military parades, their presidential palaces and their motorcades. But the more intelligent among them have realized that their power to change their people's circumstances is severely circumscribed, and that if they want to make a difference they will have no alternative but to swim in a bigger pool.

The main influence propelling Africans towards greater economic and political unity, however, is globalization. Faced with a trading system which insists on transnational capital having *carte blanche,* Africans are increasingly recognizing that they will have to stand together if they are to defend (or advance) their own interests. As individual nation-states within artificial borders they can too easily be picked off or played against each other by the corporations and the global accountants in a post-colonial version of divide-and-rule.

Both the threat and the example of the European Union have given the idea of an African Union new impetus—and an African bloc incorporating the economic weight of both South Africa and the oil states of the Maghreb would have infinitely more clout than even the most populous nation, Nigeria, let alone minnows like Togo or Cape Verde. There is little likelihood, for example, of a permanent seat on the UN Security Council being granted to any single African country—but an African Union representing a tenth of the world's people could hardly be denied it.

At the OAU summit in Abuja, Nigeria, in 1991, a treaty was signed establishing an African Economic Community and holding out the prospect of 'Africa-wide monetary union' and a continental Parliament by the year 2025.

But the long timetable led to understandable suspicion that this was mere rhetoric. As the Pan-African scholar Julius O Ihonvbere wrote in 1994: 'Given that none of the current leaders will be in office by 2025, the current decision to finalize arrangements for a regional community in 34 years appears to be an attempt to buy time and give the impression that something was being done as a response to the crisis.'[6]

The idea of a united Africa seemed to have been consigned to the past. But in the 1990s it was born again

The headlong onrush of globalization in the 1990s has, however, made it clear that this leisurely pace towards a united Africa is inadequate. Prompted by the Libyan leader, Muammar Qadhafi, the OAU held an extraordinary summit in Sirte, Libya, in September 1999, and committed itself to a fast-track process. The Sirte Declaration decided to establish an African Union, to 'ensure the speedy establishment of... the African Central Bank, the African Monetary Union, the African Court of Justice and, in particular, the Pan-African Parliament. We aim to establish that Parliament by the year 2000.'

This deadline looks to be too ambitious but it signals the new sense of urgency. As we go to press the draft treaty for an African union—which would completely replace the OAU—has been approved by an extra-ordinary ministerial council and will be put to a summit meeting in Lomé in July. Such radical moves toward international unity only ever tend to come to fruition after a cataclysm, just as painful memories of the Second World War provided the motivating force for both the United Nations and the European Union. But make no mistake: Africa has undergone a disaster equivalent in scale to a world war over the last two decades, and unity may well be the eventual result here too.

Of course there will be all kinds of obstacles in the road over the next few years. The Francophone countries, for instance, have historically been the least keen on the notion of Pan-African union. But even that negative consensus, based on a special relationship with Paris and a direct link to the French franc, has been undermined in the last few years by another example of continental unity—when France entered the European single currency it was unable to take its former African colonies' currencies along for the ride.

Ironically one of the main stumbling blocks may be a development which in itself underlines the slow death of the African nation-state: the move towards regional unity. The most notable example of this is, like Pan-Africanism, a resurrection of an old idea: Tanzania, Kenya and Uganda have agreed to form an East African Union like that first planned by Nyerere, Kenyatta and Obote in the early 1960s; an East African passport is envisaged, together with eventual expansion to include Rwanda and Burundi. This makes a great deal of sense but may detract from the momentum towards a wider union.

> # There is a chance that Pan-African institutions could help unlock some of the continent's more intractable problems

Even if it eventually comes about, a United States of Africa would not, needless to say, be a panacea for all the continent's ills. If it were to work, power would have not only to be handed upwards to a federal administration but also transferred downwards to local areas and communities, enhancing genuine participation and democracy. Realizing that greater unity carries with it the nightmare possibility of a bigger stage for the kind of despots with which too many African nations have been blighted, the OAU last year established a new rule which will from now on exclude from its table and its consultations any leader who has come to power by military coup. A simple rule—but one which betokens a dramatic sea change since the 1980s when most of the continent's leaders owed their position to the gun.

There is a chance that Pan-African institutions and sensibilities could help unlock some of the continent's more intractable problems. It is at least arguable that there would be less scope for ethnic conflict or national rivalry within an African Union than at present—countries are often chronically destabilized in part by a colonial border which divides an ethnic group in two or three. The current peace process in Burundi—a painstak-ing programme overseen first by Nyerere and now by Mandela—is in itself an encouraging example of a Pan-African initiative aimed at forestalling another genocide. On the AIDS front, meanwhile, it is much easier to imagine a coherent programme of action emerging from a federal government than from national leaders, too many of whom have adopted an ostrich mentality.

This may be over-optimistic. But we would rather err on the side of hope than of despair. And while *The Economist* and a hundred other leader-writers in the West would have us believe that Foday Sankoh's atrocities in Sierra Leone sum up the continent, we would rather focus on the resilience and resourcefulness of the millions of Africans who have never held a gun. This silent majority could shake the world if it spoke with a single voice.

In one tiny, though fascinating, way it already has. The BBC World Service recently conducted its own poll asking people from the continent who was the greatest African of the twentieth century. The result, they thought, would be a foregone conclusion: yet more eulogies to Nelson Mandela were planned. Instead they were shocked to find that Kwame Nkrumah won by a mile, an index of the rising stock of the Pan-Africanism he embodied. Nkrumah did not live to see it, but 'Africa United' may well be an idea whose time has finally come.

Notes

1. UNICEF State of the World's Children 1982, 1992, 1997 and 2000.
2. World Bank World Development Report 1982; UNICEF State of the World's Children 2000.
3. Rabia Abdelkrim-Chikh, 'African Realities' electronic digest, 2000.
4. Speech in Accra on the 40th anniversary of Ghana's independence, 6 March 1997.
5. I am indebted for this thought to Tajudeen Abdul-Raheem.
6. Keynote address to The All-African Students' Conference, University of Guelph, Canada, 27 May 1994.

Globalization and Its Discontents

An African Tragedy

If man were to begin by studying himself, he would see how incapable he is of going beyond himself. How could it be possible for a part to know the whole? But he may perhaps aspire to a knowledge of at least those parts which are on the same scale as himself. But the different parts of the world are all so closely linked and related together that I hold it to be impossible to know one without knowing the other and without knowing the whole.

—Blaise Pascal, Fragment 72

David E. Apter

ALTHOUGH PROTEST against globalization continues to mount, most of its foes' indictments are so sweeping as to render real solutions nugatory. Undiscriminating critiques undermine the opposition and distort its purposes. This is a pity. Fringe violence aside, such protests are the only way to register serious objections to a process that proceeds under its own steam, yet generates discontents of great practical and moral importance.

To blame all developmental ills on globalization is to demonize it rather than to understand its complexities. The truth is that if one can't live with globalization, one can't live without it either. Its consequences are many and diverse. But the devil is in the details. And details are not the concern of most protesters.

The same criticisms apply to most supporters of globalization. Their Olympian spirit keeps their heads in the clouds. Despite professions of pragmatism, they are swept along by hubris regarding the benign effects of the market. Their market models ignore the social overhead costs of global development and its impact in real terms: population displacement and dispersion, loss of patrimonies, marginalization of sectors of the population, social polarization, fragmenting institutions, increased corruption, and criminality. Globalization poses a host of problems for education, training, and the constructive use of human talent. All this is particularly visible in Africa.

Africa is an extreme case in degree rather than kind. It illustrates many of the negative consequences globalization has for vulnerable people, with few public or private resources at their disposal and virtually no access to economic or political redress. The continent demonstrates how people can catch it both ways. On the one hand, Africans suffer from a declining share in the globalization process. This lack of participation leaves them further and further behind economically, poorer in both relative and real terms. The result is fewer political opportunities to address social needs with any likelihood of success. On the other other hand, where globalization

has occurred, poverty and social cleavages have been exacerbated (especially in such capital cities as Abidjan, Nairobi, or Lagos). The gap between rich and poor becomes more socially complex and politically serious. There are good grounds for asserting that however poor a country is, its rich will continue to get richer.

In short, globalization has negative effects both in its presence and absence. Similar dilemmas can be seen in parts of Latin America, the Middle East, the Philippines, Indonesia, and elsewhere, often accompanied by violence and terrorism. But in Africa the dysfunctions of globalization are particularly sour, coming as they do on top of the effects of postcolonialism, rapacious one-party regimes and military dictatorships, the erosion of educational and professional standards, government decay, and dilapidated institutions. Things are very bad, and proposals for debt reduction, institution building, and other remedies won't bring real improvement.

Marginalizations

Africa has undergone what can be called double marginalization. The continent has experienced an absolute decline in relation to the global economy. Within countries, there has been social polarization between a relatively concentrated sector of functional elites (most of them working for global firms or in government or educational institutions) and a growing sector of the population that is "superfluous." The result is a kind of social distortion. With the growth of the "superfluous" sector, governments face increasingly unmanageable overload problems.

The paradox is plain enough. The African economies need to be more securely integrated into global markets, primarily through private investment, yet it is this very integration that accentuates the problems globalization is supposed to solve. If that private investment is in high technology, it will employ relatively few workers, and so the new ventures will increase rather than decrease the "marginals" in countries with strikingly asymmetrical patterns of social inequalities. There are, in other words, unintended social inefficiencies of the market. Even with economic growth, they are likely to work at cross-purposes with political democracy. For although it may be possible for globalization to promote growth and democracy simultaneously it is very difficult for it to do so in the face of expanding social inequities. This suggests an underlying logic:

- Growth depends on increasing productivity;
- Increased productivity depends on innovations that take form as capital-intensive industry at the expense of labor-intensive industry;
- This leads to redundancies in the labor-intensive sectors, especially among the unskilled and poorly educated. Their prolonged unemployment turns an economic condition (lack of a job) into a social condition of marginality;
- "Marginals," as a sector of "functionally superfluous" people, have few prospects for improvement. Life is risky for them, and their marginality individualizes risk because collective options for change seem nonexistent. Schools and training programs become useless, tending to validate failure rather than success. Something happens to people who have minimal access to facilities and resources for a sufficiently long time. Although they need the most, they receive the least and wind up as unwitting co-conspirators in their own demise.
- The result is a type of social "pathology" demanding vast state expenditures on compensatory and welfare programs that themselves are invariably inefficient.
- Rising social overhead costs reduce the state's ability to balance equity and growth.

One unhappy, dangerous result is what can be called "cleavage politics," the social consequence of dividing societies into technologically literate and technologically disadvantaged populations. Among the disadvantaged, globalization's social and economic cleavages produce a politics that highlights differences of religion, caste, race, language, and other categorical affiliations. They promote conflict rather than mediation and lead people to stress "principles" rather than interests. All this reinforces the negative rather than the positive aspects of diversity and pluralism. The institutional framework required of a democratic political system is weakened, and political parties cannot function properly. It becomes very difficult to devise strategies to promote growth without exacerbating marginality or to induce innovative reforms without intensifying social polarization.

All this may be summarized in a contrarian hypothesis: market-driven economic growth in Africa will produce marginality, social disruption, and political turbulence, expanding opportunities for the few without a corresponding enlargement of opportunities for the many. These conditions will prejudice future democratic possibilities. In other words, drawing Africa deeper into the global economy will ensure that it will remain what it has become: a graveyard of failed projects and failed democracies.[1]

Complexities

Africa's social problems are complex both in kind and in pedigree. Hopes were high at independence that the new states would finally be able to transcend externally imposed economic, political, and social limitations. But the postcolonial situation was aggravated by social and political experiments that aimed to jump-start economies. They were practically all failures, leaving behind problems

that plagued succeeding generations of political leaders. The long-term legacy of colonialism was very mixed. It is patronizing to assume that postcolonial African leaders, whatever the pressures on them, were not their own agents. Despite an understandable reluctance to hold themselves accountable [for] much of what went wrong, they were in fact not simple proxies or clients of outside powers. Many were complicit with those powers and thereby bolstered their own positions and feathered their own nests. Experience demonstrates that despite repeated calls for "transparency," globalization provides new opportunities for bad government.

Bad government includes overvalued currencies, heavy indebtedness, inefficient investment, inadequate services, poorly performing state corporations, low labor productivity, urban unemployment, declining agricultural output, inadequate (costly and minimal) social services, weak administrative and health facilities, and drastically eroding educational standards. All these lead toward increasingly fragile infrastructures, corruption, technological backwardness, a dearth of skilled workers, and poor economic diversity. Take the case of Zimbabwe. Land reform there stalled, inflation is 40 percent, the university is closed, and the organized opposition has been brutally repressed. Corruption extends to the highest places.

Africa's share of the global social product declined radically in the 1970s and 1980s. Africa accounted for only 1.6 percent of international private capital flows by the first half of the 1990s. The total gross national product for sub-Saharan African countries is less than 10 percent of the GNP of the United States. Add broken-down road, rail, and communication facilities; heavy indebtedness; AIDS; and ethnic and religious conflicts, which range from civil unrest to genocidal violence; and it is hardly surprising that African civil societies don't function and political institutions are inadequate.

Worst of all, generations of children have been sacrificed—that is, maimed psychologically, physically, and educationally. Excepting South Africa, virtually all African countries are worse off today than at independence. If one finds hopeful signs it is mostly due to diminished expectations.

The Politics of Negative Pluralism

One way to consider the scope of Africa's tragedy is in terms of the "social pathologies" to which I have pointed. It is only at the extreme, where marginality becomes a permanent condition, that it becomes a social pathology. People have little faith that work will be appropriately rewarded. Cause and effect appear to them to be something random. Daily life is so precarious that individual survival becomes the sole object, and the means may easily include violence.

If such circumstances extend sufficiently, marginals become a "class," but not in a Marxist sense. They become a category made up of people forced out of the labor market and struggling for basic survival. These people create their own ways of life and subcultures that are very different from and threatening to the "functional" elites. The latter then seek physical boundaries to protect themselves. In most parts of Africa, the elites live in worlds that are physically and socially apart from the "marginals." The elites and their children have access to education, knowledge, and expertise together with funds; their physical space is ordered and sheltered.

The marginals tend to random violence, an individualized enterprise; but when risk does get collectivized and turned into social movements, it can lead to group violence ranging from terrorism to genocide (as we have seen in Rwanda, Sierra Leone, and the Sudan). Because economic marginality begets social marginality, people turn to other affiliations—religious, ethnic, linguistic, ideological—to protect themselves. The reinforcement of collective identities, a propensity to parochial rather than citizen commitments, can sunder societies, communities, and institutions. Hence globalization, a universalizing phenomenon, generates damaging particularisms and negative pluralism.

Nᴇᴡ ꜰᴏʀᴍꜱ ᴏꜰ organization and power—with distinct types of membership, jurisdiction, and obligation—are one consequence of this untrusting world. Power becomes defined as an ability to sustain loyalty and punish betrayal. These are the preconditions for the organization of anti-state movements that encourage people to act in concert, to transcend individual limitations and engage in violent acts, to create both symbolic and moral capital (in the absence of other kinds). The tolerance and flexibility one associates with positive pluralism are absent.

In Africa, free-market efforts to remedy the economic distortions produced by government interventions and fiscal manipulations have resulted in growing polarization, cleavages, inequalities and conflict. Apart from Ghana, Senegal, Botswana, and a few other countries, democracy is mostly a façade. Few institutions—administrative, educational, health—are in practice what they pretend to be. Elites struggle not to be overwhelmed by those below. They build barriers to protect themselves, rely on guards, private networks, and personal clienteles (often based on ethnicity). Associates don't trust one another and live their lives under scrutiny. They are vulnerable to racketeers, and their children are permanent targets for kidnappers. The more powerful a political leader, the more shuttered, sheltered, and monitored his life. Political leaders become remote from reality (but they know how to plunder the treasury). Patronage and favoritism become means of access to them, rather than knowledge and capabilities. In short, power produces its own vulnerabilities. The higher someone rises in the hierarchy, the more he or she becomes an object of suspicion, jealousy, and anger. In such circumstances it is easy to justify pursuit of more money, power, and the use of secret police.

Because of these sorts of conditions, many of the most principled and best educated Africans leave to work abroad. Many of those who stay are underused, their abilities and knowledge wasted. Just when globalization demands innovative knowledge, Africa has fewer and fewer resources. In the meantime, those in the population who are superfluous to globalization and lack skills for capital-intensive industry often move into the cities. There they become large, miserable squatter populations.

Most people seem to accept these conditions. But not everyone does, and the result is an economic underworld of brigandage and violence. For those on the bottom, negative pluralism means that the world is a gambler's universe. In it, most people are losers. So a desire emerges to change the odds by changing the rules. It is one short step from parochial affiliations to local mafias that rely on clanship, clientelism, and identity politics. Boundaries are tense throughout today's Africa. Outsiders are feared, and this encourages xenophobia. With "identity" reconfigured as a power base, the path opens to ethnic civil wars and crime.[2]

These conditions favor the rise of authoritarian, capricious political figures. By ignoring its own social consequences, globalization intensifies, directly and indirectly, conditions that turn people away from the idea of common citizenship. Negative pluralism overwhelms governments whose institutionalization is already weak. Leaders become unwilling or unable to deploy political power for useful purposes. When there is enormous economic scarcity on the bottom of the social pyramid, pluralism can undo democracy. Mutual hostilities among groups intensify, narrow interests are promoted, and competition among political parties becomes war by other means. Old political conflicts are reinvented in new contexts. Institutions that should bind society and help it to transcend narrow affiliations become sites of intense competition and rallying points for the mobilization of parochialism. Social and cultural cleavages are reinforced just where mediation of differences is most needed. The sort of civil obligation, discipline, and forbearance required of public and private life in a democracy can't be sustained.

The central premise of positive pluralism—which is essential to democracy—is free choice. In contrast, negative pluralism is based on group loyalty and the explicit limitation of choice. Or, to put it another way, positive pluralism defines the terms and conditions of freedom and choice. Negative pluralism, however, defines terms and conditions of identity and affiliation.[3] In circumstances of marginality, identity becomes more important and there is less tolerance of others. When choice is limited to those who are functionally significant and identity defines the functionally superfluous, state and society will come into conflict. Negative pluralism downgrades commonalities among human beings and elevates differences, turns claims into rights, interests into principles, and maximizes cleavage politics. Difference becomes the priority and the basis for representation and accountability. The result is universal sectarianism and stalemate.

Implications

Today, most African countries are façade democracies. Most of them hold elections, even when one party dominates (often representing a dominant ethnic group), and others are prevented by various means from coming to power. Elections are not the sole gauge of democracy, but they are, of course, important, substantive milestones. As Michael Bratton and Nicolas van de Walle point out in *Democratic Experiments in Africa,* the number of sub-Saharan countries holding competitive elections quadrupled between 1990 and 1995. The number of seats held by opposition parties rose from a paltry 10 percent in 1989 to a more robust 41 percent by 1994. Twenty-nine "founding elections" were held in thirty-eight countries between 1990 and 1994. That is, elections paved a route out of authoritarian regimes. "Not a single de jure one-party state remained in Africa" by 1994, write Bratton and van de Walle. "Governments instead adopted new constitutional rules that formally guaranteed basic political liberties, placed limits on tenure and power of chief political executives, and allowed multiple parties to exist and compete in elections. To all appearances, the African one-party state is not only politically bankrupt but, at least as a legal entity, extinct."

Unfortunately this has turned out to be more wishful thinking than fact. In the last few years African political leaders in some countries—for example, Cameroon, Gabon, and Kenya—have turned democratic institutions into a charade. In Congo, Burundi, Rwanda, the Central African Republic, Angola, and elsewhere, the state itself is a precarious institution. In Liberia, Sierra Leone, and Sudan, bandit warlordism saps the institutions of civil society, and torture, sexual abuse, and abduction of child soldiers remain common. There has been a decline in military coups at the center, but a rash of other forms of military upheaval.

INDEED, IF my hypotheses about globalization and its consequences are true, even if democracy may be accepted in the long term and market principles embraced by some political leaders, these will be insufficient to ensure a steady growth of substantive democratic politics and civil societies. The problems posed by globalization must be attenuated. Platitudes, hortatory admonitions, and boiler-plate solutions proffered by such international agencies as the World Bank and the International Monetary Fund won't take Africans very far. It is fine to proclaim the need for restored self-confidence and mutual confidence, for greater transparency, and for institutional reform and human rights. But good intentions are likely to be eroded in the daily grubbing around of political and social survival.

Platitudes make it easier for protagonists and foes of globalization to confront each other in mutual incomprehension. Leaders in the industrial countries now talk about the environment, and the plight of the poor in "emerging" economies and exhibit sympathy for the effectively disenfranchised. But their chief concern is sustaining economic momentum. For them, globalization is the main road to social improvement. Their solution to the problems of globalization is more globalization. For some of globalization's more radical opponents, however, class struggle is simply being resuscitated in modern dress. They see globalization as the same old capitalism with global appurtenances. Theirs is a lonely position, for they are left with a socialism that is all critique, rather than a socialism of solutions.

If placating gestures by leaders in industrial countries are largely just that—gestures—and if the zeal of their opponents is often accompanied by knee-jerk radicalism, then what can be done? It might be useful to establish civic responsibility as a criterion for dealing with governments—in the same way investors look to the management of a company in evaluating their options.

Most important, however, is to push everyone to refocus. Much greater attention must be paid to how people are affected locally by distant globalization policies. Greater complexity must be introduced into policy calculations by Western or global development institutions. Policy makers need to go and see for themselves the consequences of their decisions; they need to spend time living in the circumstances they help create. They need to recognize the value of knowledge in detail, of specifics. What may work in one setting can be a disaster in another.

Even though democracy in Africa is often a façade, politics has in fact become more populist. People have become more vocal. In the most remote areas, life dances more and more to its own tunes and pays its own pipers.[4] Still, the African experience teaches us that there are many strategies for the use and abuse of power; it doesn't just work through familiar political channels, but it also works through countless ordinary strategies of life based on everything from blood ties to healing rituals, from the role of women to that of soothsayers. Territoriality and property disputes can be linked to ancestral rights. Self-constituted domains of meaning are, increasingly, also self-constituted domains of action. Boundaries, jurisdictions, and affiliations are the raw materials of power and conflict, not to speak of proscriptions.

So if globalization is not going to stop—and it won't—what is needed is far more attention paid to its vicissitudes. New theoretical and practical approaches are needed to deal with them. Communities, local and national, differ drastically. At a minimum, far greater care is needed in the preparation and organization of developmental projects, as well as a better sense of how to anticipate problems down the road. The whole concept of basic preparatory research needs to be enlarged, and the boundaries of what is regarded as relevant knowledge exponentially expanded. Most general theory rides roughshod over particularities of place and circumstance in ways that almost guarantee failure. Moreover, it becomes impossible to learn from mistakes. We need site-specific information in the social and cultural as well as the economic senses, rather than formulaic solutions.

If the African experience demonstrates in extreme form some of the problems of globalization, it also highlights the fact that solutions are yet to be found. They won't be until it is recognized that global knowledge requires local knowledge; and local knowledge requires adapting both market models and policies to situation, circumstance, and place. I am suggesting that globalization compels a second information revolution. The first was technological in every sense of the term. Now we must tell those who are remaking the world just what they are really doing.

Notes

1. The World Bank's *World Development Report* for 1997 speaks convincingly of the need to reinvigorate "the state's capacity" and calls for an activist rather than a minimalist state. But how to accomplish this remains something of a mystery. See *The State in a Changing World* (Oxford University Press, 1997), p. 27.

2. Not so long ago many things said today of Africa were said about China, Malaysia, and Indonesia.

3. Nigeria provides one example. North-South relations remain tense. Despite a nominally democratic regime, the country remains hostage to northern military officers. Corruption remains rampant with the concentration of wealth in the Abuja area largely a rake-off from oil revenues extracted from the south. Meanwhile, Makoko, a Lagos slum, is pitched on a swamp. Most people who manage somehow to survive do so with few of life's basic necessities. It might as well be on the other end of the earth. No wonder that the Yoruba and Ibo peoples in the south of the country began talk of secession. *New York Times* (July 20, 1998).

4. See Andrew Apter, *Black Critics and Kings: The Hermeneutics of Power in Yoruba Society* (University of Chicago Press, 1992).

DAVID E. APTER's most recent books are *The Legitimization of Violence* and *Revolutionary Discourse in Mao's Republic* (with Tony Saich).

From *Dissent*, Spring 2002, pp. 13-18. © 2002 by Dissent. Reprinted by permission.

PERSPECTIVES

An End to Africa's Wars: Rethinking International Intervention

MARINA OTTAWAY

F*rom Eritrea to Angola, a large swath of Africa has been engulfed by war for several years. The situation is unlikely to improve any time soon, because the conflicts arise from the disintegration of postcolonial states and thus from the order that was imposed on Africa by outside states. Wars will continue to flare up until a new order emerges. Such an order could either be imposed and maintained with force by the international community—that is, industrialized countries and the United Nations—or it could be based on new territorial and political arrangements reflecting the balance of power among African forces.*

No Winners

The international community has decided that it must help restore stability in Africa by promoting negotiations, providing peacekeepers to monitor the implementation of agreements, and shoring up and reconstructing crumbling states. This policy has been a failure: commitment of resources has not matched the rhetoric, and international intervention has not been sufficient to bring any African conflicts to an end. International intervention, however, has been sufficient to prevent conflicts from ending the way most conflicts do: through the victory of one side. The current policy of the international community does not help populations and does not even help the international community. The reputation of the United Nations is becoming seriously tarnished by its African failures, and there is much resentment among Africans towards the United States.

The major source of conflict in Africa at present is the political and economic decay of a growing number of postcolonial states. Political decay has created a power vacuum in many governments that have only nominal control over their territories and little power over means of coercion. Economic decay has worsened the situation because, in the absence of a viable legal-administrative structure, violence is the only way of securing access to resources—the major conflicts in Africa involve diamond-rich countries. Economic decay also provides the warring factions with an endless supply of fighters, including child-soldiers who see few other career prospects in war-torn nations. The lone exception to this scenario of decay-induced conflict is the war between Ethiopia and Eritrea, which stems from the two countries' ambition to build strong states; in other words, theirs is a classic war between states vying for power and economic advantage.

The Great Lakes region provides the most dramatic, but by no means the only, example of how decay breeds civil war, which can then expand into inter-state war. The multiple, interlocked conflicts in the area stem from the implosion of the Democratic Republic of the Congo, formerly Zaire, after decades of mismanagement. With the government no longer controlling the country, border regions have become havens for both opposition movements and the armies of neighboring countries such as Uganda and Rwanda. This, in turn, has invited political and armed intervention from all neighbors in a tangle that has been dubbed Africa's first world war. A victorious warlord, Laurent Kabila, has been installed in Kinshasa as president and enjoys diplomatic recognition as the country's leader by virtue of being there; however, two armed opposition movements, the RCD (Congolese Rally for Democracy), which is itself divided into two antagonistic factions, and the MLC (Movement for the Liberation of the Congo) combine to wage war on the government and periodically on each other with the support of Uganda and Rwanda. Troops from these two countries operate in Congolese territory against the government. Other armed groups, including the indigenous Mai-Mai, the Rwandan Hutu Interahamwe, and the remnants of the pre-1994 Rwandan army, add to the complexity by simultaneously pursuing their own goals and those of Kabila.

Conflicts in Sierra Leone, Angola, Sudan, Rwanda, Burundi, and Somalia have also sprung from the decay plaguing the continent. Other countries, such as Nigeria, are highly vulnerable. Nigeria's return to civilian government in 1998 was a positive step, but not a guarantee that the country will succeed in avoiding further decay and possibly violent conflict. Even the future of the countries of southern Africa, until recently the most stable and promising, is threatened by a political crisis in Zimbabwe and by their staggering rates of HIV infection—around a quarter of the adult population—which is bound to affect their economies and most probably their politics with unpredictable consequences.

Lukewarm Interventions

The international community's response to the spreading conflict in Africa has lacked coherence, to a large extent because the fighting does not significantly threaten the security or economic interests of any major power. The international community has responded to conflict not because of any clear self-interest but instead for humanitarian reasons and a rather vague, general interest in keeping Africa from sinking into complete chaos and becoming a breeding ground for new diseases. Humanitarian interests and hazy predictions of future threats have not elicited strong and clear policy responses to Africa's predicament. Rather, they have led to a dangerous combination of the idealistic and thus ambitious goals rooted in humanitarian considerations and the scant resources, and lukewarm commitment associated with the absence of immediate, concrete interests on the part of most industrialized countries. Such a combination is a recipe for failed interventions. A scenario that is becoming typical in Africa is the signing, under pressure from the international community, of a peace agreement to which the warring factions are not seriously committed; the implementation of the agreement is therefore dependent on a UN presence which has rarely been sufficient to do the job.

Some of the interventions have been completely ineffectual, as in Sierra Leone, where peacekeepers have been unable to protect themselves from being kidnapped by the rebels, let alone fulfill their mandate. In other cases, the international community's approach has been not only ineffectual but morally outrageous, as in Rwanda, where UN presence was reduced at the height of the genocide in 1994. Most major powers, including the United States, were unwilling to increase their symbolic commitment to the level necessary to protect the population. Finally, some international interventions aimed at ending conflict have simply made the situation worse, as in Angola, where the internationally-sponsored and monitored agreements have been used repeatedly by the rebels to rearm, reorganize, and eventually restart the fight.

The intervention in the Democratic Republic of the Congo at present also appears headed for futility or disaster—one can only hope for the former. The international community has supported negotiations and the resulting Lusaka agreement. This unrealistic pact is based on the assumptions that armed movements at war with each other can suddenly lay aside their differences, without any new developments having taken place on the ground; abide by a cease-fire agreement; and engage, together with representatives of civil society, in a national dialogue that will produce an agreement on a new political system in 45 days. The only concrete support the international community has provided for the Lusaka agreement so far is the deployment of a small number of observers with the pledge that 500 observers protected by 5,000 peacekeepers will be positioned in the country if the cease-fire is ever implemented. This is fewer than the number initially deployed in tiny Sierra Leone to support the Lome agreement; by September 2000, the United Nations was calling for an increase in the number of peacekeepers to 20,000, over three times the initial figure.

It is time to reassess the effectiveness of the international community's efforts to settle African conflicts and to rethink when and how interventions should take place, what the goal of intervention should be, and how intervention can best be managed. Only if the goals are realistic and the means are adequate can the international community have an impact on the conflicts in Africa.

Unlikely Conditions for Success

The purported goals of international intervention have matured during the last decade. In the "good old days" of the Cold War, it was enough to maintain the stability of pro-Western regimes and to deny access to the Soviet Union, and if these goals could be best attained by supporting a friendly dictator like South Africa gave Sese Seko in Zaire, so be it. Fortunately, the international community has moved away from that position, but, as often happens, it has swung too far in the opposite direction: the goals now are to stop conflict immediately through negotiations, to restore democratic and accountable government—from bullets to ballots in one smooth transition—and to do so while preserving the territorial integrity of all post-colonial states within the boundaries established by the colonial powers.

Commendable goals indeed. Unfortunately, so far there is little evidence that they can be attained except under special circumstances. Namibia and Mozambique did indeed go from conflict to stability under international supervision, but success was based on an unusual combination of factors. In Namibia, the South African government simply gave up its fight to keep control of the territory, and the major opposition movement had no difficulty accepting an electoral process where it was certain to win by a very wide margin. Furthermore, the United Nations deployed over 8,000 military and civilian personnel to administer the transition, a large number for a country of one million people.

Mozambique was also blessed with special circumstances. The conflict was two-sided, and neither side had the resources to continue fighting: the opposition movement, Renamo, lost its outside support when South Africa gave up the attempt to preserve apartheid, and the government was highly dependent on the donor community, which had no intention of financing a war. By contrast, all active conflicts at present are multi-sided. Even in small Burundi there are at present 17 parties and two armed movements that need to sign on for any agreement to be implemented. Furthermore, in Angola, Sierra Leone, and the Democratic Republic of the Congo, the conflict is self-financed, supported on all sides by the sale of diamonds.

Since the present model only appears to work under special circumstances, what can the international community do elsewhere? Although theoretically it could impose the settlement it wants on any country, if it were willing to provide a presence comparable to that in Bosnia or Kosovo for an indefinite period of time, in practice this can happen only in exceptional cases. This is not because racist attitudes prevent the commitment, as many Africans have come to believe, but because the scale of the conflict, the size of the territories involved, and the logistical problems preclude such intervention in many countries. A very simplistic calculation, based on population size, suggests that an international presence in the Democratic Republic of the Congo comparable to that in Kosovo would require the deployment of some 900,000 military and civilian personnel. The figure should not be taken literally but is a sobering reminder of what robust intervention means in large, messy countries.

Selective Intervention

The key to more effective policy is to realign goals and commitment. This means increasing commitment in those cases where it is warranted and where it could be effective, and settling for more modest goals in the other, more numerous cases. When the international community opts for the immediate cessation of conflict, the preservation of the existing states, and their reconstruction as democratic entities, it must provide real rather than symbolic resources, and it must be prepared to sustain the commitment for a long period.

In all other cases, it would be more helpful, or at least less harmful, if the international community allowed conflicts to reach a decisive turning point before becoming involved, even if this implied that some post-colonial states might not survive.

At present, there is only one conflict in Africa where the international community should increase its commitment to fit presently stated goals, and that is in Sierra Leone. The present level of commitment has proven vastly insufficient, thus the choice is to increase it or to pull out altogether. Pulling out of Sierra Leone at this point would destroy once and for all the credibility of the United Nations in Africa, while making it impossible to get support for any kind of international intervention in the future. The failure in Somalia is still casting a long shadow, and another failure would be the final blow.

On the other hand, the small size of Sierra Leone makes it conceivable to provide sufficient peacekeepers and civilian personnel to end the conflict and to reconstruct the country. While it is proving difficult for the United Nations to get member states to commit sufficient troops to Sierra Leone, success is not impossible as it would be, for example, in the Congo. Finally, Sierra Leone would provide a manageable test case for the effectiveness of the international community's prescriptions for Africa. Does the international community know how to put an end to years of chaos without using undue force and hurting civilians? Can it bring the culprits to justice? Can it demobilize the combatants? Can it build the small,

efficient, professional army it believes suitable for African countries? Can it develop a modern police force, an honest civil service, and an apolitical judiciary in a reasonable span of time? If it cannot be done in Sierra Leone, there is no point pretending it can be done elsewhere.

At the opposite extreme are the conflicts from which the international community should step back altogether until something clearly changes on the ground. Sudan is the prime example here. It has become obvious over the years that none of the parties in that war are ready for compromise and that negotiations are simply a game, an attempt by each participant to bamboozle the international community into believing in their good will and to put the blame on the other side. There is nothing to be gained by continuing this process. Stopping it will not bring the conflict to an end any faster, but it will at least send a signal to groups in other countries that the international community is not willing to participate in a game of perpetual negotiations.

Finally, there are cases, such as the Congo, where it is too early for the international community to pull back completely, but where massive intervention is neither warranted nor possible. If groups are still engaged in talks and outsiders can play a useful role, there is no reason not to continue—talk is cheap. But the international community must also make clear to all sides the limits of its willingness to engage. [The] combatant needs to be told clearly that if he does not want to abide by the Lusaka agreement, he cannot count on the rest of the world to save his country from disintegration if the fortunes of war turn against him. If rebel movements have no interest in compromise, they must risk defeat. The international community must clarify that it neither wants to, nor can, hold together a country whose leaders only want war and that it is up to them to make the choice. Even if the international community remains involved in the diplomatic effort to end the fighting, it must stop pushing for ideal frameworks and quick solutions, because they cannot work. The participants in the civil war in the Congo will never agree on a democratic political system in 45 days.

African conflicts will eventually come to an end, as conflicts always do. The example of Somalia suggests that, left to their own devices, and without the hope of getting more resources from the international community, or protection in the form of a cease-fire when the going gets too tough, the warring factions will find their own solutions. They will not be ideal solutions based on territorial integrity for all countries and democracy in our lifetime. However, it is time that the international community stepped back, lest in trying to promote ideal solutions without providing even remotely sufficient resources, it prolongs conflict in Africa.

MARINA OTTAWAY is co-director of the Democracy and Rule of Law Project at the Carnegie Endowment for International Peace.

From *Harvard International Review*, Winter 2001, pp. 20-23. © 2001 by Harvard International Review. Reprinted with permission.

Moving Africa off the Back Burner

Building on the European Model
Challenges for the African Union

BY HENRY OWUOR

AS FOR ANY other international body seeking to stake its place in the international arena, the African Union (AU) still has a long way to go. But as the Chinese say, a journey of a thousand miles starts with a single step. It may be recalled that the European Union (EU), on which the AU is modeled, was initially just a coal-and-steel-trading body. It now boasts the largest common market with 370 million consumers.

The EU's strong point has been the numerous treaties that govern its operations. The first such treaty was signed in Paris in 1951 to formalize the body's operations with an assembly to monitor them.

This is where Africa comes in. Any association of independent states can only function if all its members have a sense of belonging. None should feel left out, as was the case with East and Central African countries that weren't consulted when Nigerian, South African, Senegalese, and Algerian presidents met leaders of the G-8 nations to seek support for the New Partnership for Africa's Development (NEPAD). NEPAD is an initiative of the AU, yet not a single East African country is on its steering committee.

But these are what can be said to be teething problems, as occurred in the early stages of the EU when Britain refused to join and later had a rough time negotiating terms for entry. It was not until 1973 that Britain finally joined the community together with Ireland and Denmark.

Another area where Africa fares worse than the EU is the fact that the European body has only 15 countries to deal with. Africa starts with 53 nations loaded with a myriad of problems and needs. Even if one considers just one task, conflict prevention, a look at all the conflicts on the continent shows that this will not be an easy one to deal with.

But still, the AU comes in well armed. Under its Act, it can intervene in the affairs of member states on issues such as elections. This provision is already in use in Madagascar, where the AU has refused to recognize Marc Ravalomanana, who claims to have won disputed presidential elections, or Didier Ratsiraka, the former president who he claims to have beaten. Instead, it wants a fresh poll.

The AU plans a special summit in six months' time to review some contentious issues, especially under NEPAD, whose peer-review scheme is still being watched suspiciously by long-time dictators not used to outside scrutiny.

The NEPAD Peer Review will be run by retired presidents, judges, and other eminent persons who will police African countries in areas such as the timely holding of elections and the respect for human rights.

Still, some figures are very troubling: The World Bank forecasts that the number of poor people in Africa will rise from just over 300 million out of a total population of 659 million in 2000 to 345 million people by 2015.

Add to this the fact that as the AU was being launched in Durban early this month, more than half the members owed its precursor, the Organization of African Unity, more than US$50 million in membership fees, and you see the challenge it faces.

The task African countries are embarking on has a lot of similarities with that of the EU, says Finn Thilsted, the Danish ambassador to Kenya, whose country currently holds the presidency of the EU. He says both the EU and the AU "are out to promote trade and get rid of borders as they are a hindrance to free movement of goods and services."

Europe, like Africa, says Thilsted, has major differences in the lifestyles of its people. But still, European countries share similarities: For example, England and Denmark were both sea-faring nations and have contacts dating back to the Middle Ages.

The same may not be said for Africa, but still, some cultures did penetrate [the continent]: the Kiswahili language spread from the East African coast all the way to the Congo. Given the debate on adopting a working language for the AU, not even Kiswahili or Arabic can emerge as the main language. Arabic is the main language in northern Africa, while Kiswahili reigns in East Africa. Africa therefore finds itself using three colonial languages— English, French, and Portuguese.

The AU has given itself 10 years in a period dubbed "the age of capacity building" to radically change Africa. What it needs are treaties among African nations on various goals it wants achieved, such as human rights, with the AU setting standards for members to meet. Once this system begins, other treaty areas will be explored. Such power tactics as military coups and rigged elections should be among the first to be tackled in African treaties. The AU is moving in the right direction by creating an Economic and Social Council, through which nongovernmental organizations (NGOs) and trade unions will have the right to participate in the affairs of the continental body. At the moment, most African countries suppress NGO operations, which they see as a threat to their hold on power.

From *World Press Review*, October 2002, p. 7.

Moving Africa off the Back Burner

Africa, NEPAD, and the G-8

Turning Benefactors into Partners

BY CHINUA AKUKWE

THE RECENT G-8 conference in Canada included an Africa Day devoted to discussions between G-8 leaders and selected African leaders on the New Partnership for Africa's Development (NEPAD). At the end of this interaction, the G-8 leaders presented a G-8 Africa Action Plan, which has come under relentless criticisms from Western and African media, civil society organizations, and professional bodies for being too vague and lacking in ambition in terms of resources and commitments. It is important to ponder what's next for the triangular relationship between African leaders, NEPAD, and the G-8.

I begin by stating without reservations that Africa and the G-8 countries need each other for many reasons. The G-8 countries need Africa as a bulwark in the war against terrorism, for access to oil and gas reserves to meet the voracious needs of their industries and middle class, as a last frontier for profitable private-sector expansion, and to continue the spread of democracy.

> GUIDING LIGHTS? South Africa's Thabo Mbeki and Nigeria's Olusegun Obasanjo are movers in the African Union, but both get mixed reviews for their leadership records.

Africa, on the other hand, needs the G-8 for accelerated new financial investments, for deep and sustained

debt relief, to complement poverty-alleviation efforts, for infrastructure development, and—very importantly—to win the war against HIV/AIDS. The idea that a continent of more than 700 million people may be left in the lurch in an increasingly global political and economic system is wishful thinking.

The G-8 Action Plan for all intents and purposes can be seen as a "G-8 Intention Plan." This reading is best captured by the wily President [Olusegun] Obasanjo of Nigeria, who stated in his address to G-8 leaders in Canada, "For us in Africa, the journey is just beginning.... We are hopeful that, now that you have embarked with us on this long journey, you will continue with us to the end, and that we can therefore continue to count on your support whenever we might need it."

African and G-8 leaders are already aware that the triangular partnership with NEPAD is a long-term proposition. The key is to ensure that short-term steps are taken now to maintain the momentum of the partnership. These short-term steps should be completed within the next two years, with some as early as the end of the year.

Short-Term Action Steps for Africa and NEPAD:

1. African leaders should obtain wider support for NEPAD. Civil society organizations, professional bodies, community-based organizations, the legislative branches of government, the private sector, and students are not yet on board regarding

NEPAD. I have no doubt that the G-8 leaders, all seasoned politicians, are aware of this anomaly.

2. NEPAD must focus on a development emergency in Africa: HIV/AIDS. I fail to see how NEPAD can become sustainable in the face of a deadly onslaught from HIV/AIDS in Africa. The Joint U.N. Program on HIV/AIDS (UNAIDS) estimates that more than 19 million Africans have died of AIDS, and by 2020, 55 million more Africans, mostly men and women in the prime productive ages, will die in the absence of concerted treatment and prevention programs.

3. The new African Union should become a magnet of ideas on Africa's renewal rather than a dumping ground for retired, tired, or out-of-favor politicians and bureaucrats. The role of NEPAD in the African Union should be settled as quickly as possible. Regional integration issues should become a major focus of the African Union.

4. Democracy and peaceful transfer of power should become rote in Africa. The triangular relationship between Africa, NEPAD, and the G-8 rests on strong democratic systems and seamless transfers of political power. The ongoing crude attempts by some African leaders to perpetuate themselves in power after decades of personalized leadership will not only hurt their countries but also have ripple effects in Africa.

5. The private sector and civil society should become trusted partners of the government. Today, governments in Africa remain the most lucrative entity in their societies. But

the Africa-NEPAD-G-8 partnership requires a robust, respectful relationship between the private sector, civil society, and the public sector. It is highly unlikely that the public sector in Africa will create all the new jobs that will see Africa trade or work its way out of poverty. It is also unlikely that the public sector can serve as the ultimate watchdog of its own activities.

6. Scarce resources should be managed prudently and with accountability. The triangular relationship between Africa, NEPAD, and the G-8 may rise or fall on this premise. African governments cannot create a situation whereby their citizens or international partners will lose confidence in their capacity to manage national and international resources effectively.

7. Organize and tap the expertise of Africans living in the West. I believe that G-8 leaders are aware of the skills and expertise of Africans living in the West. Discussions on technical and even financial expertise are likely to gravitate to the potential contributions of these Africans. Leaders should see the expertise of these Africans as strategic assets to be tapped immediately.

Short-Term Action Steps for the G-8 Nations:

1. Cancel or substantially reduce the debts of African nations. The US$1 billion stated in the G-8 Action Plan toward the Heavily Indebted Poor Countries (HIPC) initiative is a good step forward, but not enough by a mile. The G-8 nations should end the 2001 situation in Africa where more than 16 African nations spent more on debt repayments than on health care, according to UNAIDS. In addition, UNAIDS indicates that several African countries with HIV prevalence above 20 percent are not eligible for the current HIPC initiative. Debt relief for Africa will be a significant test of the G-8's commitment to Africa's renewal.

2. Assist African governments to repatriate looted monies. President Obasanjo estimates that Africa has lost at least US$140 billion through corruption since the 1960s. The G-8 can accelerate the return of some of these monies to Africa for verifiable investments in health care, education, and infrastructure development.

3. Increase development assistance to Africa. The G-8 plans to devote half or more of new development assistance (about $6 billion) to Africa by 2006. Why not sooner, to help jumpstart the NEPAD initiative? The G-8 leaders say they are committed to the principle that no African country will be denied the chance to achieve United Nations Millennium Development Goals through lack of finance. Why not, in addition to accelerating the promised development assistance, increase the aggregate amount of aid by the next summit in 2003?

4. Meet the required financial outlay for the Global Fund to Fight AIDS, Tuberculosis, and Malaria. The G-8 should fully fund the Global Fund to the tune of $10 billion a year as requested by the U.N. Secretary-General Kofi Annan. So far, about $2 billion is available over a number of years, with roughly $800 million available to the fund for this year.

5. Decrease agricultural subsidies in the West. Massive agricultural subsidies inhibit African farmers from accessing European and North American markets. The G-8 leaders in Canada committed by 2005 to "reductions of all forms of export subsidies with a view to their being phased out, and substantial reductions in trade-distorting domestic support." Again, why not make strides in this vital area sooner rather than later?

In conclusion, Africa and the G-8 nations are joined in a mutually beneficial, triangular relationship with the NEPAD initiative. The 2002 G-8 meeting set the ball rolling on this long-term relationship. The road ahead will definitely have its twists and turns. In the interim, both Africa and the G-8 have some serious work to do in order to sustain the partnership.

From *World Press Review*, October 2002, pp. 8-9.

The Road to Johannesburg

Embracing Sustainability Science:
The Challenges for Africa

by Godwin O.P. Obasi

There is growing international concern about seeking a new and deeper understanding of the complex interaction between human society and the environment. This understanding is essential to the quest for sustainable development—meeting the needs of the present without compromising the ability of future generations to meet their own needs. Because it is generally recognized that the world's present development path is not sustainable—in spite of substantial efforts in the last decade—the world community is now focusing attention on what needs to be done.

It has been increasingly recognized that to meet the needs of a growing society and to avoid further undermining the Earth's essential life-support systems, a new paradigm of scientific inquiry needs to be invoked. It must address the complex interaction between the various components of the Earth system together with the impact of society on the system. This paradigm is referred to as sustainability science—that is, science and technology in service of a transition toward sustainability—and it requires in-depth exploration of goals and

trends, including consideration of environmental threats and opportunities as well as relevant actions.[1]

A crucial element of sustainability science is the recognition of multidimensional interactions (including the human dimension)—hence the necessity for the participation of not only those in the natural sciences but also those in the social sciences. A group of scientists that met at Friibergh Manor near Stockholm, Sweden, in 2000 addressed this subject and noted that progress in sustainability science requires the following: development of a research framework that integrates global and local perspectives to shape a place-based understanding of the interactions between environment and society; initiation of focused research programs on a small set of understudied questions that are central to achieving a deeper understanding of those interactions; and promotion of better use of existing tools and processes for linking knowledge to action in pursuit of a sustainable transition.[2] By its very nature, sustainability science differs in fundamen-

tal ways from most sciences, and it emphasizes action and social learning in addition to the more classical scientific approach involving hypotheses formulation, observation, verification, understanding, and prediction.

Background

When the International Council for Science (ICSU) initiated the Global Change Programme in the early 1980s, Africa had been undergoing change. Sub-Saharan Africa was experiencing a severe drought that had first manifested itself in the mid-1960s, affecting food production and resulting in famine, water shortages, malnutrition, and diseases. Overgrazing in the Sahel region and overexploitation of wood fuel promoted desertification in the semi-arid areas of western Africa, and there was an overall degradation of the environment. The experience in Africa was partly responsible for the establishment of the Brundtland Commission, which produced the now often-quoted publication *Our Common Future*, in which the concept of sustainable development in regard to the Earth system emerged.[3]

In addressing sustainable development and sustainability science, it is helpful to consider a number of trends. One major consideration is increasing population—the current world population is just more than 6 billion, and it is projected to reach 10 billion by 2050.[4] About half of all people live in cities, and rapid urbanization and the growth of megacities are causes of concern. In sub-Saharan Africa, the current population of 600 million is projected to reach 1.7 billion by 2050—the fastest rate of increase of any major region in the world. The rate of population growth is double that of food production, thus necessitating food importation. In Nigeria, the population is projected to more than double and to reach approximately 244 million by 2050; Lagos is expected to become one of the world's megacities, with a projected population of 24 million by 2020.[5]

Moreover, 340 million people in Africa—half its population—live on less than U.S. $1 per day, one measure of poverty level. Only 58 percent of the population has access to safe water. The rate of illiteracy for people older than 15 is 41 percent. There are only 18 mainline telephones per 1,000 people, compared with 146 per 1,000 for the world as a whole and 567 per 1,000 for high-income countries.[6]

Africa has constantly faced a number of other challenges that have major implications for sustainable development, including globalization, unfavorable terms of trade, increasing debt burdens, declining agricultural production, political instability, and civil strife. These are exacerbated by recurrent floods and droughts as well as by other natural disasters, which set back the economies of a number of countries—and thus sustainable development. There is a wide spectrum of concerns relating to the health and well-being of the people, and science and technology have been poorly funded. Laboratory equipment is inadequate in many schools and institutions of higher learning, and advanced research infrastructure is nonexistent in many universities.

African countries have adopted a common approach to address these concerns. The New African Initiative, approved by the Organization of African Unity Summit in July 2001, seeks to eradicate poverty and to place African countries, individually and collectively, on a path of sustainable growth and development.[7] The program identifies sectoral priorities and expands on several initiatives to address the pertinent areas of concern. These initiatives include human resources development, infrastructure development, diversification of production and exports, market access, resource mobilization, capital flow, and environmental protection. Specific areas of action also have been developed, including poverty reduction, health, education, information and communication technology, energy, transport, water and sanitation, science and technology, agriculture, desertification, wetland conservation, coastal management, and global warming.

In addition, the African Regional Round-Table of Eminent Persons, organized by the United Nations in Cairo in June 2001, brought together a select group of Africans to contribute to the development of African perspectives and concerns that will be considered at the World Summit on Sustainable Development (Rio+10) in Johannesburg in late August 2002. The summit is to assess progress that the world community has made in its path to sustainable development since the UN Conference on Environment and Development in Rio de Janeiro in 1992. Because it is generally accepted that the objectives of Rio were not fully met, the Johannesburg summit will consider a more realistic agenda whose objectives can be met in the face of current developments such as globalization and market liberalization.

The Research Challenge

If the sciences are to address sustainable development issues effectively, they will need a new vision for international partnership and cooperation as well as a new form of social arrangement with society. A major step in laying down the framework of sustainability science occurred at the groundbreaking meeting at Friibergh Manor. This framework identified an initial set of core questions that seek to draw research attention to understanding the basic character of interactions between nature and society (see the box). Addressing these questions as they relate to Africa will be challenging and will demand much creativity and resourcefulness.

The Friibergh meeting also highlighted the importance of developing appropriate research strategies. A basic challenge is to undertake wide-ranging but integrated studies of the Earth system as a whole, in its full functional and geographical complexity. At the same time, there is a need to pursue the more applied scientific understanding that can help societies develop in

Core Questions of Sustainability Science

In 2000, a group of scientists accomplished a major task in laying down the framework of sustainability science at a meeting at Friibergh Manor, near Stockholm, Sweden. They identified an initial set of core questions that seek to draw research attention to understanding the basic character of interactions between nature and society.[1] The questions are:

• How can the dynamic interactions between nature and society—including lags and inertia—be better incorporated into emerging models and conceptualizations that integrate the Earth system, human development, and sustainability?

• How are long-term trends in environment and development, including consumption and population, reshaping nature-society interactions in ways relevant to sustainability?

• What determines the vulnerability or resilience of the nature-society system in particular kinds of places and for particular types of ecosystems and human livelihoods?

• Can scientifically meaningful "limits" or "boundaries" be defined that would provide effective warning of conditions beyond which the nature-society systems incur a significantly increased risk of serious degradation?

• What systems of incentive structures—including markets, rules, norms, and scientific information—can most effectively improve social capacity to guide interactions between nature and society toward more sustainable trajectories?

• How can today's operational systems for monitoring and reporting on environmental and social conditions be integrated or extended to provide more useful guidance for efforts to navigate a transition toward sustainability?

• How can today's relatively independent activities of research planning, monitoring, assessment, and decision support be better integrated into systems for adaptive management and societal learning?

1. R. W. Kates et al., "Sustainability Science," *Science*, 27 April 2001, 641–42.

ways that sustain global life-support systems. Sustainable development requires an ongoing adaptive learning process. A strategy must be built for using scientific and technical knowledge as the basis for future actions in a number of areas.

Research carried out in recent years demonstrates clearly that the Earth system behaves as a single, self-regulating system comprised of physical, chemical, biological, and human components. The interactions and feedbacks between the components are complex, and they exhibit multiscale temporal and spatial variability. A fundamental aspect is the very significant influence of human activities on the Earth's environment. In regard to the continued emission of greenhouse gases, the Third Assessment Report of the World Meteorological Organization (WMO)/UN Environment Programme's Intergovernmental Panel on Climate Change (IPCC) concluded that "there is new and stronger evidence that most of the warming observed over the last 50 years is attributable to human activities."[8] Humankind is irrevocably modifying the Earth's land surface, cryosphere, oceans, coasts, water cycle, ecosystem, and many biogeochemical processes.

However, the changes occurring cannot be understood fully in terms of simple cause-and-effect relationships. Anthropogenic effects include multiple consequences in the highly complex, nonlinear Earth system; they interact with each other and with local and regional changes in a manner that is difficult to understand and even more difficult to predict. Moreover, the dynamics of the Earth system appear to be characterized by critical thresholds and abrupt changes, which could be triggered inadvertently by human activities.

In terms of several key environmental parameters, the Earth system already has moved outside the range of natural variability exhibited in human history. The nature of changes, their magnitude, and their rates are unprecedented. WMO—the lead UN agency concerned with meteorology, climate, hydrology, and related environmental science—is urgently tackling a number of the scientific issues involved, including:

- documenting how the global climate and the Earth's environment have changed in the past;
- understanding changes in the climate system on all time scales and improving the ability to predict climate variability and change, including

interseasonal scales as well as desertification;

- identifying the factors that influence natural disasters (such as floods and droughts) and other extreme weather events, as well as how to mitigate their effects;
- addressing fundamental issues related to water-resources development, particularly assessing the quantity and quality of water resources;
- contributing to questions on environmental quality, such as pollution levels, transport of transboundary pollution, and ozone depletion; and
- contributing to the quest for food security and poverty alleviation.

The human dimensions of global environmental change must be considered. This requires an understanding of the major human causes of change and how these causes vary over time, across space, and between economic sectors and social groups. Another human dimension that must be determined is the consequences of global environmental change on key life-support systems (such as water, air, natural ecosystems, and agriculture) as well as on economic and social systems. In addition, the underlying social processes or driving forces behind the human relationship to the global environment, such as human attitudes and behavior, population dynamics, institutions, and economic and technological transformations, need to be understood. A scientific foundation must be developed for evaluating potential human responses to global change, their effectiveness, their cost, and the basis for deciding among the options.

A Framework for Action: Core Studies in Africa

Making progress in sustainability science by meeting research challenges will require unprecedented interdisciplinary cooperation and coordination. There is a need to bring together scientists in all fields of environmental research (such as physical climate scientists, atmospheric chemists, experts in biogeochemical processes, social scientists, and economists). Relevant national, regional, and global environmental and social programs should be strengthened and developed by governments, concerned UN agencies, the scientific community, and private and public

funding institutions—indeed, by all stakeholders.

Core areas of study can help address sustainable development concerns in Africa. The continent is not homogenous; there is much variety in terms of population distribution, climate, land form, biota, culture, and socioeconomic circumstances. This is important to recall as sustainability science requires the development of place-based strategies and approaches. Eight areas are of particular concern to Africa: food security, natural resources management and biodiversity, desertification, water resources, natural disasters, health and diseases, coastal zones and sea-level rise, and climate projection. While each of these areas can be addressed separately, it is important to recognize that they interact with each other and that there are feedback loops. For instance, these concerns—individually and collectively—could trigger movement of populations, possibly resulting in alterations of social structure, cultural identity, and political stability. These, in turn, can exacerbate the problems associated with sustainable development. The comprehensive IPCC Third Assessment Report provides very useful material for addressing these concerns in relation to the impact of climate change.[9]

Food Security

Of the 49 least-developed countries of the world, Africa hosts 34. In 21 countries in sub-Saharan Africa, the undernourished have an average deficit of more than 300 kilocalories per person per day. Africa is the only region of the developing world where the regional average of food production per person has been declining over the last 40 years.[10] In Africa as a whole, food consumption exceeded domestic production by 50 percent during the mid-1980s, when severe droughts were prevalent; in the 1990s, it exceeded domestic production by more than 30 percent.

For development goals to be achieved, agricultural growth must take place in a way that safeguards the natural environment. In addition to being a significant source of food, agriculture is also the primary way of life in Africa. About 70 percent of the population lives on farming, and agricultural products amount to 40 percent of exports. Relative to other regions in the world, African regions (with the exception of North Africa) have the lowest food-security index. (The index of food security is based on trends in food production, available food as a percentage of requirements, and arable land per capita.) At the same time, African regions have the lowest ability to adapt to further changes, as indicated by the UN Development Programme's Human Development Index, a composite of measures of life expectancy, literacy, education, and income.

Africa has substantial natural and human resources that can be used to make regional gains in food security. Such gains can be achieved through methods such as applying biotechnology to develop new crop varieties. An example is the development of New Rice for Africa varieties by the West African Rice Development Authority, with support of its partners. These rice varieties can provide 50 percent more grain than current varieties when cultivated with traditional rain-fed systems and without fertilizer. Irrigation systems can be improved by researching the availability of water for distribution and cost-effective utilization, measured in terms of its quality and quantity. An interdisciplinary approach can help increase agricultural yield through science-based information, assessment, and prediction, together with extension work and good communication with the farming community. In addition, science-based climate variability forecasts and assessments can help decision makers take appropriate measures to improve food availability.

Natural Resources Management and Biodiversity

The Food and Agriculture Organization reports that in Africa, forests cover 5 million square kilometers—one-sixth of the continent's land area. Forest and woodland species provide firewood, structural timber, traditional medicines, staple food, and drought emergency food. A large part of the population lives in rural areas and depends on trees and shrubs for subsistence. Firewood and charcoal provide approximately 70 percent of the energy used in Africa.[11] Africa contains about one-fifth of all known species of plants, mammals, and birds, as well as one-sixth of amphibians and reptiles. This biodiversity provides an important resource for African people, but intensive use of forest species is bound to have an effect on indigenous biodiversity.

Within the next 10 to 20 years, biodiversity of flora and fauna is likely to be affected by environmental changes, including climate change. Africa is highly characterized by ecosystem control through disturbance. Fire and grazing regimes, for example, interact with climate change in a significant

way to control biodiversity through rapid, discontinuous "ecosystem switches." Ecosystem switches are accompanied by drastic species shifts and sometimes even extinction. Theory and application can be further developed to predict the extent and nature of future ecosystem switches as well as geographical shifts of species.

The development of mitigation and adaptation strategies for managing biodiversity can be strengthened by collaboration among countries and across disciplines, and the UN Convention on Biological Diversity (CBD) can provide a framework. With respect to declining tree resources, adaptation strategies should include natural regeneration of local species, energy-efficient cooking stoves, sustainable forest management, and community-based resource management. For instance, it has been reported that in Senegal and Kenya, new cooking stoves have been introduced with gains in energy efficiency.

Desertification

The UN Convention to Combat Desertification (UNCCD) defines desertification as "land degradation in arid, semi-arid and dry subhumid areas resulting from various factors including climatic variations and human activities."[12] In Africa, these areas cover 13 million square kilometers (km^2)—43 percent of the continent's land area—on which 270 million people (40 percent of the continent's population) live. Some 250,000 people perished in the Sahel drought of 1968–73.

Desertification in Africa has reduced by 25 percent the potential vegetative productivity of more than 7 million km^2. The relative importance of anthropogenic and climatic factors in causing desertification should be studied further. The major anthropogenic factors are unsustainable agricultural practices, overgrazing, and deforestation. Population growth is certainly another element for consideration. On the other hand, climatic factors including precipitation and temperature determine the potential distribution of terrestrial vegetation and constitute principal factors in the genesis and evolution of soil. Certainly, there are feedback mechanisms among the climatic factors as well as between them and the anthropogenic factors.

There is a need for further studies to consider the vicious cycle linking desertification to environmental change. As desertification proceeds, agricultural and livestock yields decline, thus reducing the population's survival options. People lose

vital ecosystem services such as firewood, traditional medicine species, and emergency food species, rendering them more vulnerable to future environmental change.

The diversification of the use of resources also should be studied further. For instance, in southern Kenya, Masai herders have adopted farming as a supplement to or replacement for livestock herding. In the future, seasonal climate forecasting may help farmers and herders to be aware of times when resource diversification has a higher probability of being successful. User communication and interaction also need to be improved.

Water Resources

Better management of freshwater will be key to improving health and sustainable economic development in many African countries. Water quantity and quality assessments should go hand in hand, and the planning and management of water resources must take into account the likelihood of floods and droughts to occur.

The sustainable management of available freshwater is essential for economic and social well-being. Freshwater has already become critically scarce, and the global mean per capita runoff has decreased by more than 40 percent since 1970 to 7,600 cubic meters (m^3) per year, most notably in Africa, Asia, and Europe.[13] There are now eight African countries that have renewable freshwater resources less than 1,000 m^3 per capita per year, which is a benchmark for freshwater scarcity. In these countries, competition for water for agricultural, domestic, and industrial purposes is clearly evident. Some estimates suggest that the amount of freshwater currently available for each person in Africa is about one-quarter of what it was in 1950, while in Asia and South America, it is about one-third. The situation is getting worse as a consequence of rapid population growth, expanding urbanization, and increased agricultural and industrial use. In 2000, about 300 million Africans lived in a water-scarce environment. By 2025, the number of countries in Africa experiencing water stress will rise to 18—thus affecting 600 million people.[14]

Competition for freshwater resources is expected to grow, and the pressure on these resources will continue to increase in the foreseeable future as the population rises. Shortage of freshwater is expected to be the most dominant water problem in this century and—along with water quality—it could well jeopardize all other efforts to secure sustainable development, even leading to social and political instability in some cases. As a result of difficult economic conditions, insufficient knowledge of the freshwater resources in Africa is often at the heart of many water-related problems.

Research is needed to support efforts to provide adequate water resources for Africa, and these efforts need to address several challenges, including population pressure, problems associated with land use (such as erosion and siltation), and possible ecological consequences of land-use change on the hydrological cycle. Climate change will make addressing these problems more complex. The greatest impact will continue to be felt by the poor, who have the most limited access to water resources. The issues that should be studied include affordable water purification systems; desalinization of seawater; water-use strategies—especially demand management—in industry, settlements, and agriculture; shared basin management, which necessitates international agreements; intensified monitoring to improve data reliability; intensive research into flood-control management technology; and innovation in building designs.

There is an urgent need to intensify the density of monitoring stations to improve climate change scenarios. The cost of rehabilitating stations that are in disrepair is not beyond the financial capability of African countries. Whatever strategies are adopted for optimizing water availability, their successful development is contingent on reliable meteorological and hydrological information.

Natural Disasters

Floods, droughts, storms, earthquakes, landslides, and other natural disasters all contribute to an enormous annual toll of human suffering, loss of life, and property damage in Africa. In the past 20 years, natural disasters worldwide have killed more than 3 million people (with 90 percent of deaths occurring in developing countries), inflicted injuries, facilitated the spread of diseases, and displaced more than 1 billion people. Annual economic losses related to natural disasters have been estimated at about $50 billion to $100 billion globally. In Africa in 2000 alone, more than 1,000 lives were lost in connection with natural disasters that struck four countries (Egypt, Madagascar, Mozambique, and Zimbabwe). Associated economic losses for just five countries in Africa (Botswana, Morocco, Mozambique, South Africa, and Zimbabwe) were estimated to total $1.8

billion. In Mozambique, the cost of damage due to the flooding disaster in 2000 represented 11.6 percent of its gross national product.[15]

The Mozambique floods were fed by tropical cyclone Eline, which also affected Botswana, Madagascar, Swaziland, South Africa, and Zimbabwe. It has been reported that much of the development Mozambique had achieved since the end of the civil war in 1992 was swept away by the worst flooding in southeastern Africa in the past century. Almost one-fifth of the country's only highway and large sections of railway linking Mozambique to Zimbabwe were destroyed.

The damage and costs inflicted by natural disasters often extend well beyond the affected countries or regions. No nation can afford to ignore the growing impacts of natural disasters, and many African countries are highly vulnerable to them. Technical means exist—and others are under development—to reduce losses through improved warning and preparedness systems, all-hazard risk assessments for the design of safer structures, land-use zoning in areas prone to natural disasters, and other means. Greater emphasis can be given to enhancing the use of information, research results, and technology that are already available.

One adaptive measure against extreme events relates to the ability to give adequate warning of imminent danger and to deliver relief. Facilities to broadcast timely information on developing events such as storms to rural populations remain weak and can be improved. Natural disaster-mitigation plans should be developed and/or enhanced, and financial resources should be earmarked.

Health and Diseases

Africa is facing an enormous health crisis that contributes to and is exacerbated by poverty.[16] People subject to hunger and malnutrition are especially vulnerable to diseases. The seriousness of this health crisis, particularly the devastation due to HIV/AIDS, has been recognized by African governments—the first Organization of African Unity Summit on HIV/AIDS and other infectious diseases was held in Abuja, Nigeria, in April 2001. Twenty-four African countries figure among those most affected by HIV/AIDS.

Malaria is also a serious problem. It has been estimated that malaria slows economic growth in Africa by up to 1.3 percent each year.[17] It has been reported that malaria-

free countries average three times higher per capita gross domestic product than other countries with malaria epidemics, even after adjusting for relevant factors such as government policy and geography. There is increasing evidence that climate plays a significant role in malaria epidemics.[18] In a highland area of Rwanda, for example, malaria incidence increased by 337 percent in 1987, and 80 percent of this variation could be explained by rainfall and temperature.[19] A similar association has been reported in Zimbabwe. Other epidemics in East Africa have been associated largely with El Niño. Understanding how climate affects the transmission of these diseases will lead to enhanced preparedness.

Communities that are exposed to waterborne diseases such as cholera could reduce the risk of infections by using safe drinking-water technologies. Several simple and inexpensive techniques have been found to be effective in reducing the risk of cholera infection from contaminated water. A simple filtration procedure using domestic sari material can reduce the number of cholera vibrio attached to plankton in raw water from ponds and rivers commonly used as sources of drinking water. The use of 5-percent calcium hypochlorite to disinfect water and the subsequent storage of treated water in narrow-mouthed jars have produced drinking water from nonpotable sources that has met the standards of the World Health Organization for microbiologic quality. In many cases, boiling water is not possible because firewood and charcoal are scarce, particularly in flooded conditions. These low-cost technologies should be studied further and made widely available.

Furthermore, there is a need to strengthen health and medical research, including research into traditional medicines, which should be considered a priority. Here, the Convention on Biological Diversity can come into play, as many traditional medicines come from biological species that are now under threat of extinction.

Coastal Zones and Sea-Level Rise

In Africa, as elsewhere, human settlement is heavily concentrated within 100 kilometers of coastal zones. More than one-quarter of the population of Africa resides within these areas, thus rendering a significant number of people vulnerable to rises in sea level as a result of climate change.

Changes in climatic conditions would have severe impacts not only on the distribution of human settlements but also on the quality of life in particular areas. By modeling the effects of a projected 38-centimeter mean global sea-level rise between 1990 and 2080, it was estimated that the average annual number of people in Africa affected by flooding could increase from 1 million in 1990 to 70 million in 2080. It was also indicated that Banjul, the capital of The Gambia, could disappear in 50-60 years through coastal erosion and sea-level rise, putting more than 42,000 people at risk.[20] Studies should be undertaken relating to response strategies to rising sea levels (retreat, adaptation, or defense) and their physical impacts. An ordered, planned program could minimize losses from rising sea levels.

Studies on coastal defense work, planned on a long-term scale, will help to develop such defenses well before the crisis occurs and to spread total capital costs over many years. As the problem of coastal management is basically regional, further studies should include regional integration among coastal-zone states, recognition by all governments of regional vulnerability to climate change impacts, and political and institutional stability that allows intergenerational projects to be sustained without interruption from political upheavals.

African Climate Projection

Observational records show that the African continent is warmer than it was 100 years ago. Warming through the twentieth century has been at the rate of about 0.05 degrees Celsius per decade, with slightly greater warming from June to August and from September to November than at other times. The five warmest years in Africa have all occurred since 1988, with 1988 and 1995 being the two warmest years. Africa's rate of warming is not dissimilar to the global rate, and the periods of most rapid warming—the 1910s to 1930s, and the post-1970 period—occurred simultaneously in Africa and the rest of the world.[21]

Studies have produced comprehensive characterizations of regional climate change projections for the twenty-first century. An analysis conducted specifically for Africa has projected future annual warming to be from 0.2 to more than 0.5°C per decade—ten times the rate during the twentieth century. This warming is expected to be greatest over the interior of semi-arid margins of the Sahara and central southern Africa.

Future changes in mean seasonal rainfall in Africa are not as well defined. Analyses were conducted of future rainfall changes for three African regions—the Sahel, east Africa, and southeast Africa—to illustrate the extent of inter-model differences for these regions and to put future modeled changes in the context of past observed changes. Although results vary, there is a general consensus for wetting in east Africa, drying in southeast Africa, and a poorly specified outcome for the Sahel. Under the lowest warming scenario, few areas will experience changes from December to February or from June to August that exceed two standard deviations of natural variability by 2050. The exception is parts of equatorial east Africa, where rainfall will increase by 5–20 percent from December to February and decrease by 5–10 percent from June to August. Under the two intermediate warming scenarios, significant decreases (10–20 percent) in rainfall during March to November, which includes the critical grain-filling period, will be apparent in northern Africa by 2050, as will be 5–15 percent decreases in rainfall during the growing season (November to May) in southern Africa. Under the most rapid scenario of global warming, increasing areas of Africa will experience significant changes in summer or winter rainfall.

One further need in formulating adaptive strategies is more refined regional climate change scenarios—especially a better understanding of extreme events. In southern Africa, for example, most of the regional climate change scenarios are rather ambivalent with regard to precipitation. In 1998–99, the city of Harare, Zimbabwe, suffered damage to its roads because sewer transport could not cope with entrained storm water. If it were accepted that the frequency of such seasons will likely increase, future designs of infrastructure could take cognizance of that prediction.

Moreover, it is important to incorporate the human dimension and cross-linkages in climate change and its variability. Significant steps are already being taken by many groups of scientists and through international programs. For example, the World Climate Research Programme—which WMO jointly sponsors with ICSU and the Intergovernmental Oceanographic Commission of the UN Education, Scientific and Cultural Organization (UNESCO)—is joining forces with the International Geosphere-Biosphere Programme and the International Human Dimensions Programme of Global Environmental Change in studies of food systems and water resources. The impact of global environmental change on food production and food availability throughout the world, as well as vulnerability in different regions and among different social groups, are being evaluated. Key supporting themes question how different societies and different categories of producers might adapt their food systems to cope with global change and changing demands, and what the environmental and socioeconomic consequences of adaptation to these changes are.

Capacity and Institution Building

The agenda for sustainability science needs to be expanded and deepened, and the involvement of—and interactions among—development agencies, scientists, and environmental experts must be reinforced. There is also a need to plan and promote a set of specific and focused research, development, and application programs, which could spring from regional workshops and could be the basis for attracting development resources and investments.

The existing infrastructure and capacity to ensure sustainable development are inadequate in Africa. Virtually no country, region, or sector is well placed to meet the unique needs of integrating across many different scientific disciplines or of fostering the interaction necessary for making progress in sustainability science. An appropriate enabling environment must be developed. Generating adequate scientific capacity and institutional support is particularly urgent. In this respect, the development of centers of excellence in Africa is a sine qua non. There are some centers of excellence that already exist and can be further developed, such as the ones for mathematics in Abuja, for biotechnology in Kampala and Harare, and for solar energy in Dar-es-Salaam. Such centers can provide the needed foci in their specialized fields for which other centers in Africa—as well as elsewhere—could be networked for specialized work. Networking arrangements could help bridge the gap in the level of development in relevant fields, particularly in the so-called digital divide between the north and south.

Advances in information and communication technology, notably the Internet, offer an excellent infrastructure that can link interdisciplinary research teams and support the activities involved in sustainability science. However, there are still barriers to this form of communication in

Africa, and progress in sustainability science will depend on remedying the lack of a networking infrastructure that embraces all countries. Multilateral, government, private foundation, and private sector sources of funding should be sought to redress the situation.

African national science academies can play a crucial role in furthering sustainability science in Africa. Their establishment and/or strengthening should be promoted. The range of disciplines in such science academies should be sufficiently broad to address African areas of concern within the realm of sustainability science. In particular, the interaction among social and natural scientists is imperative. The Third World Academy of Sciences and the African Academy of Sciences should provide the necessary encouragement and support in this endeavor. The African Academy of Sciences is known to be preparing a directory of African scientists and their specializations. Such a directory can help identify who is doing what and where it is being done, and it can help further interdisciplinary programs.

Conclusion

The World Summit on Sustainable Development provides an opportunity to connect science to the political agenda for sustainable development. Science must be directed at further exploring the character of nature-society interactions and guiding these interactions along sustainable pathways. Wide discussions in the scientific community in both the developing and developed worlds must be encouraged regarding key questions, appropriate methodologies, and institutional needs. A strategic framework is needed for the further development of sustainability science at the international, regional, and national levels, and it should include a strengthening of the needed scientific base and capacity in the relevant disciplines.

It is important to emphasize that enhanced partnership and cooperation should be fostered among the various stakeholders to facilitate the development of appropriate strategies to ensure integrated activities and the provision of related services, including the necessary monitoring of Earth-system components. Such cooperation should lead to support for sustainable development activities through,

among other things, the provision of advance warnings on the occurrence of natural disasters and improved predictions for El Niño and climate change as well as predictions relevant to the management of water and other natural resources and the protection of environmental quality. This should result in more effective contribution to the regional climate change assessments of IPCC and to the implementation of international environmental instruments such as the UN Framework Convention on Climate Change, UNCCD, and CBD.

Appropriate information, public affairs, and advocacy activities should ensure that policy makers and the public recognize the contributions of sustainability science to sustainable development and encourage them to provide institutional, policy, and resource support. Sustainability science truly can be in the full service of present and future generations of humankind, leading to peace, security, and prosperity—our desired common future, our common journey.

NOTES

1. G. O. P. Obasi, "Sustainability Science in Africa" (keynote address given at the African Regional Workshop on Sustainability Science, Abuja, Nigeria. 13 November 2001).

2. R. W. Kates et al., "Sustainability Science," *Science*, 27 April 2001, 641–42.

3. World Commission on Environment and Development, *Our Common Future* (Oxford: Oxford University Press, 1987).

4. United Nations Population Fund, *The Status of World Population 1998—The New Generation* (New York. 1998).

5. United Nations. *The Role of the United Nations in Support of the Efforts of the African Countries to Achieve Sustainable Development*, Report of the UN Secretary-General to the High-Level Segment of ECOSOC (Economic and Social Council), July 2001.

6. Organization of African Unity. *A New African Initiative* (approved by the Organization of African Unity Summit on 11 July 2001).

7. Ibid.

8. Intergovernmental Panel on Climate Change (IPCC), *Climate Change 2001: The Scientific Basis*, Third Assessment Report (Cambridge, U.K.: Cambridge University Press, 2001).

9. IPCC. *Climate Change 2001: Impacts, Adaptation and Vulnerability*, Third Assessment Report (Cambridge, U.K.: Cambridge University Press, 2001).

10. UN, note 5 above.

11. IPCC, note 9 above.

12. United Nations Convention to Combat Desertification. UN document number A/AC.241/27 (New York: UN, 1994).

13. G. O. P. Obasi, "Freshwater Management in Africa—Some issues and Challenges" (lecture given at the Fifth General Conference of the African Academy of Sciences, Hammamet, Tunisia, 1999).

14. World Bank, *Towards Environmentally Sustainable Development in Sub-Saharan Africa: A Framework for Integrated Coastal Zone Management Building Blocks for Africa 2025*, Paper 4, Post UNCED (UN Conference on Environment and Development) Series, Environmentally Sustainable Development Division, Africa Technical Department (Washington, D.C., 1995).

15. S. G. Cornford, "Human and Economic Impacts of Weather Events in 2000," *WMO Bulletin* 50, no. 4 (October 2001): 284–300.

16. UN, note 5 above.

17. Ibid.

18. World Health Organization, "The State of World Health, 1997 Report," *World Health Forum* 18 (1998): 248–60.

19. M. E. Loevinsohn, "Climate Warming and Increased Malaria in Rwanda," *Lancet* 343 (t994): 714–48.

20. B. P. Jallow et al., "Coastal Zone of The Gambia and the Abidjan Region in Cote d'Ivoire: Sea Level Rise Vulnerability Response Strategies, and Adaptation Options," in N. Mimura, ed., *National Assessment Results of Climate Change: Impacts and Responses* (Oldendorf/Luhe, Germany: Inter-Research, 1999) 129–36.

21. IPCC, note 9 above.

Godwin O. P. Obasi is secretary-general of the World Meteorological Organization. This article is drawn from the keynote address he delivered at the African Regional Workshop on Sustainability Science in Abuja, Nigeria, in November 2001 (see note 1). He can be reached at the World Meteorological Organization, 7 bis, avenue de la Paix, Case postale No. 2300, CH-1211, Geneva 2, Switzerland.

From *Environment*, May 2002, pp. 8-19. Reprinted with permission of the Helen Dwight Reid Educational Foundation. Published by Heldref Publications, 1319 Eighteenth St., NW, Washington, DC 20036-1802. © 2002.

WORLD IN REVIEW

Disconnected Continent

The Difficulties of the Internet in Africa

MAGDA KOWALCZYKOWSKI, Staff Writer, *Harvard International Review*

While the Internet has become an integral part of the Western world, it has only just arrived in Africa. In the United States and northern Europe, an average of one out of three people uses the Internet. In Africa, as in much of the developing world, Internet usage rates below one percent are the norm. Fortunately, this situation is changing rapidly, as initiatives by the United Nations and private corporations attempt to bridge the global digital divide.

According to official statistics, the future of connectivity in African countries is bright. The number of telephone lines is growing at a rate of 10 percent per year, and all of the main lines in Botswana and Rwanda are digital, compared to just under half of all lines in the United States. Cellular phone service, limited to six African countries a decade ago, is available in 42 countries today. Columbia Technology's Africa-One project is expected to complete an optical fiber network for the entire continent this year at a cost of US$1.6 billion.

Support from other private corporations looks promising as well. In cooperation with the Harvard Center for International Development, commercial technology giants such as Sun Microsystems, AOL-Time Warner, and Hewlett-Packard have pledged US$10 million over the next two years toward technology designed to improve the quality of life in 12 developing nations. With figures like these, it is easy to gloss over the real problems in the implementation of this vast network.

WorldTeacher: Namibia

In July 2001, I stepped into a fully equipped computer lab at the teacher resource center in Ongwediva, northern Namibia, beginning my part in a project to help spread computer and Internet literacy to the developing world as a WorldTeach volunteer. With video cameras, CD burners, and high-speed connections, the lab could easily have been in the United States. Although this teacher resource center in Ongwediva is state-of-the-art, it must serve the technological needs of all of northern Namibia. Because

of large distances between towns, a lack of vehicles, and limited awareness, many teachers do not exploit this resource to the fullest extent possible.

WorldTeach is designed to eliminate these problems. The 16-teacher contingency traveled to Namibia to teach computer and Internet literacy to students at various primary and secondary schools throughout the country An experiment of sorts, it was the first time that a program of this magnitude had been implemented in the developing world. A partnership with Schoolnet, an England-based company that equips schools with computers and technical support, made the project possible: Schoolnet provided the network, WorldTeach provided the teachers. After a period of training, each WorldTeacher was sent to a Schoolnet-sponsored site where he or she gave lessons on computer and Internet use.

Various development reports and meetings in the past few years have pointed to the explosive potential of information technology in Africa. Statistics on the increasing numbers of telephone lines, Internet Service Providers, and Internet connections have shed light on the vast potential of expanding markets and opportunities in this sector. Expectations are not reality, however. From my viewpoint on the ground in Namibia, I found the actual situation to be much more complex than the large companies and idealistic groups wishing to help these countries would like to pretend.

The enormous project at hand faces many glitches that must be smoothed out before the Internet can become a regular and integral part of the Namibian culture. One difficulty has been financial. Although the first school I visited in Edundja had a special budget allotted for the use of the Internet, it simply was not enough to sustain the program after I had left. I was there five days a week for three weeks, leaving the dial-up connections on all day in the hopes of getting as many students as possible to explore cyberspace. I taught the eighth, ninth, and tenth graders how to open email accounts, how to search the web, and even how to make their own website. The

phone bill amassed during this time proved too hefty for the administration to allow Internet use to continue after my departure. At my second school in Uis, I made sure to use the Internet more sparingly. Even so, this school was poorer than the first and the Internet did not thrive there either.

Another issue tied closely to the monetary situation is technical support and resources. Both of the schools in which I taught had many technical problems with the computers. With problems ranging from missing files to insufficient memory, the computers needed to be maintained and fixed on a regular basis, an expensive and time-consuming proposition. When students and teachers have not had a chance to familiarize themselves with computers, they can potentially damage them. Some problems are relatively small and can be fixed quickly and easily, but when the caretakers responsible for the computers do not understand even the basics of troubleshooting, salvageable computers are left to gather dust. Because the teachers and students do not want to break the computers, they are often reluctant to use them without supervision from an expert. At both schools, the teachers were initially very nervous about touching the keyboard. They wanted and needed someone to help them navigate through this technological endeavor.

Once the teachers and students overcame their initial hesitations, they were very enthusiastic about the possibilities of how the Internet could serve them. I attended the students' other classes to see what was being taught, and I supplemented their lessons with extra information and media from the Internet. The students were enthralled by an animation of blood flow in the heart after learning about blood cells in their biology class. The teachers were excited to find lesson plans and activity ideas online and wanted to learn more about this teaching tool.

Barriers to Change

Despite the initial interest of the teachers and students, the situation soon returned to the status quo. The allotted time of three weeks spent at each school was just not enough to change everyone's thinking and integrate the Internet into the education of Namibian students. That transformation would have required many more resources and a concerted effort over a longer period of time. Admitting the Internet into their community entails a recognition of the outside world. The Internet is an innovation of global proportions, but to introduce the ideas behind it to someone who has never worked on a computer before is quite a formidable task.

Even if the students were to understand the power and potential of the Internet, it would not do them much good in the short run. In the rural areas, socio-economic problems prevented students from making the most of their education. Only the rich can attend the University of Namibia or the Namibian Polytechnic University. Many young girls drop out of school due to teenage pregnancy. Parents often want their children at home, tending to the oxen and harvesting the mahangu, a crop used to make the porridge Namibians eat daily. When even the best students end up working as cashiers in the local markets after graduation, the benefits of a good education can be hard to see. Students cannot understand the link between studying hard and a bright future.

Political shortcomings also obstruct the integration of the Internet into society. The Namibian government does not have computers for all of its offices and workers. It is difficult to see how a government can fully embrace the technological movement when it does not understand how to use the Internet in its own affairs. The few schools with computers obtain them from nongovernmental organizations. Without a push from the government for these types of programs, technology cannot be fully integrated into schools and communities. The government is responsible for phone lines, roads, and the other infrastructure vital to the fledgling Internet movement. It alone has the power to break up the monopoly of Telecom, the only communications company in the country, which would make dial-up Internet access much more affordable. In a place where the public treasury is a precious resource that must be parceled out based on national priorities, the Internet takes a back seat to more immediately pressing matters such as controlling the AIDS epidemic.

The programs that are implemented at the local level must be thoroughly researched and carefully planned while leaving enough flexibility to adapt to special conditions. The 2001 WorldTeach program was structured for breadth rather than depth in its first year. To make sure as many schools as possible were connected during one short summer, each

teacher covered two schools in a six-week period. The schools spanned the geographic and demographic spectrum from urban areas where students were aware of the Internet and possessed basic computer skills to the deepest rural parts of Namibia where students had never seen a computer. Because each school was different, each WorldTeacher had to adapt lessons to the environment and resources that were available. My first school at Edundja in the north had a lab of 23 computers that students used on a weekly basis. It required a completely different teaching approach than my second school in Uis, which had three computers, only one of which could access the Internet.

Another factor that should have been considered in the implementation of the program was the different schedules of the schools. Toward the end of the summer, the students were studying for their trimester exams. Trying to fit another class into their schedules was a daunting task, especially given only three weeks at each school. I worked closely with the principals and maximized my time at each school by staying after class and working with the teachers in small groups. The age of students was another factor that affected teaching styles. WorldTeachers placed in primary schools could not teach things the same way as WorldTeachers placed in secondary schools taught them. Even in secondary schools, ages ranged from 13 to 21.

Financing the Future

Because money is the single most important factor in bringing the Internet to the African continent, securing funding will be vital to developing technology there. Already, advances are being made at an extraordinary rate. The Africa Bureau of the United Nations Development Programme has already agreed to a US$6 million fund to improve Internet connectivity in Africa with a project called the Internet Initiative for Africa. The United Nations has also announced the beginning of a US$11.5 million program called Harnessing Information Technology for Development, which provides funding for various information and communications technology (ICT) projects throughout Africa.

As a result of this push for ICT, the number of Internet users in Africa will soon rise to four million, according to Mike Jensen, an independent Internet development researcher. With Eritrea finally getting permanent Internet connectivity in 2001, every country on the continent has some degree of Internet access. In 1996, only 11 countries were connected. I personally saw how Internet use grew more common as I moved from rural towns to the Namibian capital of Windhoek.

Benefiting from the influx of capital and technology investment by the United Nations, Namibia is poised to make good on the promise of Internet growth.

Jensen's report on African technological advances also discusses the launching of another program to spur technological growth. The United Nations Economic, Social and Cultural Organization (UNESCO) has recently established the Creating Learning Networks for African Teachers project to connect teachers to the Internet and to assist teaching colleges in using ICTs in literacy and other education programs. The project has already been implemented in Zimbabwe, is being initiated in Senegal, and will eventually reach 20 countries with further outside support.

Along with the US$10 million donation to digital-divide initiatives, the private sector is contributing to the technological growth of Africa in many capacities. Hewlett-Packard has begun a global technology outreach program that has spurred the economic growth of countries such as Senegal and Ghana. Its new project, dubbed "HP e-inclusion," is expanding the possibilities for Internet technology in these countries. Other ICT companies, if not directly involved, are advising countries on how to best integrate technology into their development processes. Academics and executives from top ICT companies met with South African President Thabo Mbeki in October 2001 to begin this process. This was the first meeting of Mbeki's International Task Force on Information, Society, and Development, which was set up to advise him on how South Africa can develop its ICT capacity to promote investment, economic growth, and job creation. "The private sector must respond to the goals, objectives, programs, and mission as defined by the government and its people," said Hewlett-Packard CEO Carly Fiorina, who attended the conference. "In responding to the programs we see an opportunity not only to do well, but also an opportunity to do good."

This historic meeting in South Africa has opened the way for the rest of the continent. With more companies looking to diversify into new markets, Africa is looking attractive to investors. As nations that gained independence in recent decades have grown into fledgling democracies, they have solidified property rights laws and done much to improve the business environment. While the overall outlook may appear positive, many issues could thwart the development of this technological revolution. At the level of ideas, concepts, and statistics, the Internet enjoys enthusiastic support, but it is the practical matters that will determine the pace of technological advancement.

DEATH STALKS A CONTINENT

In the dry timber of African societies, AIDS was a spark. The conflagration it set off continues to kill millions. Here's why

By Johanna McGeary

IMAGINE YOUR LIFE THIS WAY. You get up in the morning and breakfast with your three kids. One is already doomed to die in infancy. Your husband works 200 miles away, comes home twice a year and sleeps around in between. You risk your life in every act of sexual intercourse. You go to work past a house where a teenager lives alone tending young siblings without any source of income. At another house, the wife was branded a whore when she asked her husband to use a condom, beaten silly and thrown into the streets. Over there lies a man desperately sick without access to a doctor or clinic or medicine or food or blankets or even a kind word. At work you eat with colleagues, and every third one is already fatally ill. You whisper about a friend who admitted she had the plague and whose neighbors stoned her to death. Your leisure is occupied by the funerals you attend every Saturday. You go to bed fearing adults your age will not live into their 40s. You and your neighbors and your political and popular leaders act as if nothing is happening.

Across the southern quadrant of Africa, this nightmare is real. The word not spoken is AIDS, and here at ground zero of humanity's deadliest cataclysm, the ultimate tragedy is that so many people don't know—or don't want to know—what is happening.

As the HIV virus sweeps mercilessly through these lands—the fiercest trial Africa has yet endured—a few try to address the terrible depredation. The rest of society looks away. Flesh and muscle melt from the bones of the sick in packed hospital wards and lonely bush kraals. Corpses stack up in morgues until those on top crush the identity from the faces underneath. Raw earth mounds scar the landscape, grave after grave without name or number. Bereft children grieve for parents lost in their prime, for siblings scattered to the winds.

The victims don't cry out. Doctors and obituaries do not give the killer its name. Families recoil in shame. Leaders shirk responsibility. The stubborn silence heralds victory for the disease: denial cannot keep the virus at bay.

The developed world is largely silent too. AIDS in Africa has never commanded the full-bore response the West has brought to other, sometimes lesser, travails. We pay sporadic attention, turning on the spotlight when an international conference occurs, then turning it off. Good-hearted donors donate; governments acknowledge that more needs to be done. But think how different the effort would be if what is happening here were happening in the West.

By now you've seen pictures of the sick, the dead, the orphans. You've heard appalling numbers: the number of new infections, the number of the dead, the number who are sick without care, the number walking around already fated to die.

But to comprehend the full horror AIDS has visited on Africa, listen to the woman we have dubbed Laetitia Hambahlane in

Durban or the boy Tsepho Phale in Francistown or the woman who calls herself Thandiwe in Bulawayo or Louis Chikoka, a long-distance trucker. You begin to understand how AIDS has struck Africa—with a biblical virulence that will claim tens of millions of lives—when you hear about shame and stigma and ignorance and poverty and sexual violence and migrant labor and promiscuity and political paralysis and the terrible silence that surrounds all this dying. It is a measure of the silence that some asked us not to print their real names to protect their privacy.

HALF A MILLION AFRICAN CHILDREN WERE INFECTED WITH HIV LAST YEAR

Theirs is a story about what happens when a disease leaps the confines of medicine to invade the body politic, infecting not just individuals but an entire society. As AIDS migrated to man in Africa, it mutated into a complex plague with confounding social, economic and political mechanics that locked together to accelerate the virus' progress. The region's social dynamics colluded to spread the disease and help block effective intervention.

We have come to three countries abutting one another at the bottom of Africa—Botswana, South Africa, Zimbabwe—the heart of the heart of the epidemic. For nearly a decade, these nations suffered a hidden invasion of infection that concealed the dimension of the coming calamity. Now the omnipresent dying reveals the shocking scale of the devastation.

AIDS in Africa bears little resemblance to the American epidemic, limited to specific high-risk groups and brought under control through intensive education, vigorous political action and expensive drug therapy. Here the disease has bred a Darwinian perversion. Society's fittest, not its frailest, are the ones who die—adults spirited away, leaving the old and the children behind. You cannot define risk groups: everyone who is sexually active is at risk. Babies too, unwittingly infected by mothers. Barely a single family remains untouched. Most do not know how or when they caught the virus, many never know they have it, many who do know don't tell anyone as they lie dying. Africa can provide no treatment for those with AIDS.

They will all die, of tuberculosis, pneumonia, meningitis, diarrhea, whatever overcomes their ruined immune systems first. And the statistics, grim as they are, may be too low. There is no broad-scale AIDS testing: infection rates are calculated mainly from the presence of HIV in pregnant women. Death certificates in these countries do not record AIDS as the cause. "Whatever stats we have are not reliable," warns Mary Crewe of the University of Pretoria's Center for the Study of AIDS. "Everybody's guessing."

THE TB PATIENT

CASE NO. 309 IN THE TUGELA FERRY HOME-CARE PROGRAM shivers violently on the wooden planks someone has knocked into a bed, a frayed blanket pulled right up to his nose. He has the flushed skin, overbright eyes and careful breathing of the tubercular. He is alone, and it is chilly within the crumbling mud walls of his hut at Msinga Top, a windswept outcrop high above the Tugela River in South Africa's KwaZulu-Natal province. The spectacular view of hills and veld would gladden a well man, but the 22-year-old we will call Fundisi Khumalo, though he does not know it, has AIDS, and his eyes seem to focus inward on his simple fear.

Before he can speak, his throat clutches in gasping spasms. Sharp pains rack his chest; his breath comes in shallow gasps. The vomiting is better today. But constipation has doubled up his knees, and he is too weak to go outside to relieve himself. He can't remember when he last ate. He can't remember how long he's been sick—"a long time, maybe since six months ago." Khumalo knows he has TB, and he believes it is just TB. "I am only thinking of that," he answers when we ask why he is so ill.

But the fear never leaves his eyes. He worked in a hair salon in Johannesburg, lived in a men's hostel in one of the cheap townships, had "a few" girlfriends. He knew other young men in the hostel who were on-and-off sick. When they fell too ill to work anymore, like him, they straggled home to rural villages like Msinga Top. But where Khumalo would not go is the hospital. "Why?" he says. "You are sick there, you die there."

"He's right, you know," says Dr. Tony Moll, who has driven us up the dirt track from the 350-bed hospital he heads in Tugela Ferry. "We have no medicines for AIDS. So many hospitals tell them, 'You've got AIDS. We can't help you. Go home and die.'" No one wants to be tested either, he adds, unless treatment is available. "If the choice is to know and get nothing," he says, "they don't want to know."

Here and in scattered homesteads all over rural Africa, the dying people say the sickness afflicting their families and neighbors is just the familiar consequence of their eternal poverty. Or it is the work of witchcraft. You have done something bad and have been bewitched. Your neighbor's jealousy has invaded you. You have not appeased the spirits of your ancestors, and they have cursed you. Some in South Africa believe the disease was introduced by the white population as a way to control black Africans after the end of apartheid.

Ignorance about AIDS remains profound. But because of the funerals, southern Africans can't help seeing that something more systematic and sinister lurks out there. Every Saturday and often Sundays too, neighbors trudge to the cemeteries for costly burial rites for the young and the middle-aged who are suddenly dying so much faster than the old. Families say it was pneumonia, TB, malaria that killed their son, their wife, their baby. "But you starting to hear the truth," says Durban home-care volunteer Busi Magwazi. "In the church, in the graveyard, they saying, 'Yes, she died of AIDS.' Oh, people talking about it even if the families don't admit it." Ignorance is the crucial reason the epidemic has run out of control. Surveys say many Africans here are becoming aware there is a sexually transmitted disease called AIDS that is incurable. But they don't think the risk applies to them. And their vague knowledge does not translate into changes in their sexual behavior. It's easy to see why so many don't yet sense the danger when few talk openly about the disease. And Africans are beset by so plentiful a roster of perils—famine, war, the violence of desperation or ethnic hatred, the regular illnesses of poverty, the dangers inside mines or on the roads—that the delayed risk of AIDS ranks low.

A CONTINENT IN PERIL

17 million Africans have died since the AIDS epidemic began in the late 1970s, more than 3.7 million of them children. An additional 12 million children have been orphaned by AIDS. An estimated 8.8% of adults in Africa are infected with HIV/AIDS, and in the following seven countries, at least 1 adult in 5 is living with HIV

1. Botswana

Though it has the highest per capita GDP, it also has the highest estimated adult infection rate—**36%**. 24,000 die each year. 66,000 children have lost their mother or both parents to the disease.

2. Swaziland

More than **25%** of adults have HIV/AIDS in this small country. 12,000 children have been orphaned, and 7,100 adults and children die each year.

3. Zimbabwe

One-quarter of the adult population is infected here. 160,000 adults and children died in 1999, and 900,000 children have been orphaned. Because of AIDS, life expectancy is 43.

4. Lesotho

24% of the adults are infected with HIV/AIDS. 35,000 children have been orphaned, and 16,000 adults and children die each year.

5. Zambia

20% of the adult population is infected, 1 in 4 adults in the cities. 650,000 children have been orphaned, and 99,000 Zambians died in 1999.

6. South Africa

This country has the largest number of people living with HIV/AIDS, about **20%** of its adult population, up from 13% in 1997. 420,000 children have been orphaned, and 250,000 people die each year from the disease.

7. Namibia

19.5% of the adult population is living with HIV. 57% of the infected are women. 67,000 children are AIDS orphans, and 18,000 adults and children die each year.

Source: UNAIDS

THE OUTCAST

TO ACKNOWLEDGE AIDS IN YOURSELF IS TO BE BRANDED AS monstrous. Laetitia Hambahlane (not her real name) is 51 and sick with AIDS. So is her brother. She admits it; he doesn't. In her mother's broken-down house in the mean streets of Umlazi township, though, Laetitia's mother hovers over her son, nursing him, protecting him, resolutely denying he has anything but TB, though his sister claims the sure symptoms of AIDS mark him. Laetitia is the outcast, first from her family, then from her society.

For years Laetitia worked as a domestic servant in Durban and dutifully sent all her wages home to her mother. She fell in love a number of times and bore four children. "I loved that last man," she recalls. "After he left, I had no one, no sex." That was 1992, but Laetitia already had HIV.

She fell sick in 1996, and her employers sent her to a private doctor who couldn't diagnose an illness. He tested her blood and found she was HIV positive. "I wish I'd died right then," she says, as tears spill down her sunken cheeks. "I asked the doctor, 'Have you got medicine?' He said no. I said, 'Can't you keep me alive?' " The doctor could do nothing and sent her away. "I couldn't face the word," she says. "I couldn't sleep at night. I sat on my bed, thinking, praying. I did not see anyone day or night. I ask God, Why?"

Laetitia's employers fired her without asking her exact diagnosis. For weeks she could not muster the courage to tell anyone. Then she told her children, and they were ashamed and frightened. Then, harder still, she told her mother. Her mother raged about the loss of money if Laetitia could not work again. She was so angry she ordered Laetitia out of the house. When her daughter wouldn't leave, the mother threatened to sell the house to get rid of her daughter. Then she walled off her daughter's room with plywood partitions, leaving the daughter a pariah, alone in a cramped, dark space without windows and only a flimsy door opening into the alley. Laetitia must earn the pennies to feed herself and her children by peddling beer, cigarettes and candy from a shopping cart in her room, when people are brave enough to stop by her door. "Sometimes they buy, sometimes not," she says. "That is how I'm surviving."

Her mother will not talk to her. "If you are not even accepted by your own family," says Magwazi, the volunteer home-care giver from Durban's Sinoziso project who visits Laetitia, "then others will not accept you." When Laetitia ventures outdoors, neighbors snub her, tough boys snatch her purse, children taunt her. Her own kids are tired of the sickness and don't like to help her anymore. "When I can't get up, they don't bring me food," she laments. One day local youths barged into her room, cursed her as a witch and a whore and beat her. When she told the police, the youths returned, threatening to burn down the house.

But it is her mother's rejection that wounds Laetitia most. "She is hiding it about my brother," she cries. "Why will she do nothing for me?" Her hands pick restlessly at the quilt covering her paper-thin frame. "I know my mother will not bury me properly. I know she will not take care of my kids when I am gone."

Jabulani Syabusi would use his real name, but he needs to protect his brother. He teaches school in a red, dusty district of KwaZulu-Natal. People here know the disease is all around them, but no one speaks of it. He eyes the scattered huts that make up his little settlement on an arid bluff. "We can count 20 who died just here as far as we can see. I personally don't remember any family that told it was AIDS," he says. "They hide it if they do know."

Syabusi's own family is no different. His younger brother is also a teacher who has just come home from Durban too sick to work anymore. He says he has tuberculosis, but after six months the tablets he is taking have done nothing to cure him. Syabusi's wife Nomsange, a nurse, is concerned that her 36-year-old brother-in-law may have something worse. Syabusi finally asked the doctor tending his brother what is wrong. The doctor said the information is confidential and will not tell him. Neither will his brother. "My brother is not brave enough to tell me," says Syabusi, as he stares sadly toward the house next door, where his only sibling lies ill. "And I am not brave enough to ask him."

Kennedy Fugewane, a cheerful, elderly volunteer counselor, sits in an empty U.S.-funded clinic that offers fast, pinprick blood tests in Francistown, Botswana, pondering how to break through the silence. This city suffers one of the world's highest infection rates, but people deny the disease because HIV is linked with sex. "We don't reveal anything," he says. "But people are so stigmatized even if they walk in the door." Africans feel they must keep private anything to do with sex. "If a man comes here, people will say he is running around," says Fugewane, though he acknowledges that men never do come. "If a woman comes, people will say she is loose. If anyone says they got HIV, they will be despised."

Pretoria University's Mary Crewe says, "It is presumed if you get AIDS, you have done something wrong." HIV labels you as living an immoral life. Embarrassment about sexuality looms more important than future health risks. "We have no language to talk candidly about sex," she says, "so we have no civil language to talk about AIDS." Volunteers like Fugewane try to reach out with flyers, workshops, youth meetings and free condoms, but they are frustrated by a culture that values its dignity over saving lives. "People here don't have the courage to come forward and say, 'Let me know my HIV status,'" he sighs, much less the courage to do something about it. "Maybe one day…"

Doctors bow to social pressure and legal strictures not to record AIDS on death certificates. "I write TB or meningitis or diarrhea but never AIDS," says South Africa's Dr. Moll. "It's a public document, and families would hate it if anyone knew." Several years ago, doctors were barred even from recording compromised immunity or HIV status on a medical file; now they can record the results of blood tests for AIDS on patient charts to protect other health workers. Doctors like Moll have long agitated to apply the same openness to death certificates.

THE TRUCK DRIVER

HERE, MEN HAVE TO MIGRATE TO WORK, INSIDE THEIR COUN-tries or across borders. All that mobility sows HIV far and wide, as Louis Chikoka is the first to recognize. He regularly drives the highway that is Botswana's economic lifeline and its curse. The road runs for 350 miles through desolate bush that is the Texas-size country's sole strip of habitable land, home to a large majority of its 1.5 million people. It once brought prospectors to Botswana's rich diamond reefs. Now it's the link for transcontinental truckers like Chikoka who haul goods from South Africa to markets in the continent's center. And now the road brings AIDS.

Chikoka brakes his dusty, diesel-belching Kabwe Transport 18-wheeler to a stop at the dark roadside rest on the edge of Francistown, where the international trade routes converge and at least 43% of adults are HIV-positive. He is a cheerful man even after 12 hard hours behind the wheel freighting rice from Durban. He's been on the road for two weeks and will reach his destination in Congo next Thursday. At 39, he is married, the father of three and a long-haul trucker for 12 years. He's used to it.

Lighting up a cigarette, the jaunty driver is unusually loquacious about sex as he eyes the dim figures circling the rest stop. Chikoka has parked here for a quickie. See that one over there, he points with his cigarette. "Those local ones we call bitches. They always waiting here for short service." Short service? "It's according to how long it takes you to ejaculate," he explains. "We go to the 'bush bedroom' over there [waving at a clump of trees 100 yds. away] or sometimes in the truck. Short service, that costs you 20 rands [$2.84]. They know we drivers always got money."

Chikoka nods his head toward another woman sitting beside a stack of cardboard cartons. "We like better to go to them," he says. They are the "businesswomen," smugglers with gray-market cases of fruit and toilet paper and toys that they need to transport somewhere up the road. "They come to us, and we negotiate privately about carrying their goods." It's a no-cash deal, he says. "They pay their bodies to us." Chikoka shrugs at a suggestion that the practice may be unhealthy. "I been away two weeks, madam. I'm human. I'm a man. I have to have sex."

What he likes best is dry sex. In parts of sub-Saharan Africa, to please men, women sit in basins of bleach or saltwater or stuff astringent herbs, tobacco or fertilizer inside their vagina. The tissue of the lining swells up and natural lubricants dry out. The resulting dry sex is painful and dangerous for women. The drying agents suppress natural bacteria, and friction easily lacerates the tender walls of the vagina. Dry sex increases the risk of HIV infection for women, already two times as likely as men to contract the virus from a single encounter. The women, adds

Chikoka, can charge more for dry sex, 50 or 60 rands ($6.46 to $7.75), enough to pay a child's school fees or to eat for a week.

UNVANQUISHED

A Fighter in a Land of Orphans

Silence and the ignorance it promotes have fed the AIDS epidemic in Africa perhaps more than any other factors. In Malawi, where until the end of dictator Hastings Banda's rule in 1994 women were barred from wearing short skirts and men could be jailed for having long hair, public discussion of AIDS was forbidden. According to the government, AIDS didn't exist inside Malawi. Catherine Phiri, 38, knew otherwise. She tested positive in 1990, after her husband had died of the disease. Forced to quit her job as a nurse when colleagues began to gossip, she sought refuge with relatives in the capital, Lilongwe. But they shunned her and eventually forced her to move, this time to Salima on beautiful Lake Malawi. "Even here people gossiped," says Phiri, whose brave, open face is fringed by a head of closely cropped graying hair.

Determined to educate her countrymen, Phiri set up a group that offers counseling, helps place orphans and takes blood that can then be tested in the local hospital. "The community began to see the problem, but it was very difficult to communicate to the government. They didn't want to know."

They do now. According to a lawmaker, AIDS has killed dozens of members of Parliament in the past decade. And Malawi's government has begun to move. President Bakili Muluzi incorporates AIDS education into every public rally. In 1999 he launched a five-year plan to fight the disease, and last July he ordered a crackdown on prostitution (though the government is now thinking of legalizing it). At the least, his awareness campaign appears to be working: 90% of Malawians know about the dangers of AIDS. But that knowledge comes too late for the estimated 8% of HIV-positive citizens— 800,000 people in 1999—or the 276,000 children under 15 orphaned by the disease.

Last October, Phiri picked up an award for her efforts from the U.N. But, she says, "I still have people who look at me like trash…" Her voice trails off. "Sometimes when I go to sleep I fear for the future of my children. But I will not run away now. Talking about it: that's what's brave."

—By Simon Robinson/Salima

Chikoka knows his predilection for commercial sex spreads AIDS; he knows his promiscuity could carry the disease home to his wife; he knows people die if they get it. "Yes, HIV is terrible, madam," he says as he crooks a finger toward the businesswoman whose favors he will enjoy that night. "But, madam, sex is natural. Sex is not like beer or smoking. You can stop them.

But unless you castrate the men, you can't stop sex—and then we all die anyway."

Millions of men share Chikoka's sexually active lifestyle, fostered by the region's dependence on migrant labor. Men desperate to earn a few dollars leave their women at hardscrabble rural homesteads to go where the work is: the mines, the cities, the road. They're housed together in isolated males-only hostels but have easy access to prostitutes or a "town wife" with whom they soon pick up a second family and an ordinary STD and HIV. Then they go home to wives and girlfriends a few times a year, carrying the virus they do not know they have. The pattern is so dominant that rates of infection in many rural areas across the southern cone match urban numbers.

IN SOME AFRICAN COUNTRIES, THE INFECTION RATE OF TEEN GIRLS IS FOUR TIMES THAT OF BOYS

If HIV zeros in disproportionately on poor migrants, it does not skip over the educated or the well paid. Soldiers, doctors, policemen, teachers, district administrators are also routinely separated from families by a civil-service system that sends them alone to remote rural posts, where they have money and women have no men. A regular paycheck procures more access to extramarital sex. Result: the vital professions are being devastated.

Schoolmaster Syabusi is afraid there will soon be no more teachers in his rural zone. He has just come home from a memorial for six colleagues who died over the past few months, though no one spoke the word AIDS at the service. "The rate here—they're so many," he says, shaking his head. "They keep on passing it at school." Teachers in southern Africa have one of the highest group infection rates, but they hide their status until the telltale symptoms find them out.

Before then, the men—teachers are mostly men here—can take their pick of sexual partners. Plenty of women in bush villages need extra cash, often to pay school fees, and female students know they can profit from a teacher's favor. So the schoolmasters buy a bit of sex with lonely wives and trade a bit of sex with willing pupils for A's. Some students consider it an honor to sleep with the teacher, a badge of superiority. The girls brag about it to their peers, preening in their ability to snag an older man. "The teachers are the worst," says Jabulani Siwela, an AIDS worker in Zimbabwe who saw frequent teacher-student sex in his Bulawayo high school. They see a girl they like; they ask her to stay after class; they have a nice time. "It's dead easy," he says. "These are men who know better, but they still do it all the time."

THE PROSTITUTE

THE WORKINGWOMAN WE MEET DIRECTS OUR CAR TO A reedy field fringing the gritty eastern townships of Bulawayo, Zimbabwe. She doesn't want neighbors to see her being interviewed. She is afraid her family will find out she is a prostitute, so we will call her Thandiwe. She looked quite prim and proper in her green calf-length dress as she waited for johns outside 109 Tongogaro Street in the center of downtown. So, for that matter, do the dozens of other women cruising the city's dim street corners: not a mini or bustier or bared navel in sight. Zimbabwe is in many ways a prim and proper society that frowns on commercial sex work and the public display of too much skin.

FINANCIAL AID

A Lending Tree

Getting ahead in Africa is tough. Banks lend money only to the middle class and the wealthy. Poor Africans—meaning most Africans—stay poor. It's even harder if you're sick. Without savings to fall back on, many HIV-positive parents pull their kids out of school. They can't afford the fees and end up selling their few possessions to feed the family. When they die, their kids are left with nothing.

Though not directly targeted at people with AIDS, microcredit schemes go some way toward fixing that problem. The schemes work like minibanks, lending small amounts—often as little as $100—to traders or farmers. Because they lack the infrastructure of banks and don't charge fees, most charge an interest rate of as much as 1% a week and repayment rates of over 99%—much better than that for banks in Africa, or in most places.

Many microcredit schemes encourage clients to set aside some of the extra income generated by the loan as savings. This can be used for medical bills or to pay school fees if the parents get sick. "Without the loans I would have had to look for another way to make money," says Florence Muriungi, 40, who sings in a Kampala jazz band and whose husband died of AIDS four years ago. Muriungi, who cares for eight children—five of her own and three her sister left when she too died of AIDS—uses the money to pay school fees in advance and fix her band's equipment. Her singing generates enough money for her to repay the loans and save a bit.

Seventeen of the 21 women at a weekly meeting of regular borrowers in Uganda care for AIDS orphans. Five are AIDS widows. "I used to buy just one or two bunches of bananas to sell. Now I buy 40, 50, 60," says Elizabeth Baluka, 47, the group's secretary. "Every week I put aside a little bit of money to help my children slowly by slowly."

—By Simon Robinson/Kampala

That doesn't stop Thandiwe from earning a better living turning tricks than she ever could doing honest work. Desperate for a job, she slipped illegally into South Africa in 1992. She cleaned floors in a Johannesburg restaurant, where she met a cook from back home who was also illegal. They had two daughters, and they got married; he was gunned down one night at work.

She brought his body home for burial and was sent to her in-laws to be "cleansed." This common practice gives a dead husband's brother the right, even the duty, to sleep with the widow. Thandiwe tested negative for HIV in 1998, but if she were positive, the ritual cleansing would have served only to pass on the disease. Then her in-laws wanted to keep her two daughters because their own children had died, and marry her off to an old uncle who lived far out in the bush. She fled.

Alone, Thandiwe grew desperate. "I couldn't let my babies starve." One day she met a friend from school. "She told me she was a sex worker. She said, 'Why you suffer? Let's go to a place where we can get quick bucks.'" Thandiwe hangs her head. "I went. I was afraid. But now I go every night."

She goes to Tongogaro Street, where the rich clients are, tucking a few condoms in her handbag every evening as the sun sets and returning home strictly by 10 so that she won't have to service a taxi-van driver to get a ride back. Thandiwe tells her family she works an evening shift, just not at what. "I get 200 zim [$5] for sex," she says, more for special services. She uses two condoms per client, sometimes three. "If they say no, I say no." But then sometimes resentful johns hit her. It's pay-and-go until she has pocketed 1,000 or 1,500 Zimbabwe dollars and can go home—with more cash than her impoverished neighbors ever see in their roughneck shantytown, flush enough to buy a TV and fleece jammies for her girls and meat for their supper.

"I am ashamed," she murmurs. She has stopped going to church. "Every day I ask myself, 'When will I stop this business?' The answer is, 'If I could get a job'…" Her voice trails off hopelessly. "At the present moment, I have no option, no other option." As trucker Chikoka bluntly puts it, "They give sex to eat. They got no man; they got no work; but they got kids, and they got to eat." Two of Thandiwe's friends in the sex trade are dying of AIDS, but what can she do? "I just hope I won't get it."

In fact, casual sex of every kind is commonplace here. Prostitutes are just the ones who admit they do it for cash. Everywhere there's premarital sex, sex as recreation. Obligatory sex and its abusive counterpart, coercive sex. Transactional sex: sex as a gift, sugar-daddy sex. Extramarital sex, second families, multiple partners. The nature of AIDS is to feast on promiscuity.

79% OF THOSE WHO DIED OF AIDS LAST YEAR WERE AFRICAN

Rare is the man who even knows his HIV status: males widely refuse testing even when they fall ill. And many men who suspect they are HIV positive embrace a flawed logic: if I'm already infected, I can sleep around because I can't get it again. But women are the ones who progress to full-blown AIDS first and die fastest, and the underlying cause is not just sex but power. Wives and girlfriends and even prostitutes in this part of the world can't easily say no to sex on a man's terms. It matters little what comes into play, whether it is culture or tradition or the pathology of violence or issues of male identity or the sub-servient status of women.

Beneath a translucent scalp, the plates of Gertrude Dhlamini's cranium etch a geography of pain. Her illness is obvious in the thin, stretched skin under which veins throb with the shingles that have blinded her left eye and scarred that side of her face. At 39, she looks 70. The agonizing thrush, a kind of fungus, that paralyzed her throat has ebbed enough to enable her to swallow a spoon or two of warm gruel, but most of the nourishment flows away in constant diarrhea. She struggles to keep her hand from scratching restlessly at the scaly rash flushing her other cheek. She is not ashamed to proclaim her illness to the world. "It must be told," she says.

Gertrude is thrice rejected. At 19 she bore a son to a boyfriend who soon left her, taking away the child. A second boyfriend got her pregnant in 1994 but disappeared in anger when their daughter was born sickly with HIV. A doctor told Gertrude it was her fault, so she blamed herself that little Noluthando was never well in the two years she survived. Gertrude never told the doctor the baby's father had slept with other women. "I was afraid to," she says, "though I sincerely believe he gave the sickness to me." Now, she says, "I have rent him from my heart. And I will never have another man in my life."

Gertrude begged her relatives to take her in, but when she revealed the name of her illness, they berated her. They made her the household drudge, telling her never to touch their food or their cooking pots. They gave her a bowl and a spoon strictly for her own use. After a few months, they threw her out.

Gertrude sits upright on a donated bed in a cardboard shack in a rough Durban township that is now the compass of her world. Perhaps 10 ft. square, the little windowless room contains a bed, one sheet and blanket, a change of clothes and a tiny cooking ring, but she has no money for paraffin to heat the food that a home-care worker brings. She must fetch water and use a toilet down the hill. "Everything I have," she says, "is a gift." Now the school that owns the land under her hut wants to turn it into a playground and she worries about where she will go. Gertrude rubs and rubs at her raw cheek. "I pray and pray to God," she says, "not to take my soul while I am alone in this room."

Women like Gertrude were brought up to be subservient to men. Especially in matters of sex, the man is always in charge. Women feel powerless to change sexual behavior. Even when a woman wants to protect herself, she usually can't: it is not uncommon for men to beat partners who refuse intercourse or request a condom. "Real men" don't use them, so women who want their partners to must fight deeply ingrained taboos. Talk to him about donning a rubber sheath and be prepared for accusations, abuse or abandonment.

A nurse in Durban, coming home from an AIDS training class, suggested that her mate should put on a condom, as a kind of homework exercise. He grabbed a pot and banged loudly on it with a knife, calling all the neighbors into his house. He pointed the knife at his wife and demanded: "Where was she between 4 p.m. and now? Why is she suddenly suggesting this? What has changed after 20 years that she wants a condom?"

Schoolteacher Syabusi is an educated man, fully cognizant of the AIDS threat. Yet even he bristles when asked if he uses a condom. "Humph," he says with a fine snort. "That question is nonnegotiable." So despite extensive distribution of free condoms, they often go unused. Astonishing myths have sprung up. If you don one, your erection can't grow. Free condoms must be too cheap to be safe: they have been stored too long, kept too hot, kept too cold. Condoms fill up with germs, so they spread AIDS. Condoms from overseas bring the disease with them. Foreign governments that donate condoms put holes in them so that Africans will die. Education programs find it hard to compete with the power of the grapevine.

THE CHILD IN NO. 17

IN CRIB NO. 17 OF THE SPARTAN BUT CROWDED CHILDREN'S ward at the Church of Scotland Hospital in KwaZulu-Natal, a tiny, staring child lies dying. She is three and has hardly known a day of good health. Now her skin wrinkles around her body like an oversize suit, and her twig-size bones can barely hold her vertical as nurses search for a vein to take blood. In the frail arms hooked up to transfusion tubes, her veins have collapsed. The nurses palpate a threadlike vessel on the child's forehead. She mews like a wounded animal as one tightens a rubber band around her head to raise the vein. Tears pour unnoticed from her mother's eyes as she watches the needle tap-tap at her daughter's temple. Each time the whimpering child lifts a wan hand to brush away the pain, her mother gently lowers it. Drop by drop, the nurses manage to collect 1 cc of blood in five minutes.

The child in crib No. 17 has had TB, oral thrush, chronic diarrhea, malnutrition, severe vomiting. The vial of blood reveals her real ailment, AIDS, but the disease is not listed on her chart, and her mother says she has no idea why her child is so ill. She breast-fed her for two years, but once the little girl was weaned, she could not keep solid food down. For a long time, her mother thought something was wrong with the food. Now the child is afflicted with so many symptoms that her mother had to bring her to the hospital, from which sick babies rarely return.

VIRGINITY TESTING IS BACK The practice of virginity testing used to be part of traditional Zulu rites. It is regaining popularity among anxious mothers who believe that if their daughters remain virgins, they won't get AIDS.

She hopes, she prays her child will get better, and like all the mothers who stay with their children at the hospital, she tends her lovingly, constantly changing filthy diapers, smoothing sheets, pressing a little nourishment between listless lips, trying to tease a smile from the vacant, staring face. Her husband works in Johannesburg, where he lives in a men's squatter camp. He comes home twice a year. She is 25. She has heard of

AIDS but does not know it is transmitted by sex, does not know if she or her husband has it. She is afraid this child will die soon, and she is afraid to have more babies. But she is afraid too to raise the subject with her husband. "He would not agree to that," she says shyly. "He would never agree to have no more babies."

Dr. Annick DeBaets, 32, is a volunteer from Belgium. In the two years she has spent here in Tugela Ferry, she has learned all about how hard it is to break the cycle of HIV transmission from mother to infant. The door to this 48-cot ward is literally a revolving one: sick babies come in, receive doses of rudimentary antibiotics, vitamins, food; go home for a week or a month; then come back as ill as ever. Most, she says, die in the first or second year. If she could just follow up with really intensive care, believes Dr. DeBaets, many of the wizened infants crowding three to a crib could live longer, healthier lives. "But it's very discouraging. We simply don't have the time, money or facilities for anything but minimal care."

Much has been written about what South African Judge Edwin Cameron, himself HIV positive, calls his country's "grievous ineptitude" in the face of the burgeoning epidemic. Nowhere has that been more evident than in the government's failure to provide drugs that could prevent pregnant women from passing HIV to their babies. The government has said it can't afford the 300-rand-per-dose, 28-dose regimen of AZT that neighboring nations like Botswana dole out, using funds and drugs from foreign donors. The late South African presidential spokesman Parks Mankahlana even suggested publicly that it was not cost effective to save these children when their mothers were already doomed to die: "We don't want a generation of orphans."

Yet these children—70,000 are born HIV positive in South Africa alone every year—could be protected from the disease for about $4 each with another simple, cheap drug called nevirapine. Until last month, the South African government steadfastly refused to license or finance the use of nevirapine despite the manufacturer's promise to donate the drug for five years, claiming that its "toxic" side effects are not yet known. This spring, however, the drug will finally be distributed to leading public hospitals in the country, though only on a limited basis at first.

The mother at crib No. 17 is not concerned with potential side effects. She sits on the floor cradling her daughter, crooning over and over, "Get well, my child, get well." The baby stares back without blinking. "It's sad, so sad, so sad," the mother says. The child died three days later.

The children who are left when parents die only add another complex dimension to Africa's epidemic. At 17, Tsepho Phale has been head of an indigent household of three young boys in the dusty township of Monarch, outside Francistown, for two years. He never met his father, his mother died of AIDS, and the grieving children possess only a raw concrete shell of a house. The doorways have no doors; the window frames no glass. There is not a stick of furniture. The boys sleep on piled-up blankets, their few clothes dangling from nails. In the room that passes for a kitchen, two paraffin burners sit on the dirt floor alongside the month's food: four cabbages, a bag of oranges and one of potatoes, three sacks of flour, some yeast, two jars of oil

and two cartons of milk. Next to a dirty stack of plastic pans lies the mealy meal and rice that will provide their main sustenance for the month. A couple of bars of soap and two rolls of toilet paper also have to last the month. Tsepho has just brought these rations home from the social-service center where the "orphan grants" are doled out.

Tsepho has been robbed of a childhood that was grim even before his mother fell sick. She supported the family by "buying and selling things," he says, but she never earned more than a pittance. When his middle brother was knocked down by a car and left physically and mentally disabled, Tsepho's mother used the insurance money to build this house, so she would have one thing of value to leave her children. As the walls went up, she fell sick. Tsepho had to nurse her, bathe her, attend to her bodily functions, try to feed her. Her one fear as she lay dying was that her rural relatives would try to steal the house. She wrote a letter bequeathing it to her sons and bade Tsepho hide it.

As her body lay on the concrete floor awaiting burial, the relatives argued openly about how they would divide up the profits when they sold her dwelling. Tsepho gave the district commissioner's office the letter, preventing his mother's family from grabbing the house. Fine, said his relations; if you think you're a man, you look after your brothers. They have contributed nothing to the boys' welfare since. "It's as if we don't exist anymore either," says Tsepho. Now he struggles to keep house for the others, doing the cooking, cleaning, laundry and shopping.

The boys look at the future with despair. "It is very bleak," says Tsepho, kicking aimlessly at a bare wall. He had to quit school, has no job, will probably never get one. "I've given up my dreams. I have no hope."

Orphans have traditionally been cared for the African way: relatives absorb the children of the dead into their extended families. Some still try, but communities like Tsepho's are becoming saturated with orphans, and families can't afford to take on another kid, leaving thousands alone.

Now many must fend for themselves, struggling to survive. The trauma of losing parents is compounded by the burden of becoming a breadwinner. Most orphans sink into penury, drop out of school, suffer malnutrition, ostracism, psychic distress. Their makeshift households scramble to live on pitiful handouts—from overstretched relatives, a kind neighbor, a state grant—or they beg and steal in the streets. The orphans' present desperation forecloses a brighter future. "They hardly ever succeed in having a life," says Siphelile Kaseke, 22, a counselor at an AIDS orphans' camp near Bulawayo. Without education, girls fall into prostitution, and older boys migrate illegally to South Africa, leaving the younger ones to go on the streets.

1 IN 4 SOUTH AFRICAN WOMEN AGES 20 TO 29 IS INFECTED WITH HIV

EVERY DAY SPENT IN THIS PART OF AFRICA IS ACUTELY DEPRESSING: there is so little countervailing hope to all the stories of the dead and the doomed. "More than anywhere else in the world, AIDS in Africa was met with apathy," says Suzanne LeClerc-

Madlala, a lecturer at the University of Natal. The consequences of the silence march on: infection soars, stigma hardens, denial hastens death, and the chasm between knowledge and behavior widens. The present disaster could be dwarfed by the woes that loom if Africa's epidemic rages on. The human losses could wreck the region's frail economies, break down civil societies and incite political instability.

In the face of that, every day good people are doing good things. Like Dr. Moll, who uses his after-job time and his own fund raising to run an extensive volunteer home-care program in KwaZulu-Natal. And Busi Magwazi, who, along with dozens of others, tends the sick for nothing in the Durban-based Sinoziso project. And Patricia Bakwinya, who started her Shining Stars orphan-care program in Francistown with her own zeal and no money, to help youngsters like Tsepho Phale. And countless individuals who give their time and devotion to ease southern Africa's plight.

But these efforts can help only thousands; they cannot turn the tide. The region is caught in a double bind. Without treatment, those with HIV will sicken and die; without prevention, the spread of infection cannot be checked. Southern Africa has no other means available to break the vicious cycle, except to change everyone's sexual behavior—and that isn't happening.

The essential missing ingredient is leadership. Neither the countries of the region nor those of the wealthy world have been able or willing to provide it.

South Africa, comparatively well off, comparatively well educated, has blundered tragically for years. AIDS invaded just when apartheid ended, and a government absorbed in massive transition relegated the disease to a back page. An attempt at a national education campaign wasted millions on a farcical musical. The premature release of a local wonder drug ended in scandal when the drug turned out to be made of industrial solvent. Those fiascoes left the government skittish about embracing expensive programs, inspiring a 1998 decision not to provide AZT to HIV-positive pregnant women. Zimbabwe too suffers savagely from feckless leadership. Even in Botswana, where the will to act is gathering strength, the resources to follow through have to come from foreign hands.

AIDS' grip here is so pervasive and so complex that all societies—theirs and ours—must rally round to break it. These countries are too poor to doctor themselves. The drugs that could begin to break the cycle will not be available here until global pharmaceutical companies find ways to provide them inexpensively. The health-care systems required to prescribe and monitor complicated triple-cocktail regimens won't exist unless rich countries help foot the bill. If there is ever to be a vaccine, the West will have to finance its discovery and provide it to the poor. The cure for this epidemic is not national but international.

The deep silence that makes African leaders and societies want to deny the problem, the corruption and incompetence that render them helpless is something the West cannot fix. But the fact that they are poor is not. The wealthy world must help with its zeal and its cash if southern Africa is ever to be freed of the AIDS plague.

A UGANDAN TALE

Not Afraid to Speak Out

Major Rubaramira Ruranga knows something about fighting. During Idi Amin's reign of terror in Uganda in the 1970s, Ruranga worked as a spy for rebels fighting the dictator. After Amin's ouster, the military man studied political intelligence in Cuba before returning to find a new dictator at the helm and a blood war raging. Hoping for change, Ruranga supplied his old rebel friends with more secrets, this time from within the President's office. When he was discovered, he fled to the bush to "fight the struggle with guns."

The turmoil in Uganda was fueling the spread of another enemy—AIDS. Like many rebel soldiers, Ruranga was on the move constantly to avoid detection. "You never see your wife, and so you get to a new place and meet someone else," he says. "I had sex without protection with a few women." Doctors found he was HIV positive in 1989. "They told me I would die in two to three years, so I started preparing for when I was away. I told my kids, my wife. Worked on finishing the house for them. I gave up hope." But as he learned about AIDS, his attitude changed. After talking to American and European AIDS activists—some had lived with the disease for 15 years or more—"I realized I was not going to die in a few years. I was reborn, determined to live."

He began fighting again. After announcing his HIV status at a rally on World AIDS Day in 1993—an extraordinarily brave act in Africa, where few activists, let alone army officers, ever admit to having HIV—he set up a network for those living with HIV/AIDS in Uganda, "so that people had somewhere to go to talk to friends." And while Uganda has done more to slow the spread of AIDS than any other country—in some places the rate of infection has dropped by half—"we can always do better," says Ruranga. "Why are we able to buy guns and bullets to kill people and we are not able to buy drugs to save people?" The fight continues.

—By Simon Robinson/Kampala

The Seeds of Hunger

AIDS, drought and the king's denial of a problem raise the risk of famine in Swaziland

By Michael Grunwald
Washington Post Staff Writer

SIGWE, Swaziland

There is a time to reap and a time to sow, and now is the time to sow. The first rains have fallen on the parched lowlands, softening the soil for the plow. But Saraphina Simelane has no seeds to plant. Joseph Dlamini has seeds, but no money to hire oxen. Julia Gwebu has no seeds, no money, no oxen and no time; she spends her days in her thatched stone hut tending a daughter with AIDS.

Halfway through the planting season for maize, the traditional staple for this land-locked kingdom in southern Africa, hardly anyone is plowing, much less sowing. Droughts ravaged Swaziland's last two harvests, and relief agencies are handing out seeds to 12,000 subsistence farmers. But another 38,000 households have nothing to put in the ground, and while foreign aid organizations are feeding them for now, they can't reap what they don't sow.

"It looks like we're in for another disaster," says Ben Nsibandze, chairman of Swaziland's National Disaster Task Force. "It's becoming almost endemic. We're just hand-to-mouth, hand-to-mouth."

The World Food Program estimates that about one-fourth of the kingdom's 1.1 million citizens are now at risk of starvation. And aid groups report the worsening situation in tiny Swaziland—rated a middle-income country by the United Nations—looks mild compared to looming catastrophes in larger, poorer nations such as Angola, Malawi, Zambia and Zimbabwe in southern Africa, or Ethiopia and Eritrea in the Horn of Africa.

More than 30 million Africans are threatened by famine, and the situation is getting worse. In Zambia and Malawi, 70 percent of households have no seeds. In Zimbabwe, the figure is 94 percent. Prices are skyrocketing, oxen are too thin to haul plows, and an El Niño could bring a third year of drought.

The crisis has roots in bad weather, bad policies and bad economies, but it's the AIDS epidemic—which has slashed average life expectancy to 45 years or less in every southern African country but one—that has prompted humanitarian agencies to call this a "new variant famine." Simelane, 65, cares for five orphaned grandchildren. Dlamini, 21, is responsible for four orphaned siblings. Breadwinners are dying or growing too sick to work or are selling off farm supplies to pay for health care at a time when bread is in short supply. In the lingo of foreign aid, millions of potential drought victims already have "diminished coping abilities."

"The numbers are staggering," says Judith Lewis, the U.N. coordinator for southern Africa, a region where AIDS has flourished amid poverty, substandard health care and a culture that traditionally frowns more on discussion of sex than the practice of it. "When we look at the vulnerability—the world's highest malnutrition rates, the world's highest HIV rates—we're bracing for the worst."

Swaziland's people are not yet starving to death, thanks to emergency aid that began arriving in July. The nation is not reeling from decades of war, like Angola, or expelling many of its productive farmers, like Zimbabwe, or refusing to accept thousands of tons of genetically modified food, like Zambia. It will receive a scant fraction of the $500 million the United Nations hopes to spend in southern Africa through March.

BUT SWAZILAND PROVIDES A WINdow into Africa's unfolding food crisis. It is smaller than New Jersey, and hunger is limited to the rural south and east of the country. It is a functioning country, with well-paved roads, well-regarded schools and developed urban areas. It is Africa's only absolute monarchy, but it is a relatively transparent society. World Bank data show that since declaring independence in 1968, Swaziland has been less reliant on foreign aid than any other country in Africa.

Nsibandze says that Swaziland is now in danger of becoming a perpetual welfare state, "constantly appealing to our international friends." The country is producing less than one-third of its own food, and more than one-third of its adults are HIV-positive. The first rains this year were again a month late, and the second rains did little good because there were few seeds in the ground. Mostafa Imam, a Swazi who runs the U.N. Food and Agriculture Organization program here, swells with emotion as he points out the untilled fields and dusty pastures that dominate the landscape of the southern lowlands.

"I weep for my country," he says. "When you can plow, you can hope. Right now there is no hope."

Grace Mkhabela is in charge of distributing U.N. maize and seeds at the Sigwe police station, where hundreds of villagers come for help.

She is a manager for a local nonprofit group, Swaziland Farmer Development, and she decides who is hungry enough to eat and who is strong enough to plant.

Simon Mamba, 61, lost his entire crop last year. He has 20 mouths to feed, and he had to take three grandchildren out of school because he could not afford the fees. Since Mamba has a low-paying mining job, he is ineligible for food aid.

"We have to focus on the most vulnerable," Mkhabela explains. And seeds are being distributed only to food recipients because hungry people tend to eat seeds instead of plant them.

"I'm too rich for food but too poor for seeds!" Mamba said. "It's better for me to die."

Simelane is unemployed, and she is certainly poor, crammed in a dingy home with five orphans who sleep on her floor. The two youngest, Pati and Pendulili, have bellies distended from malnutrition. Their sweatpants have more holes than cloth; their shoes are caked with mud and ripped to shreds. Although Simelane is eligible for food aid, she does not get seeds. "We cannot serve the old and weak," Mkhabela says. "We must give seeds to strong people who will plant them with success."

Then there is Irene Dlamini, no relation to Joseph. She is only 40 years old but just as poor. She has 11 children who live on two spoonfuls of porridge a day. None of them go to school anymore. Yet she does not get seeds

Famine Threatens

Drought, floods and misgovernment have led to failed crops in southern Africa, putting more than 14 million people at risk of starvation, the World Food Program reports. High rates of AIDS have exacerbated the misery.

Countries most affected by famine

Southern African countries most at risk of famine:

Country and population (in millions)		Percentage of population likely in need of food aid by next month
Lesotho	2.2	29%
Malawi	10.5	31%
Mozambique	19.4	3%
Swaziland	1.1	24%
Zambia	9.9	29%
Zimbabwe	11.4	60%

SOURCES: WFP, UNAIDS, CIA

Countries that are among those with the highest HIV/AIDS rates:

	Percentage of people who are HIV infected or have AIDS	Life expectancy at birth, in years
Zimbabwe	33.7%	37
Swaziland	33.4	38
Lesotho	31	47
Zambia	21.6	37
South Africa	19.9	45
Malawi	16	37
Mozambique	13	35

THE WASHINGTON POST

either. Mkhabela could not say why Dlamini was unlucky. There just aren't enough seeds to go around.

"It's a very painful feeling," Dlamini says. "My children think I neglect them."

Even those who receive seeds are by no means assured of a crop. Joseph Dlamini, the young man trying to feed four orphans, has been unable to hire oxen or a tractor to plow his fields. And it may be too late for him to plant maize, which should be a foot high by now. Still, he is trying to build a fence around his bone-dry fields out of acacia branches, just in case. "It's all I can do," he says.

Drought is particularly lethal to maize crops, but maize is central to Swazi culture, so aid agencies have struggled to persuade Swazis to diversify into more drought-resistant crops. But the darkest shadow over Swaziland's nutrition problems is the AIDS epidemic.

It is rarely spoken by name; Gwebu, for example, says her daughter has "a blood problem." But it is hard not to notice that the vast majority of the villagers at the food lines in Sigwe are either children or seniors.

Health officials say life expectancy in Swaziland has fallen by 25 years since the AIDS epidemic began.

In the countryside, teenage Swazi girls are selling sex—and spreading HIV—for $5 an encounter, exactly what it costs to hire oxen for a day of plowing.

"People just can't cope," Lewis said. "This isn't sustainable."

The international community, led by the United States, has rushed in enough food to stave off famine in Swaziland this year, and by all accounts the food is being directed to people in need. But donors have been reluctant to invest in longer-term solutions—or even medium-term fixes such as seed kits, which cost $31—in part because Swaziland has failed to make those investments itself. Politics is also at the heart of this crisis.

"It's probably more important than weather," a Western diplomat says. "Make that definitely more important."

"It seems like we're losing our direction," Sibonelo Mngomezulu says. "We need to rechallenge our priorities. We have to think about our people and what they need. The king needs to be enlightened."

THOSE ARE BOLD WORDS IN SWAZIland, where political parties are banned and criticism of the popular king, Mswati III, can be tantamount to sedition. They are especially bold from Mngomezulu, who happens to be the third of the king's 10 wives. She is also one of his key advisers; many others, the queen says, are "selfish and corrupt," and she blames them for the global notoriety he attracted earlier this month after aides tried to intimidate judges into rejecting a lawsuit accusing him of abducting his 10th wife.

It is no secret what the queen means by priorities. A $900,000 proposal for emergency aid has languished for months; a $500,000 proposal to upgrade the royal fleet

of luxury cars was swiftly approved. The government has no irrigation projects for maize—only for sugar plantations controlled by the king. Rural farmers have no way to finance their own irrigation projects because the king holds title to their land. And even though the government has stopped giving away seeds to farmers, an on-again, off-again agreement to buy the king a jet for $60 million—twice the country's health budget—appears to be on again.

"Obviously, it would make better sense to spend money elsewhere," the queen says in an interview in her palace.

The king has not even declared a state of emergency, which would have released more foreign money to buy seeds in time for planting season.

He may consider it too much of an admission of failure, or he may not even realize the seriousness of the situation. At a recent ceremony, the U.S. ambassador, James McGee, showed the king photographs of hungry Swazis that the ambassador had taken. The king's response, according to a witness: "Oh, that's nice."

In January, Mswati will oversee Swaziland's most sacred ceremony, the annual Ncwala, and he will give his people permission to eat the year's first maize. "I just hope there is maize to eat," says Imam, the U.N. agriculture official. "We should not be a nation that depends on the world."

Lessons from The Fastest-Growing Nation: Botswana?

GDP in the diamond-rich economy has skyrocketed thanks to free-market principles and prudent management. And its AIDS offensive is showing Africa the way

By John Koppisch

Quick—name the country that's grown the fastest over the last few decades. It's not any of the four Asian Tigers, though they come close. It's not China—its economy didn't get going until the 1980s. Surely it wouldn't be any country in Africa, where every decade is a lost decade.

Wrong. The country posting the highest gross-domestic-product growth since 1966 is Botswana, a landlocked nation in southern Africa that's two-thirds desert. Its economy has expanded an average of 7% a year since it won independence from Britain that year, according to the World Bank. And it continues to expand—growth is projected at 5% this year—despite a rampant AIDS epidemic that is now the country's biggest challenge. On a continent where virtually every country is worse off than it was at independence, Botswana offers some lessons in economic management for its neighbors.

What accounts for Botswana's sterling achievement? The easy answer would be diamonds, discovered under the desert in 1967. And diamond mining is indeed the keystone of the economy. But odd as it may seem, natural resources like diamonds, gold, or oil are often more of a curse than a blessing. Examples are chronically unstable Venezuela (oil) and the war-torn former Zaire (gold and diamonds). The better answer is that this peaceful democracy has largely stuck to free-market principles, even when its vast diamond wealth could have led it astray. Botswana keeps taxes low. It respects property rights and hasn't nationalized any businesses. It avoids the corrup-

tion that easy money encourages, and doesn't waste much money on grandiose air forces or white-elephant industrial projects. Indeed, Botswana is rated Africa's freest economy by the Washington-based Cato Institute in its annual report, *Economic Freedom of the World*.

Back in 1966, Botswana wasn't a model for anything. It boasted just three-and-a-half miles of paved roads and only three high schools in a country of 550,000 people. The per-capita income was $80 a year. Water was so scarce, and valuable, that the currency would later be named the pula, meaning rain.

Today Botswana is one of Africa's few enclaves of prosperity. Per-capita income has jumped to $6,600. In a continent of weak currencies, the pula is strong, backed by one of the world's highest per-capita reserves ($6.2 billion). Budget surpluses are the norm, and Moody's Investors Service rates Botswana's government bonds higher than Japan's. Botswana's world-class game parks are so free of poachers that they must cull their elephant herds. As a result the government is a leader in the fight to expand legal ivory trading.

Any developing country could have adopted Botswana's sensible policies. But in Africa, most newly independent nations pursued the fashion of the day—socialist or nationalist agendas that called for a strongman at the head of a one-party state, state-owned industries, barriers to foreign investment and trade, and, inevitably, lots of debt. But Botswana was so poor there was nothing

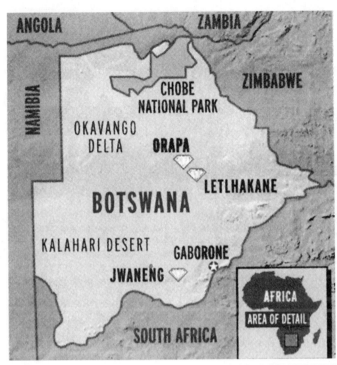

MAP BY ROGER KENNY/BW

U.N. "The one man every civil servant fears is the auditor-general."

Can Botswana keep it up? That depends on how well it meets two challenges. First, the government has become far too big, critics say, consuming half of the country's GDP, up from just 21% in 1970. So while Botswana may have the freest economy in Africa, it ranks only 38th overall in Cato's study, on a par with famously overregulated France and South Korea. "More and more of the diamond money is going into expanding the government," says Eustace Davie, director of the Johannesburg-based Free Market Foundation. "I'm very concerned that in the long run, Botswana won't be able to avoid the curse-of-resources problem, as more and more politicians, bureaucrats, and businessmen hold out their hands for government money." Ambassador Dube worries, too. "I honestly pray we never succumb to that temptation; if we do, the legacy for our children is gone."

That problem, however, pales in comparison to the catastrophe of AIDS. Fully 38.5% of the country's 15-to-40-year-olds are HIV-positive. AIDS is killing so many young people that the average life expectancy dropped from 65 in 1991 to just 39 last year. But here again, Botswana's culture of prudent management is crucial. It's mounting perhaps the most aggressive counterattack in the developing world, including a massive education campaign and a crash program of upgrading and building clinics and hospitals. Already the infection rate for 15-18-year-olds is starting to decline. The goal is to "get the problem fully under control by 2016," says Dube. As in economics, Botswana's AIDS offensive is a model for Africa. If only the rest of Africa would take note.

to nationalize and no foreign investment to block. And not having any money meant keeping strict controls on government spending. When diamonds were discovered, a culture of prudent management was already in place, so the money didn't end up in Swiss bank accounts. "The Botswana people frown on profligate spending," says Alfred Dube, the nation's permanent representative to the

Reprinted with special permission from *Business Week*, August 26, 2002, pp. 116, 118. © 2002 by The McGraw-Hill Companies, Inc.

Makeshift 'Cuisinart' Makes a Lot Possible In Impoverished Mali

It Can Do Work in a Flash, Leaving Time for Literacy And Entrepreneurship

BY ROGER THUROW

SANANKORONI, Mali—"Thump-thump-thump" is the trademark sound of the African bush. It is the dreary rhythm of village women pounding grains and nuts into breakfast, lunch and dinner with their heavy wooden pestles.

But in this village of simple mud-brick huts, the melody of daily life goes "chug-chug-chug."

"Isn't it wonderful?" marveled Biutou Doumbia, talking above the din of a diesel engine kicking into high gear. Balancing a baby on her back and cradling a large sack of peanuts in her arms, she approached a contraption that looks to have sprung from a Rube Goldberg blueprint—a most unlikely weapon in this country's war on poverty.

After paying the equivalent of 25 cents for machine time, she emptied her 15 pounds of peanuts into a funnel leading to a grinder and blender connected to another funnel, and an ooze of thick peanut butter emerged from its spout. The job was finished in 10 minutes. All that was left for Mrs. Doumbia was to scoop the peanut butter into a dozen jars and sell it on the market. Then, she said with a laugh, she might take a nap. "Before, it would take a whole day to pound and grind the peanuts by hand, and the butter still wouldn't be as fine as this."

Not only is the peanut butter better—and Mrs. Doumbia's selling easier—so is the quality of life in the 300 Mali villages that have the machine. Girls who were kept home to help with the domestic work from dawn to dusk are now going to school. Mothers and grandmothers who would have spent a lifetime pounding and grinding now have the free time to take literacy courses and start up small businesses, or to expand family farming plots and nurture a cash crop such as rice.

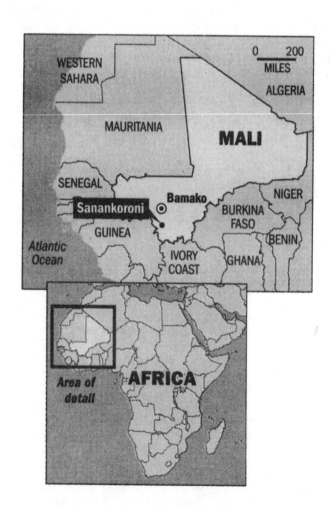

They have dubbed the durable, uncomplaining machine "the daughter-in-law who doesn't speak."

"It's changing our lives," said Mineta Keita, the 46-year-old president of the Sanankoroni women's association, which manages the machine and the flourishing business that has sprouted around it. Before it arrived a year ago, only nine women in this village of 460 people were able to read and write. Since then, she said, more than 40 have attended literacy courses. The training to prepare the women to manage the machine usually takes four to six months, and it gives them the basics in reading, writing and arithmetic. Most then continue with other courses to get better and better.

Known blandly as the "multifunctional platform" in United Nations parlance, the contraption was invented in the mid-1990s by a Swiss development worker in Mali who believed that easing the domestic load of African women would unleash their entrepreneurial zeal. The machine, simple and sturdy, was tailored for rural Africa.

A 10-horsepower motor is the centerpiece, sitting on two metal rails about 9 or 10 feet long, anchored to the floor of a small mud-brick shed. Rubber belts connect the motor to various tools: funnels that channel grain and nuts into grinders, whirring blenders that husk rice, pistons that pump water, saws that cut wood, cables that recharge batteries. It is an industrial-sized Cuisinart.

"It's not just about milling and grinding," says Laurent Coche, a Frenchman who has been deploying the machine in Mali for the U.N. Development Program and is now introducing it to neighboring countries. "The biggest impact has been to empower women."

The UNDP insists that the women who use the machine also manage it. Once the women's association in a village can scrape together about 50% of the machine's $4,000 cost, the U.N. and other donors kick in the rest. The Mali government, one of the poorest in the world, would like to see one machine in every village, and it is funneling some of its savings from international debt relief into the project.

Farma Traore, a real daughter-in-law, remembers that it used to take "three whole days" to manually grind a 100-pound bag of corn. "It's unthinkable that we would even do that anymore," she says. The machine does the job in 15 minutes.

Her brother-in-law, Sekou Traore, leaned back in a chair outside his one-room house and smiled. "Our wives aren't so tired anymore," he said. "And their hands are smoother. We like that."

Mr. Traore and several of his brothers had just returned from the fields where they cultivate their crops by hand. One of their wives served up lunch: a big bowl of rice with spicy peanut sauce. Since the women don't spend all day wielding the pestles anymore, the men say, meals are rarely late and families are spending more time together. "We're eating on time," said Mr. Traore. "There's fewer arguments."

Still, the social changes take some getting used to. "Working for women isn't an easy thing. They talk too much and are bossy," said Lassine Traore, a 19-year-old relative who has been trained to maintain the machine. He warily glanced behind his back and tended to a balky fuel injector. "But I'm happy to have this job. It beats farming."

Inside the shed housing the machine, the women's new literacy skills were on display. Two big blackboards hanging on the walls presented a full accounting of the operation. One board gave a daily reckoning of when the machine was turned on and off, what tasks it performed, how much fuel was consumed and how much money was earned. The second board listed who worked, for how long and how much they were paid for their labors. The workers—usually several women and the maintenance man—share 30% of the day's revenues. On a particularly active day, the machine may take in $10 to $15.

In nine months of operation, through March of this year, the Sanankoroni machine took in about $1,600. Of that, the women's committee paid out about $500 in salaries to the workers who rotate on part-time shifts. The committee has also managed to build up bank savings in a city nearby of more than $200 and cash reserves of $180 to cover operating expenses. That is big money in a land where average annual per capita income is less than $300, and it is nurturing even bigger ambitions. "We would like to branch out into other businesses, like dyeing clothes and making soap," said Ms. Keita, the committee president. "And we would like to dig a well to get clean water."

This past spring, in the village of Mountougoula, just outside the capital of Bamako, the women raised additional money to connect a generator to the machine and rigged up a lighting network. For the first time ever, the village of 1,580 had lights, with 280 bulbs burning brightly from dusk to midnight.

"The dark is gone," said a wide-eyed Tieoule Dembele, the village secretary. As the lights came on one recent night, he finished a bottle of Coca-Cola at an outdoor bar and sauntered back to the one-room city hall to continue his paperwork. A bare bulb shone above his desk where once hung a kerosene lamp. "We do work for 16 villages in the area," he said, "and I can't get it all done during the day."

At the maternity clinic, where 200 babies are born each year, the midwife reports healthier births under the lights. Across the dirt road, the proprietor of the general store said nightly sales are up $25 since the bulb above his counter began burning.

The chug-chug-chug of the daughter-in-law who doesn't speak pierced the stillness of the night. Soleba Doumbia, the machine's mechanic in Mountougoula, closed the door of the shed and headed home to his own bulb. "Every night, I'm teaching my two daughters how to read," he said.

One day, he figures, he may be working for them.

From *Wall Street Journal*, July 26, 2002, pp. A1, A6, by Roger Thurow. © 2002 by Dow Jones & Company, Inc. Reproduced with permission of Dow Jones & Company, Inc.

MANAGING AFRICA'S RESOURCES

Oil, Diamonds and Death

Diamonds, oil and other natural resources fuel conflict—especially in Africa. So goes one view of the situation. But is the relationship more complex and can multinational companies help stop the killing?

Alex Vines

IN THE LAST FEW YEARS THERE HAS BEEN A GROWING FOCUS on the links between natural resources and conflict, especially in Africa. The United Nations has set up a panel of experts looking at the use of natural resources and conflict in the Congo; policy institutes including the International Institute of Strategic Studies in London, the International Peace Information Service in Antwerp and the International Peace Academy in New York have commissioned studies; and non-governmental organisations such as the UK's Global Witness have campaigned on oil, diamonds and timber extraction. The World Bank's Paul Collier has been a keen advocate of the idea of resources driving disputes.

Although there is more scrutiny and funding for such studies, there is nothing new or even exclusive about natural resources fuelling conflicts on the continent. The nineteenth century scramble for Africa by the European powers included the search for cheap labour to exploit the natural resources. Africa's importance as a supplier of primary commodities continues to be one of its major attractions.

During the Cold War, parts of Africa were seen as sources of strategic minerals for the west that had to be protected from Soviet expansionism. Apartheid South Africa played on this with the proxy wars in Angola and Mozambique. Throughout the Cold War, African states found themselves wooed by the west, the Soviet Union and the Warsaw Pact and China as part of this larger struggle.

In Angola and Mozambique, Marxist-Leninist regimes came to power at independence in 1975. Protracted civil wars followed. Angola benefited from Soviet weapons and military advisers but supplied oil to Cuba and the Soviet Union in exchange. US and South African support of proxies such as the rebel UNITA fighters were also far from altruistic. Not only was there military support and training, but South African Defence Force officials made private fortunes from precious woods, ivory and diamonds.

Interestingly, Angolans who lived in Jamba, UNITA's Cold War headquarters, remember with fondness the mid-1980s. Despite the authoritarianism of the rebels, it was bloated with covert food and health aid. In this period the rebels conducted some successful hearts and minds operations. By the 1990s the aid had dried up and the rebels had to become more predatory and human rights abuse widespread.

CONTROLLING ASSETS

The US academic William Reno was one of the first to write about this post-Cold War change in behaviour. He noticed that non-state actors now needed to secure control of economic assets to survive. UNITA in Angola, the Revolutionary United Front (RUF) in Sierra Leone and the Ugandan and Rwandan proxies in the Congo all fall under this category. Two UN reports on the Congo show that, like Liberia's involvement with the RUF in Sierra Leone, Rwanda and Uganda have benefited economically from their engagement in the Congo: coffee, timber, alluvial gold, coltan or diamonds.

In Angola, access to oil and diamonds might explain how the government and rebels could afford to buy plentiful supplies of surplus Eastern European weaponry, but there are many other factors involved in the conflict itself. Resource management is only part of it—there are political, social and ethnic aspects too.

Take the insecurity in the Mano River Union—Guinea, Sierra Leone and Liberia. The war in Liberia's Lofa County is intensified by hoards of desperados fighting for rebel or government militias and security forces. They get a couple of hundred US dollars down payment and all they can loot. Many of the fighters are poor youngsters either seeking riches or forcibly conscripted; a hardened small core of professional fighters is known for gross human rights abuses. They fear the setting up of a UN Criminal Court in Sierra Leone or the erosion of their status by peace.

BUYING SECURITY

The growing tendency to focus on one commodity as a root cause or contributor to these conflicts is often misplaced. Mozambique is not mineral rich. During its civil war, government and rebel forces looted. The rebels sold some timber, semi-precious stones and ivory but the main revenue source was protection money from companies such as Lonrho, and the Cahora Bassa hydroelectric plant.

In June 1982, a Lonrho subsidiary signed a secret protection agreement with Renamo rebel leaders covering the Beira oil pipeline. It stipulated that payments of $500,000 would be made to Renamo each month from June to August, to be continued indefinitely thereafter, unless either party gave one month's notice to terminate the arrangement. These payments were made into foreign bank accounts controlled by a number of Renamo's senior internal leadership. Later, additional agreements were reached covering Lonrho's tea estates in northern Mozambique. These continued up to the signing of the General Peace Agreement in Rome in October 1992 and amounted to some $5 million.

Extortion for not attacking economic assets is not unique to the Mozambique's conflict. This also occurs in the delta in Nigeria and in Angola. In Angola's rich oil enclave of Cabinda, separatists have specialised in the abduction of expatriate oil, construction and timber workers. Over the last decade this has made Cabinda one of the most hazardous working environments outside Colombia. Separatists levy a tax that, if promptly paid, guarantees against kidnapping.

The truth is that rebels and governments in Africa will use all available resources to further their causes. A country or region does not need to be mineral rich to sustain conflict. Nor will a solution be found by targeting one commodity. Sierra Leone has one of the most diamond-dependent economies in the world. Since the late 1970s the diamond trade has become dysfunctional and dominated by corrupt practice and poor policy. By the 1990s diamonds contributed more to the destabilisation of the country and its collapse into civil war than to its wealth.

Private military companies such as Executive Outcomes and Sandline International gladly provided ring-fenced security in mineral rich areas, allowing a respite from conflict. They were, however, unable to couple it with the longer-term investment in poverty reduction and security sector reform that bilateral and multilateral assistance can.

Although the majority of the easily exploited diamond resources have been worked out, there still remain significant reserves, particularly the Kimberlite deposits of Tongo and Kono. Diamonds could generate $180 million by 2006, $40–80 million of this from the artisan sector.

CORRUPTION

The key challenge for the Sierra Leone diamond industry is endemic and high-level corruption. The Anti-Corruption Commission in Freetown recently moved against one senior government official, but there are many more widely talked about offenders. Sierra Leone is not unique, and corruption is not, of course, confined to the diamond industry, but it is in that industry that it has its 'most deleterious effect,' according to a study on Sierra Leone commissioned by the Department for International Development.

Initiatives like the Kimberly diamond process are welcome developments that add pressure for greater transparency on the origin and transfers of diamonds from southern and western Africa. The Certificate of Origins, however, cannot guarantee 'conflict-free' diamonds. There are regular reports of mixed parcels of gems from Angola, Congo and Sierra Leone.

Best mining and trading policies can help. Multinational mining and oil companies can at least ensure they are not harming the environment. But they also need to take an interest in what happens to the revenue they send to governments. In Angola one bonus payment on signing the contract with BP for an offshore oil block was earmarked for 'defence', according to the Angolan Minister of Foreign Affairs. Defence included the continued purchase of antipersonnel mines, a weapon the British government has banned.

TRANSPARENCY

The debate over the management of Angola's oil wealth continues. An oil diagnostic study by KPMG for the World Bank and the Angolan government has run into trouble over transparency. The government appears highly reluctant to show just how much oil money is diverted to off-budget expenditure.

A recent UN report on Liberia showed funds from its flag of convenience, the US-based Liberia International Shipping and Corporate Registry (LISCR), being diverted directly for off-budget payments involving sanctions busting arms and transport. The company denied any wrongdoing, claiming its role was not to police what the government did with its money—LISCR was only the agent.

Happily, there are now attempts to set up an independent audit of what happens to these funds in Liberia so that they at least do not get used again for nefarious off-budget expenditures. LISCR knows that if this is not done, clients may transfer to other shipping registries: commercial self-interest in not losing business is a strong incentive.

Clearly, greater scrutiny on how revenues are generated and organised in Africa is a good thing. The continent needs investment, and well-managed and transparent commercial operations are important. Best mining and trading policies can help. So can independent auditing and whistle-blowing by civil society, and quality investigative journalism into bad practice. Corruption takes two to tango. The more companies that refuse to dance, the better the chances of meaningful economic progress in Africa.

Alex Vines is an Associate Fellow of Chatham House. He was a member of the UN panel of experts on Liberia.

This article first appeared in the March 2002 issue of *The World Today*, pp. 19-20, published by the Royal Institute of International Affairs in London. www.theworldtoday.org Reprinted by permission.

Glossary of Terms and Abbreviations

Acquired Immune Deficiency Syndrome (AIDS) A disease of immune system dysfunction widely believed to be caused by the human immunodeficiency virus (HIV), which allows opportunistic infections to take over the body.

African Development Bank Founded in 1963 under the auspices of the United Nations Economic Commission on Africa, the bank, located in Côte d'Ivoire, makes loans to African countries, although other nations can apply.

African National Congress (ANC) Founded in 1912, the group's goal is to achieve equal rights for blacks in South Africa through nonviolent action. "Spear of the Nation," the ANC wing dedicated to armed struggle, was organized after the Sharpeville massacre in 1960.

African Party for the Independence of Guinea-Bissau and Cape Verde (PAICG) An independence movement that fought during the 1960s and 1970s for the liberation of present-day Guinea-Bissau and Cape Verde from Portuguese rule. The two territories were ruled separately by a united PAIGC until a 1981 coup in Guinea-Bissau caused the party to split along national lines. In 1981 the Cape Verdean PAIGC formally renounced its Guinea links and became the PAICV.

African Socialism A term applied to a variety of ideas (including those of Nkrumah and Senghor) about communal and shared production in Africa's past and present. The concept of African socialism was especially popular in the early 1960s. Adherence to it has not meant governments' exclusion of private-capitalist ventures.

Afrikaners South Africans of European descent who speak Afrikaans and are often referred to as *Boers* (Afrikaans for "farmers").

Algiers Agreement The 1979 peace agreement when Mauritania made peace with the Polisario and abandoned claims to Western Sahara.

Aouzou Strip A barren strip of land between Libya and Chad contested by both countries.

Apartheid Literally, "separatehood," a South African policy that segregated the races socially, legally, and politically.

Arusha Declaration A document issued in 1967 by Tanzanian President Julius Nyerere, committing the country to socialism based on peasant farming, democracy under one party, and self-reliance.

Assimilado The Portuguese term for Africans who became "assimilated" to Western ways. Assimilados enjoyed equal rights under Portuguese law.

Azanian People's Organization (AZAPO) Founded in 1978 at the time of the Black Consciousness Movement and revitalized in the 1980s, the movement works to develop chapters and bring together black organizations in a national forum.

Bantu A major linguistic classification for many Central, Southern, and East African languages.

Bantustans Areas, or "homelands," to which black South Africans were assigned "citizenship" as part of the policy of apartheid.

Basarawa Peoples of Botswana who have historically been hunters and gatherers.

Berber The collective term for the indigenous languages and peoples of North Africa.

Black Consciousness Movement A South African student movement founded by Steve Biko and others in the 1970s to promote pride and empowerment of blacks.

Boers See *Afrikaners*.

Brotherhoods Islamic organizations based on specific religious beliefs and practices. In many areas, brotherhood leaders and their spiritual followers gain political influence.

Cabinda A small, oil-rich portion of Angola separated from the main body of that country by a coastal strip of the Democratic Republic of the Congo.

Caliphate The office or dominion of a caliph, the spiritual head of Islam.

Cassava A tropical plant with a fleshy, edible rootstock; one of the staples of the African diet. Also known as manioc.

Chimurenga A Shona term meaning "fighting in which everyone joins," used to refer to Zimbabwe's fight for independence.

Committee for the Struggle against Drought in the Sahel (CILSS) A grouping of eight West African countries, formed to fight the effects of drought in the region.

Commonwealth of Nations An association of nations and dependencies loosely joined by the common tie of having been part of the British Empire.

Congress of South African Trade Unions (COSATU) Established in 1985 to form a coalition of trade unions to press for workers' rights and an end to apartheid.

Copperbelt A section of Zambia with a high concentration of copper-mining concessions.

Creole A person or language of mixed African and European descent.

Dergue From the Amheric word for "committee," the ruling body of Ethiopia following the Revolution in 1974 to the 1991 Revolution (it was overthrown by the Ethiopian People's Revolutionary Democratic Front).

East African Community (EAC) Established in 1967, this organization grew out of the East African Common Services Organization begun under British rule. The EAC included Kenya, Tanzania, and Uganda in a customs union and involved common currency and development of infrastructure. It was disbanded in 1977, and the final division of assets was completed in 1983.

Economic Commission for Africa (ECA) Founded in 1958 by the Economic and Social Committee of the United Nations to aid African development through regional centers, field agents, and the encouragement of regional efforts, food self-sufficiency, transport, and communications development.

Economic Community of Central African States (CEEAC, also known as ECCA) An organization of all of the Central African states, as well as Rwanda and Burundi, whose goal is to promote economic and social cooperation among its members.

Economic Community of West Africa (CEAO) An economic organization of former French colonies that was formed to promote trade and regional economic cooperation.

Economic Organization of West African States (ECOWAS) Established in 1975 by the Treaty of Lagos, the organization includes all of the West African states except Western Sahara. The organization's goals are to promote trade, cooperation, and self-reliance among its members.

Enclave Industry An industry run by a foreign company that uses imported technology and machinery and exports the product to industrialized countries; often described as a "state within a state."

Eritrean People's Liberation Front (EPLF) The major group fighting the Ethiopian government for the independence of Eritrea.

European Community See *European Union.*

European Union (EU) Known as the European Community until 1994, this is the collective designation of three organizations with common membership—the European Economic Community, the European Coal and Steel Community, and the European Atomic Energy Community. Sometimes also referred to as the Common Market.

Evolués A term used in colonial Zaire (the Democratic Republic of the Congo) to refer to Western-educated Congolese.

Fokonolas Indigenous village management bodies.

Food and Agricultural Organization of the United Nations (FAO) Established in 1945 to oversee good nutrition and agricultural development. **Franc Zone** (Commonly known as the CFA [*le franc des Colonies Françaises d'Afrique*] franc zone.) This organization includes members of the West African Monetary Union and the monetary organizations of Central Africa that have currencies linked to the French franc. Reserves are managed by the French treasury and guaranteed by the French franc.

Free French Conference A 1944 conference of French-speaking territories, which proposed a union of all the territories in which Africans would be represented and their development furthered.

Freedom Charter Established in 1955, this charter proclaimed equal rights for all South Africans and has been a foundation for almost all groups in the resistance against apartheid.

Frelimo See *Mozambique Liberation Front.*

French Equatorial Africa (FEA) The French colonial federation that included present-day Democratic Republic of the Congo, Central African Republic, Chad, and Gabon.

French West Africa The administrative division of the former French colonial empire that included the current independent countries of Senegal, Côte d'Ivoire, Guinea, Mali, Niger, Burkina Faso, Benin, and Mauritania.

Frontline States A caucus supported by the Organization of African Unity (consisting of Tanzania, Zambia, Mozambique, Botswana, Zimbabwe, and Angola) whose goal is to achieve black majority rule in all of Southern Africa.

Green Revolution Use of Western technology and agricultural practices to increase food production and agricultural yields.

Griots Professional bards of West Africa, some of whom tell history and are accompanied by the playing of the kora or harp-lute.

Gross Domestic Product (GDP) The value of production attributable to the factors of production in a given country regardless of their ownership. GDP equals GNP minus the product of a country's residents originating in the rest of the world.

Gross National Product (GNP) The sum of the values of all goods and services produced by a country's residents at home and abroad in any given year, less income earned by foreign residents and remitted abroad.

Guerrilla A member of a small force of irregular soldiers. Generally, guerrilla forces are made up of volunteers who make surprise raids against the incumbent military or political force.

Harmattan In West Africa, the dry wind that blows in from the Sahara Desert during January and February, which now reaches many parts of the West African coast. Its dust and haze are a sign of the new year and of new agricultural problems.

Homelands See *Bantustans.*

Horn of Africa A section of northeastern Africa including the countries of Djibouti, Ethiopia, Somalia, and the Sudan.

Hut Tax Instituted by the colonial governments in Africa, this measure required families to pay taxes on each building in the village.

International Monetary Fund (IMF) Established in 1945 to promote international monetary cooperation.

Irredentism An effort to unite certain people and territory in one state with another, on the grounds that they belong together.

Islam A religious faith started in Arabia during the seventh century A.D. by the Prophet Muhammad and spread in Africa through African Muslim leaders, migrations, and wars.

Jihad A struggle, or "holy war," waged as a religious duty on behalf of Islam to rid the world of disbelief and error.

Koran Writings accepted by Muslims as the word of God, as revealed to the Prophet Mohammed.

Lagos Plan of Action Adopted by the Organization of African Unity in 1980, this agreement calls for self-reliance, regional economic cooperation, and the creation of a pan-African economic community and common market by the year 2000.

League of Nations Established at the Paris Peace Conference in 1919, this forerunner of the modern-day United Nations had 52 member nations at its peak (the United States never joined the organization) and mediated in international affairs. The league was dissolved in 1945 after the creation of the United Nations.

Least Developed Countries (LDCs) A term used to refer to the poorest countries of the world, including many African countries.

Maghrib An Arabic term, meaning "land of the setting sun," that is often used to refer to the former French colonies of Morocco, Algeria, and Tunisia.

Mahdi The expected messiah of Islamic tradition; or a Muslim leader who plays a messianic role.

Malinke (Mandinka, or Mandinga) One of the major groups of people speaking Mande languages. The original homeland of the Malinke was Mali, but the people are now found in Mali, Guinea-Bissau, The Gambia, and other areas, where they are sometimes called Mandingoes. Some trading groups are called Dyoula.

Marabout A dervish Muslim in Africa believed to have supernatural power.

Marxist-Leninism Sometimes called "scientific socialism," this doctrine derived from the ideas of Karl Marx as modified by Vladimir Lenin; it was the ideology of the Communist Party of the Soviet Union and has been modified in many ways by other persons and groups who still use the term. In Africa, some political parties or movements have claimed to be Marxist-Leninist but have often followed policies that conflict in practice with the ideology; these governments have usually not stressed Marx's philosophy of class struggle.

Mfecane The movement of people in the nineteenth century in the eastern areas of present-day South Africa to the west and north as the result of wars led by the Zulus.

Movement for the Liberation of Angola (MPLA) A major Angolan liberation movement that has its strongest following among assimilados and Kimbundu speakers, who are predominant in Luanda, the capital, and the interior to the west of the city.

Mozambique Liberation Front (Frelimo) Mozambique's single ruling party following a 10-year struggle against Portuguese colonial rule, which ended in 1974.

Mozambique National Resistance See *Renamo.*

Muslim A follower of the Islamic faith.

Glossary of Terms and Abbreviations

Naam A traditional work cooperative in Burkina Faso.

National Front for the Liberation of Angola (FNLA) One of the major Angolan liberation movements; its original focus was limited to the northern Kongo-speaking population.

National Union for the Total Independence of Angola (UNITA) One of three groups that fought the Portuguese during the colonial period in Angola, later backed by South Africa and the U.S. CIA in fighting the independent government of Angola.

National Youth Service Service to the state required of youth after completing education, a common practice in many African countries.

Nkomati Accords An agreement signed in 1984 between South Africa and Mozambique, pledging that both sides would no longer support opponents of the other.

Nonaligned Movement (NAM) A group of nations that chose not to be politically or militarily associated with either the West or the former communist bloc.

Nongovernmental Organizations (NGO) A private voluntary organization or agency working in relief and development programs.

Organization for the Development of the Senegal River (OMVS) A regional grouping of countries bordering the Senegal River that sponsors joint research and projects.

Organization of African Unity (OAU) An association of all the independent states of Africa (except South Africa) whose goal is to promote peace and security as well as economic and social development.

Organization of Petroleum Exporting Countries (OPEC) Established in 1960, this association of some of the world's major oil-producing countries seeks to coordinate the petroleum policies of its members.

Pan Africanist Congress (PAC) A liberation organization of black South Africans that broke away from the ANC in the 1950s.

Parastatals Agencies for production or public service that are established by law and that are, in some measure, government organized and controlled. Private enterprise may be involved, and the management of the parastatal may be in private hands.

Pastoralist A person, usually a nomad, who raises livestock for a living.

Polisario Front Originally a liberation group in Western Sahara seeking independence from Spanish rule. Today, it is battling Morocco, which claims control over the Western Sahara. See Saharawi Arab Democratic Republic (SADR).

Popular Movement for the Liberation of Angola (MPLA) A Marxist liberation movement in Angola during the resistance to Portuguese rule; now the governing party in Angola.

Renamo A South African-backed rebel movement that attacked civilians in an attempt to overthrow the government of Mozambique.

Rinderpest A cattle disease that periodically decimates herds in savanna regions.

Saharawi Arab Democratic Republic (SADR) The Polisario Front name for Western Sahara, declared in 1976 in the struggle for independence from Morocco.

Sahel In West Africa, the borderlands between savanna and desert.

Sanctions Coercive measures, usually economic, adopted by nations acting together against a nation violating international law.

Savanna Tropical or subtropical grassland with scattered trees and undergrowth.

Shari'a The Islamic code of law.

Sharpeville Massacre The 1960 demonstration in South Africa in which 60 people were killed when police fired into the crowd; it became a rallying point for many antiapartheid forces.

Sorghum A tropical grain that is a traditional staple in the savanna regions.

Southern African Development Community (SADC) (Formerly the Southern African Development Coordination Conference. Its name was changed in 1992.) An organization of nine African states (Angola, Zambia, Malawi, Mozambique, Zimbabwe, Lesotho, Botswana, Swaziland, and Tanzania) whose goal is to free themselves from dependence on South Africa and to cooperate on projects of economic development.

South-West Africa People's Organization (SWAPO) Angola-based freedom fighters who had been waging guerrilla warfare against the presence of South Africa in Namibia since the 1960s. The United Nations and the Organization of African Unity now recognize SWAPO as the only authentic representative of the Namibian people.

Structural Adjustment Program (SAP) Economic reforms encouraged by the International Monetary Fund, which include devaluation of currency, cutting government subsidies on commodities, and reducing government expenditures.

Swahili A trade and government Bantu language that covers much of East Africa and Congo region.

Tsetse Fly An insect that transmits sleeping sickness to cattle and humans. It is usually found in the scrub-tree and forest regions of Central Africa.

Ujaama In Swahili, "familyhood"; government-sponsored cooperative villages in Tanzania.

Unicameral A political structure with a single legislative branch.

Unilateral Declaration of Independence (UDI) A declaration of white minority settlers in Rhodesia, claiming independence from the United Kingdom in 1965.

United Democratic Front (UDF) A multiracial, black-led group in South Africa that gained prominence during the 1983 campaign to defeat the government's Constitution, which gave only limited political rights to Asians and Coloureds.

United Nations (UN) An international organization established on June 26, 1945, through official approval of the charter by delegates of 50 nations at a conference in San Francisco, California. The charter went into effect on October 24, 1945.

United Nations Development Program (UNDP) Established to create local organizations for increasing wealth through better use of human and natural resources.

United Nations Educational, Scientific, and Cultural Organization (UNESCO) Established on November 4, 1946, to promote international collaboration in education, science, and culture.

United Nations High Commission for Refugees (UNHCR) Established in 1951 to provide international protection for people with refugee status.

Villagization A policy whereby a government relocates rural dwellers to create newer, more concentrated communities.

West African Monetary Union (WAMU) A regional association of member countries in West Africa (Benin, Burkina Faso, Côte d'Ivoire, Mali, Niger, Senegal, and Togo) that have vested authority to conduct monetary policy in a common central bank.

World Bank A closely integrated group of international institutions providing financial and technical assistance to developing countries.

World Health Organization (WHO) Established by the United Nations in 1948, this organization promotes the highest possible state of health in countries throughout the world.

Bibliography

RESOURCE CENTERS

African Studies Centers provide special services for schools, libraries, and community groups. Contact the center nearest you for further information about resources available.

African Studies Center
Boston University
270 Bay State Road
Boston, MA 02215

African Studies Program
Indiana University
Woodburn Hall 221
Bloomington, IN 47405

African Studies Educational Resource Center
100 International Center
Michigan State University
East Lansing, MI 49923

African Studies Program
630 Dartmouth
Northwestern University
Evanston, IL 60201

Africa Project
Lou Henry Hoover Room 223
Stanford University
Stanford, CA 94305

African Studies Center
University of California
Los Angeles, CA 90024

Center for African Studies
470 Grinter Hall
University of Florida
Gainesville, FL 32611

African Studies Program
University of Illinois
1208 W. California, Room 101
Urbana, IL 61801

African Studies Program
1450 Van Hise Hall
University of Wisconsin
Madison, WI 53706

Council on African Studies
Yale University
New Haven, CT 06520

Foreign Area Studies
The American University
5010 Wisconsin Avenue, NW
Washington, DC 20016

African Studies Program
Center for Strategic and International Studies
Georgetown University
1800 K Street, NW
Washington, DC 20006

REFERENCE WORKS, BIBLIOGRAPHIES, AND OTHER SOURCES

Africa Research Bulletin (Political Series), Africa Research Ltd., Exeter, Devon, England (monthly).

Africa South of the Sahara (updated yearly) (Detroit: Gale Research).

Africa Today: An Atlas of Reproductible Pages, rev. ed. (Wellesley: World Eagle, 1990).

Rosalid Baucham, *African-American Organizations: A Selective Bibliography* (Organizations and Institutional Groups) (New York: Garland, 1997).

Chris Cook and David Killingray, *African Political Facts Since 1945* (New York: Facts on File, 1990).

David E. Gardinier, *Africana Journal* notes, Volume xvii. A Bibliographic Library Guide and Review Forum (New York: Holmes & Meier, 1997).

Colin Legum, ed., *Africa Contemporary Record* (New York: Holmes & Meier) (annual).

MAGAZINES AND PERIODICALS

Africa News, P.O. Box 3851, Durham, NC 27702.

Africa Now, 212 Fifth Avenue, Suite 1409, New York, NY 10010.

Africa Recovery, DPI, Room S-1061, United Nations, New York, NY 10017.

Africa Today, 64 Washburn Avenue, Wellesley, MA 02181.

African Arts, University of California, Los Angeles, CA.

African Concord, 5–15 Cromer Street, London WCIH 8LS, England.

The Economist, 122 E. 42nd Street, 14th Floor, New York, NY 10168.

Newswatch, 62 Oregun Road, P.M.B. 21499, Ikeja, Nigeria.

The UNESCO Courier, 31, Rue François Bonvin, 75732, Paris CEDEX 15, France.

The Weekly Review, P.O. Box 42271, Nairobi, Kenya.

West Africa, Graybourne House, 52/54 Gray Inn Road, London WCIX 8LT, England.

NOVELS AND AUTOBIOGRAPHICAL WRITINGS

Chinua Achebe, *Things Fall Apart* (Portsmouth: Heinemann, 1965).
This is the story of the life and values of residents of a traditional Igbo village in the nineteenth century and of its first contacts with the West.

___, *No Longer at Ease* (Portsmouth: Heinemann, 1963).
The grandson of the major character of *Things Fall Apart* lives an entirely different life in the modern city of Lagos.

Ayi Kwei Armah, *The Beautyful Ones Are Not Yet Born* (London: Heinemann, 1992).

André Brink, *A Dry White Season* (New York: Penguin, 1989).

Syl Cheney-Choker, *The Last Harmattan of Alusine Dunba* (London: Heinemann, 1991).

Tsitsi Dangarembga, *Nervous Conditions* (Seal Press Feminist Publishing, 2002).

Buchi Emecheta, *The Joys of Motherhood* (New York: G. Braziller, 1979).
The story of a Nigerian woman who overcomes great obstacles to raise a large family and then finds that the meaning of motherhood has changed.

Nadine Gordimer, *Burgher's Daughter* (New York: Viking, 1980).

___, *A Soldier's Embrace* (New York: Viking, 1982).
These short stories treat the effects of apartheid on people's relations with each other. Films made from some of these stories are available at the University of Illinois Film Library, Urbana-Champaign, IL and the Boston University Film Library, Boston, MA.

Bessie Head, *Question of Power* (London: Heinemann, 1974).

Cheik Amadou Kane, *Ambiguous Adventure* (Portsmouth: Heinemann, 1972).
This autobiographical novel of a young man coming of age in Senegal, in a Muslim society, and, later, in a French school.

F. Kietseng, *Comrade Fish: Memoirs of a Motswana in the ANC Underground* (Pula Publishing, 1999).

Alex LaGuma, *Time of the Butcherbird* (Portsmouth: Heinemann, 1979).
The people of a long-standing black community in South Africa's countryside are to be removed to a Bantustan.

Camara Laye, *The Dark Child* (Farrar Straus and Giroux, 1954).
This autobiographical novel gives a loving and nostalgic picture of a Malinke family of Guinea.

Nelson Mandela, *Long Walk to Freedom: The Autobiography of Nelson Mandela* (New York: Little, Brown, 1995).

Okot p'Bitek, *Song of Lawino* (Portsmouth: Heinemann, 1983).
A traditional Ugandan wife comments on the practices of her Western-educated husband and reveals her own life-style and values.

Alexander McCall Smith, *The No. 1 Ladies' Detective Agency* (New York: Anchor Books, 2003).
The first in a popular series of novels set in Botswana.

Wole Soyinka, *Ake: The Years of Childhood* (New York: Random House, 1983).
Soyinka's account of his first 11 years is full of the sights, tastes, smells, sounds, and personal encounters of a headmaster's home and a busy Yoruba town.

___, *Death and the King's Horsemen* (New York: W. W. Norton, 2002).

Ngugi wa Thiong'o, *A Grain of Wheat* (Portsmouth: Heinemann, 1968).
A story of how the Mau-Mau movement and the coming of independence affected several individuals after independence as well as during the struggle that preceded it.

Amos Tutuola, *The Palm-Wine Drinkard* (Grove Press, 1994).

Yvonne Vera, *Butterfly Burning* (Baobab Books, 2000).

INTRODUCTORY BOOKS

Philip G. Altbach, *Muse of Modernity: Essays on Culture as Development in Africa* (Lawrenceville, NJ: Africa World, 1997).

Tony Binns, *People and Environment in Africa* (New York: Wiley, 1995).

Raymond Bonner, *At the Hand of Man: Peril and Hope for Africa's Wildlife* (New York: Random House, 1994).

Gwendolen Carter and Patrick O'Meara, eds., *African Independence: The First Twenty-Five Years* (Midland Books, 1986).
Collected essays surrounding issues such as political structures, military rule, and economics.

John Chiasson, *African Journey* (Upland, CT: Bradbury Press, 1987).
An examination into Africa's social life and customs.

Basil Davidson, *Africa in History* (New York: Macmillan, 1991).
A fine discussion of African history.

___, *The African Genius* (Boston: Little, Brown, 1979). Also published as *The Africans*.
Davidson discusses the complex political, social, and economic systems of traditional African societies, translating scholarly works into a popular mode without distorting complex material.

___, *The Black Man's Burden: Africa and the Curse of the Nation State* (New York: Random House, 1992).
A discussion on Africa's government and the status of the nation-state.

___, *A History of Africa*, 2nd ed. (Unwin Hyman, 1989).
A comprehensive look at the historical evolution of Africa.

Christopher Ehret, *An African Classical Age: Eastern and Southern Africa in World History, 1000 B.C. to A.D. 400* (Charlottesville: University of Virginia Press, 2001).

Clementine M. Faik-Nzuji, *Tracing Memory: Glossary of Graphic Signs and Symbols in African Art & Culture* (Seattle, WA: University of Washington Press, 1997).

Timothy J. Keegan, *Colonial South Africa and the Origins of the Racial Order* (Charlottesville: University of Virginia Press, 1997).

Anthony Appiah Kwame, Henry Louis Gates Jr., eds., *Africana: The Encyclopedia of the African and African American Experience* (BasicCivitas Books, 1999).

Paul E. Lovejoy, *Transformations in Slavery: A History of Slavery in Africa* (Cambridge: Cambridge University Press, 2000).

Amina Mama, *Beyond the Mask: Race, Gender and Identity* (London: Routledge, 1995).

John Mbiti, *African Religions and Philosophy* (Portsmouth: Heinemann, 1982).

This work by a Ugandan scholar is the standard introduction to the rich variety of religious beliefs and rituals of African peoples.

V. Y. Mudimbe, *The Invention of Africa* (Bloomington: Indiana University Press, 1988).

Joseph M. Murphy, *Working the Spirit: Ceremonies of the African Diaspora* (Boston: Beacon Press, 1994).

J. H. Kwabena Nketia, *The Music of Africa* (New York: Norton, 1974).

The author, a Ghanaian by birth, is Africa's best-known ethnomusicologist.

Gladson I. Nwanna, *Do's and Don'ts around the World: A Country Guide to Cultural and Social Taboos and Etiquette in Africa* (Baltimore: World Travel Institute, 1998).

Keith R. Richburg, *Out of Africa: A Black Man Confronts Africa* (New York: Basic Books, 1997).

Kevin Shillington, *History of Africa* (New York: Macmillan, 1995).

Bengt Sundkler and Christopher Steed, *A History of the Church in Africa* (Cambridge: Cambridge University Press, 2000).

John Thornton, *Africa and Africans in the Making of the Atlantic World, 1400–1800* (Cambridge: Cambridge University Press, 1998).

J. B. Webster, A. A. Boahen, and M. Tidy, *The Revolutionary Years: West Africa Since 1800* (London: Longman, 1980).

An interesting, enjoyable, and competent introduction.

Frank Willett, *African Art* (New York: Oxford University Press, 1971).

A work to read for both reference and pleasure.

COUNTRY AND REGIONAL STUDIES

Howard Adelman and John Sorenson, eds., *African Refugees* (Boulder: Westview, 1993).

Howard Adelman and Astri Suhrke, eds., *The Path of a Genocide: The Rwanda Crisis From Uganda to Zaire* (Transaction Publishing, 1999).

Allan R. Booth, *Swaziland: Tradition and Change in a Southern African Kingdom* (Boulder: Westview Press, 1984).

Thomas Borstelmann, *Apartheid, Colonialism, and the Cold War: The United States and Southern Africa* (New York: Oxford University Press, 1993).

Louis Brenner, ed., *Muslim Identity and Social Change in Sub-Saharan Africa* (Bloomington: Indiana University Press, 1993).

Mike Brogden and Clifford Shearing, *Policing for a New South Africa* (New York: Routledge, 1993).

Marcia M. Burdette, *Zambia: Between Two Worlds* (Boulder: Westview Press, 1988).

Amilcar Cabral, *Unity and Struggle* (Monthly Review Press, 1981).

Joao M. Cabrit, *Mozambique: The Tortuous Road to Democracy* (Palgrave Macmillan, 2001).

Thomas Callaghy and John Ravenhill, eds., *Hemmed In: Global Responses to Africa's Economic Decline* (New York: Columbia University Press, 1994).

W. Joesph Campbell, *The Emergent Independent Press in Benin and Coté d'Ivoire: From Voice of the State to Advocate of Democracy* (New York: Praeger, 1998).

Robin Cohen and Harry Goulbourne, eds., *Democracy and Socialism in Africa* (Boulder: Westview Press, 1991).

Maureen Covell, *Madagascar: Politics, Economy, and Society* (London and New York: F. Pinter, 1987).

W. A. Edge and M. H. Lekorwe, *Botswana, Politics and Society* (J. L. van Schaik, 1998).

Norman Etherington, *The Great Treks: The Transformation of Southern Africa, 1815–1854* (London: Longman, 2001).

Toyin Falola and Julius Ihonvbere, *The Rise and Fall of Nigeria's Second Republic, 1979–1984* (London: Zed Press, 1985).

Robert Fatton, *The Making of a Liberal Democracy: Senegal's Passive Revolution, 1975–85* (Boulder: L. Rienner, 1987).

Foreign Area Studies (Washington, DC: Government Printing Office).

Pumla Gobodo-Madikizela, *A Human Being Died That Night: A South African Story of Forgiveness* (New York: Houghton-Mifflin, 2003).

April A. Gordon and Donald L. Gordon, *Understanding Contemporary Africa* (Boulder: L. Rienner Publishers, 1996).

Phillip Gourevitch, *We Wish to Inform You That Tomorrow We Will Be Killed With Our Families: Stories From Rwanda* (Picador, 1999).

Joseph Hanlon, *Mozambique: The Revolution Under Fire* (London: Zed Press, 1984).

Angelique Haugerud, *The Culture of Politics in Modern Kenya* (Cambridge: Cambridge University Press, 1997).

Adam Hochschild, *King Leopold's Ghost* (New York: Houghton-Mifflin, 1999).

A fascinating and wrenching account of how Belgium's King Leopold ravaged Congo.

Tony Hodges, *Western Sahara: The Roots of a Desert War* (Westport: Laurence Hill & Co., 1983).

Gaim Kibreab, *Refugees and Development in Africa: The Case of Eritrea* (Trenton: Red Sea Press, 1987).

Gerhard Kraus, *Human Development from an African Ancestry* (London: Karnak House, 1990).

David D. Laitin and Said S. Samatar, *Somalia: Nation in Search of a State* (Boulder: Westview Press, 1987).

Karl Maier, *Angola: Promises and Lies* (Independence Educational Publishers, 2002).

Mahmoud Mamdani, *Citizen and Subject* (Princeton: Princeton University Press, 1996).

David Martin and Phyllis Johnson, *The Struggle for Zimbabwe: The Chimurenga War* (Boston: Faber & Faber, 1981).

Georges Nzongola-Ntalaja, *The Congo: From Leopold to Kabila: A People's History* (Zed Books, 2002).

Adebayo O. Olukoshi and Liisa Laakso, *Challenges to the Nation-State in Africa* (Uppsala: Nordiska Afrikainstitutel, in cooperation with Institute of Development Studies, University of Helsinki, 1996).

Eghosa E. Osaghae, *Crippled Giant: Nigeria Since Independence* (Bloomington: Indiana University Press, 1998).

Thomas O'Toole, *The Central African Republic: The Continent's Hidden Heart* (Boulder: Westview Press, 1986).

Richard Pankhurst, *The Ethiopians: A History* (London: Blackwell, 2001).

Deborah Pellow and Naomi Chazan, *Ghana: Coping With Uncertainty* (Boulder: Westview Press, 1986).

F. Jeffress Ramsay, Barry Morton, and Themba Mgadla, *Building a Nation: A History of Botswana* (Gaborone: Longman Botswana, 1996).

Richard Sandbrook, *The Politics of Africa's Economic Recovery* (Cambridge: Cambridge University Press, 1993).

Wisdom J. Tettey, ed., et al., *Critical Perspectives on Politics and Socio-Economic Development in Ghana* (African Social Studies Series, 6) (Brill Adademic Publishers, 2003).

Teun Voeten, Roz Vatter-Buck, trans., *How de Body? One Man's Terrifying Journey Through an African War* (Thomas Dunne Books, 2002).

C. W. Wigwe, *Language, Culture, and Society in West Africa* (Elms Court, UK: Arthur H. Stockwell, 1990).

Gabriel Williams, *Liberia: The Heart of Darkness* (Trafford, 2002).

Edwin Wilmsen, *Land Filled With Flies, A Political Economy of the Kalahari* (Chicago: University of Chicago Press, 1989).

Index

Index

Index